Zero-Valent Iron Reactive Materials for Hazardous Waste and Inorganics Removal

SPONSORED BY
Hazardous, Toxic, and Radioactive Waste Management Committee

Environmental and Water Resources Institute (EWRI)
of the American Society of Civil Engineers

EDITED BY
Irene M. C. Lo
Rao Y. Surampalli
Keith C. K. Lai

ASCE

EWRI

Published by the American Society of Civil Engineers

Library of Congress Cataloging-in-Publication Data

Zero-valent iron reactive materials for hazardous waste and inorganics removal / sponsored by Hazardous, Toxic, and Radioactive Waste Management Committee, Environmental and Water Resources Institute (EWRI) of the American Society of Civil Engineers ; edited by Irene M. C. Lo, Rao Y. Surampalli, Keith C. K. Lai

p. cm.

Includes bibliographical references and index.

ISBN-13: 978-0-7844-0881-0

ISBN-10: 0-7844-0881-5

1. Groundwater--Purification. 2. Hazardous waste site remediation. 3. Membrane separation. 4. Iron--Oxidation. 5. Ionic solutions. I. Lo, Irene Man-Chi. II. Surampalli, Rao Y. III. Lai, Keith C. K. IV. Environmental and Water Resources Institute (U.S.). Hazardous, Toxic, and Radioactive Waste Management Committee

TD426.Z42 2006

628.1'68—dc22 2006028834

American Society of Civil Engineers
1801 Alexander Bell Drive
Reston, Virginia, 20191-4400

www.pubs.asce.org

ISBN 13: 978-0-7844-0881-0 ISBN 10: 0-7844-0881-5
Manufactured in the United States of America.

Preface

Zero-valent iron (Fe^0) was first proposed as a reactive material for the removal of chlorinated aliphatic hydrocarbons (CAHs) in groundwater in early 1990s. The success of the corresponding laboratory experiments and the promising results from the first field demonstration in mid 1990s showed the great potential of using Fe^0 for groundwater remediation. This finding thereafter triggered the publication of a raft of research papers related to the application of Fe^0 for the removal of other recalcitrant hazardous wastes and inorganic contaminants, such as hexavalent chromium, nitrate and arsenic etc, in water. Generally, Fe^0 is used as reactive media in permeable reactive barrier (PRB) technology for groundwater remediation in which Fe^0 PRBs are installed into subsurface to intercept the flow path of the contaminant plume. Once the contaminated groundwater passively flows into the Fe^0 reactive media of the PRBs, the contaminants are removed from the groundwater. After a decade of scientific study and development, the maturity, credibility and applicability of the Fe^0 PRB technology for hazardous waste and inorganic removal have been increased substantially. Nowadays, Fe^0 PRBs have been widely recognized as an alternative to groundwater remediation in addition to the conventional pump-and-treat system.

Despite of the fact that the operating principle of Fe^0 PRB technology is simple, its design methodology is complicated. Furthermore, thorough understanding of the hydraulic characteristic, geochemical behavior and reactivity of Fe^0, and the involved removal mechanisms is required so as to obtain good performance of Fe^0 for contaminant removals. Primarily, this book is written to describe the mechanisms for the removal of various hazardous wastes and inorganics by conventional Fe^0 and innovative Fe^0-based reactive materials, which aims to provide the engineers and decision-makers an enough background to decide whether Fe^0 is cost-effective to be applied in their sites for contaminant removals. Furthermore, this book discusses some practical issues related to the design, construction and performance monitoring of Fe^0 PRBs for providing engineers some technical solutions about the engineering application of Fe^0 for hazardous waste and inorganic removal.

Chapter 1 is the introductory chapter which describes the historical development of Fe^0 and the advantages of Fe^0 PRBs over conventional pump-and-treat systems for removal of hazardous wastes in groundwater, and the maximum contaminant level (MCL) of various contaminants in groundwater and surface water. Section 1 (chapters 2 to 5) focuses on the removal of CAHs and hexavalent chromium using Fe^0 in which chapter 2 compares the Fe^0 reactivity in the removal of CAHs between laboratory Fe^0 column and a full-scale Fe^0 PRB. Chapters 3 and 4 address the simultaneous removal of CAHs and hexavalent chromium in water. The former chapter studies the application of a mixture Fe^0 and organo-clay for the removal of mixed contaminants, whereas the latter chapter evaluates the competitive effect from other CAHs on the removal of trichloroethylene (TCE) or hexavalent chromium by Fe^0. Chapter 5 gives an in-depth review of the mechanism and reaction kinetics involved in the removal of hexavalent chromium.

Section 2 (chapters 6 to 9) discusses the removal of aqueous nitrate and arsenic using Fe^0. Chapter 6 covers the fundamental study of the nitrate removal process by Fe^0 including the effect of pH, Fe^0 dosage, nitrate concentration and humic acid etc on the nitrate removal efficiency. Chapter 7 focuses on the application of clinoptilolite to adsorb the byproduct (i.e., ammonium ion) generated during the nitrate removal process by Fe^0. Chapters 8 and 9 present the arsenic removal mechanism and kinetics induced by Fe^0, evaluate the influence of the presence of competing inorganic anions on the arsenic removal efficiency, and discuss the mathematical model which is valid to describe the arsenic removal process by Fe^0.

Unlike sections 1 and 2 focusing on the contaminant removal processes using conventional granular Fe^0 as reactive materials, section 3 (chapters 10 to 12) describes the performance of various innovative Fe^0-based reactive materials on the hazardous waste removal, and the underlying theories and mechanisms involved. Chapter 10 compares the reactivity of conventional granular Fe^0 to that of the palladized granular Fe^0 in the removal of TCE. Chapter 11 presents the complicated pathways involved in the removal of halogenated methanes using nanoscale bimetallic palladized Fe^0. Chapter 12 evaluates the role of noble metal and Fe^0 in a bimetallic Fe^0 in the removal of chlorinated hydrocarbons, and investigates the factors affecting the reactivity of the bimetallic Fe^0.

Section 4 (chapters 13 to 17) concerns the practical issues pertained to the design, installation and performance monitoring of Fe^0 PRBs for groundwater remediation. Chapter 13 presents the conventional and innovative configurations, and the construction methods available for Fe^0 PRBs. On the other hand, chapter 14 describes the methodology, such as treatability testing and hydrogeologic modeling, used to design Fe^0 PRBs for hazardous waste and inorganic removal from groundwater. Chapter 15 addresses the hydraulic issues of the Fe^0 PRBs associated with the hydraulic characteristics of granular Fe^0, inadequate characterization of plume hydrogeology, construction methods and geochemistry behavior of the Fe^0 PRBs. Both chapters 16 and 17 focus on the evaluation of the performance of the full-scale Fe^0 PRBs using the tracer experiment. The former chapter discusses the application of reactive and conservative tracers in both natural gradient and forced gradient tracer experiments to evaluate the hydraulic behavior and reactivity of the Fe^0 PRBs. Chapter 17 gives a comprehensive description of the experimental method and theory required for conducting a natural gradient tracer experiment to evaluate the flow pattern of groundwater and groundwater velocity inside the Fe^0 reactive medium of the PRBs.

The efficacy and efficiency of the Fe^0 in the PRBs for contaminant removal are enhanced because of the continuous advent of innovative Fe^0-based reactive materials and the increasing understanding of the factors affecting Fe^0 reactivity and the relevant contaminant removal mechanisms. We hope that this book can give the engineers, decision-makers and researchers an in-depth understanding of the fundamental and environmental application of Fe^0 reactive materials for hazardous waste and inorganic removal.

Table of Contents

Chapter 1	**Introduction**	**1**
1.1	Historical Development of Zero-Valent Iron for Hazardous Waste Removal	1
1.2	Groundwater and Surface Water Standards	2
1.3	Comparison of the Fe^0-Based Permeable Reactive Barriers and Pump-and-Treat Systems in Hazardous Waste Removal	5
1.4	References	6

Section I Removals of Chlorinated Aliphatic Hydrocarbons and Hexavalent Chromium Using Zero-Valent Iron

Chapter 2	**Removals of Chlorinated Aliphatic Hydrocarbons by Fe^0: Full-Scale PRB vs Column Study**	**9**
2.1	Introduction	9
2.2	Experimental Section	12
2.2.1	Full-scale Fe^0 PRB Installed at Vapokon Site, Denmark	12
2.2.2	Laboratory Column Experiment	12
2.2.3	Groundwater Sample Measurements	16
2.3	Data Analysis	16
2.3.1	Determination of Longitudinal Dispersivity	16
2.3.2	Calculation of the Observed First-order Dechlorination Rate Constant	17
2.4	Results and Discussion	19
2.4.1	Longitudinal Dispersivity of the Fe^0 Packed Media	19
2.4.2	Influence of the Longitudinal Dispersivity on the CAH Concentration along Fe^0 Packed Media	21
2.4.3	The Performance on CAH Dechlorination	23
2.4.4	A Factor of Safety for the Designed Fe^0 PRB Thickness	27
2.5	Conclusions	29
2.6	References	30

Chapter 3	**Zero-Valent Iron and Organo-Clay for Chromate Removal in the Presence of Trichloroethylene**	**35**
3.1	Introduction	35
3.2	Experimental Section	36
3.2.1	Materials and Their Characterization Methods	36
3.2.2	Preparation of Organo-bentonite	36
3.2.3	Column Experiments	37
3.2.4	Data Analysis	38
3.3	Results and Discussion	39
3.4	Conclusions	45
3.5	References	45
Chapter 4	**Competitive Effects on the Dechlorination of Chlorinated Aliphatic Hydrocarbons by Zero-Valent Iron**	**47**
4.1	Introduction	48
4.2	Materials and Methods	49
4.2.1	Materials	49
4.2.2	Experimental Methods	50
4.2.3	Data Analysis	51
4.3	Results and Discussion	52
4.3.1	Dechlorination of CAHs by Fe^0	52
4.3.2	Competition between TCE and 1,1,1-TCA	52
4.3.3	Competition among TCE, 1,1,1-TCA and TCM at Various Temperatures	54
4.3.4	Competition between TCE and Cr(VI)	55
4.4	Conclusions	58
4.5	References	58
Chapter 5	**Removal of Hexavalent Chromium from Groundwater Using Zero-Valent Iron Media**	**61**
5.1	Introduction	61
5.2	Removal Mechanisms	62
5.3	Reaction Kinetics	63
5.4	Other *In Situ* Cr(VI) Removal Methods	69
5.5	Case Studies	70
5.5.1	Elizabeth City, North Carolina	70

5.5.2	Kolding, Denmark	72
5.6	Conclusions	72
5.7	References	73

Section II Removals of Nitrate and Arsenic using Zero-valent Iron

Chapter 6	**Aqueous Nitrate Reduction by Zero-Valent Iron Powder**	**77**
6.1	Introduction	77
6.2	Experimental Section	79
6.2.1	Material and Reagents	79
6.2.2	Reaction Systems and Operation	80
6.2.3	Instrumental Analyses	80
6.3	Results and Discussion	81
6.3.1	Fe^0/H_2SO_4 System	81
6.3.1.1	Effect of pH	81
6.3.1.2	Effect of Fe^0 Dosage	82
6.3.1.3	Effect of Species with Hydroxyl Group	83
6.3.2	Fe^0/CO_2 System	84
6.3.2.1	Effect of CO_2 Bubbling	84
6.3.2.2	Effect of Initial Nitrate Concentration	85
6.3.2.3	Effect of Humic Acid	86
6.3.2.4	Effect of Cations and Anions	87
6.3.2.5	Effect of Operating Modes	89
6.3.3	Issue of Undesired Byproducts and its Resolution	90
6.4	Conclusions and Recommendations	92
6.5	References	93
Chapter 7	**Removal of Nitrate from Water by a Combination of Metallic Iron Reduction and Clinoptilolite Ion Exchange Process**	**95**
7.1	Introduction	95
7.2	Materials and Methods	98
7.2.1	Chemicals	98
7.2.2	Nitrate Reduction Experiments	98
7.2.3	Ion Exchange Experiments	99
7.3	Results and Discussions	100

7.3.1	Effect of pH	100
7.3.2	Effect of Nitrate Loading	101
7.3.3	Overall Nitrate Reduction Efficiency	103
7.3.4	Removal of Ammonia in the Presence of Fe(II) Ions	104
7.4	Summary	107
7.5	References	107
Chapter 8	**Utilization of Zero-Valent Iron for Arsenic Removal from Groundwater and Wastewater**	**111**
8.1	Introduction	111
8.2	Batch Tests with Non Mine-Impacted Waters	114
8.3	Batch Test with Acid Mine Drainage	121
8.4	Effects of Competing Inorganic Anions on Arsenic Removal by Zero-Valent Iron	122
8.5	Column Tests and Field Applications	127
8.6	Mechanisms of Arsenic Removal by Zero-Valent Iron	131
8.7	Alternative Materials of Iron and Aluminum Oxides for Arsenic Removal	140
8.8	Knowledge Gaps and Research Needs	141
8.9	Conclusions	142
8.10	References	143
Chapter 9	**Removal of Arsenic from Groundwater—Mechanisms, Kinetics, Field/Pilot and Modeling Studies**	**151**
9.1	Introduction	151
9.2	Mechanism of Removal and Competing Ion Effects	154
9.3	Field/Pilot Studies and Modeling	157
9.4	Design Considerations	163
9.5	Conclusions	164
9.6	References	164

Section III Innovative Iron-based Reactive Materials

Chapter 10	**The Performance of Palladized Granular Iron: Enhancement and Deactivation**	**172**
10.1	Introduction	173

10.2	Experimental Section	174
10.2.1	Chemicals and Materials	174
10.2.2	Column Tests	175
10.2.3	Chemical Analyses	175
10.2.4	Auger Electron Spectroscopy (AES)	176
10.2.5	X-Ray Photoelectron Spectroscopy (XPS)	176
10.3	Results and Discussion	176
10.3.1	Pd Plating Efficiency	176
10.3.2	Column Tests	177
10.3.3	Surface Analysis	180
10.4	Conclusions	184
10.5	References	185
Chapter 11	**Nanoscale Bimetallic Pd/Fe Particles for Remediation of Halogenated Methanes**	**187**
11.1	Introduction	187
11.2	Experimental Section	191
11.2.1	Batch Experiments	191
11.2.2	Headspace Analysis	191
11.2.3	Kinetic Analysis	191
11.2.4	Materials and Chemicals	192
11.3	Results	192
11.3.1	Product Distributions and Reaction Rates	192
11.3.2	Correlation Analysis	194
11.3.3	Kinetic Simulation	196
11.4	Discussion	201
11.5	Conclusions	203
11.6	References	203
Chapter 12	**Reduction by Bimetallic Reactive Materials Containing Zero-Valent Iron**	**206**
12.1	Introduction	206
12.2	Noble Metals as Reduction Catalysts	207
12.3	Preparation of Bimetallic Reductants	209
12.4	Reduction Reactions of Bimetallic Materials	210
12.4.1	Direct Adsorption of Atomic Hydrogen	211

12.4.2 Dissociative Adsorption of Diatomic Hydrogen 213
12.4.3 Roles of Iron as the Primary Metal 213
12.5 Factors Affecting Reaction of Bimetallic Reductants 215
12.5.1 Effect of the Type of Noble Metal 215
12.5.2 Effect of Noble Metal Content and Surface Area 215
12.5.3 Effect of Solution Chemistry 216
12.6 Deactivation of Bimetallic Reductants 216
12.7 Nano-sized Bimetallic Reductants 217
12.8 Conclusions 218
12.9 References 218

Section IV Zero-Valent Iron Reactive Barrier: Configuration, Construction, Design Methodology, and Hydraulic Performance

Chapter 13 Configuration and Construction of Zero-Valent Iron Reactive Barriers 224
13.1 Introduction 224
13.2 Permeable Reactive Barrier Configurations 225
13.2.1 Continuous Permeable Reactive Barriers 226
13.2.2 Funnel-and-gate Permeable Reactive Barriers 226
13.2.3 Caisson Permeable Reactive Barriers 227
13.2.4 Trench Permeable Reactive Barriers 228
13.2.5 GeoSiphon™/GeoFlow 228
13.3 Emplacement Techniques for Permeable Reactive Barriers 229
13.3.1 Emplacement Techniques for Permeable Treatment Zone 230
13.3.1.1 Trench Excavation 230
13.3.1.2 Caisson-Based Emplacement 231
13.3.1.3 Mandrel-Based Emplacement 231
13.3.1.4 Continuous Trenching 232
13.3.2 Emplacement Techniques for Impermeable Funnels 232
13.3.2.1 Steel Sheet Piling 232
13.3.2.2 Slurry Wall 233
13.3.3 Innovative Technologies for the Emplacement of Permeable Treatment Zone and Impermeable Funnels 233

13.3.3.1	Jetting	234
13.3.3.2	Hydraulic Fracturing	234
13.3.3.3	Deep Soil Mixing	234
13.4	Case Studies	235
13.4.1	Configuration and Construction of a Permeable Reactive Barrier at Vapokon Site, Denmark	235
13.4.2	Configuration and Construction of a Permeable Reactive Barrier at United States Coast Guard (USCG) Support Centre in Elizabeth City, North Carolina	237
13.4.3	Configuration and Construction of a Permeable Reactive Barrier at an Industrial Site in Belfast, Northern Ireland	238
13.5	Summary	239
13.6	References	239
Chapter 14	**Design Methodology for the Application of a Permeable Reactive Barrier for Groundwater Remediation**	**243**
14.1	Introduction	243
14.2	Preliminary Assessment	245
14.3	Site Characterization	247
14.4	Reactive Media Selection	247
14.5	Treatability Testing	250
14.6	Hydrogeologic and Geochemical Modelings	254
14.7	Monitoring Plan	257
14.8	Permeable Reactive Barrier Economics	259
14.9	Summary	261
14.10	References	261
Chapter 15	**Hydraulic Issues Related to Granular Iron Permeable Reactive Barriers**	**267**
15.1	Introduction	267
15.2	Hydraulic Characteristics of Granular Iron and Impact on PRB Design	268
15.3	Influence of Inadequate Characterization of Plume Hydrogeology on Hydraulic Performance	271
15.4	Influence of Construction Methods on Hydraulic Performance	272

15.5	Influence of Long-Term Geochemical Changes on Hydraulic Performance	275
15.6	Summary	278
15.7	References	278
Chapter 16	**Tracer Experiments in Zero-Valent Iron Permeable Reactive Barriers**	**282**
16.1	Introduction	282
16.2	Tracer Experiments in Laboratory Columns	285
16.3	Tracer Experiments at PRB Sites	289
16.3.1	Rheine Site PRB	291
16.3.2	Tübingen Site PRB	294
16.4	Conclusions	300
16.5	References	301
Chapter 17	**Hydraulic Studies of Zero-Valent Iron in Permeable Reactive Barriers Using Tracer Experiment**	**309**
17.1	Introduction	309
17.2	Vapokon Site Description and Fe^0 PRB Emplacement	311
17.3	Natural Gradient Tracer Experiment for the Hydraulic Performance Monitoring of the Fe^0 PRB at Vapokon Site	312
17.3.1	Selection of Suitable Tracer Materials	312
17.3.2	Groundwater Modeling	314
17.3.2.1	Design Parameters for the Tracer Injection System	314
17.3.2.2	Design Parameters for the Groundwater Sampling System	315
17.3.2.3	Groundwater Flow Model and Solute Transport Program	316
17.3.3	Injection Wells and System	317
17.3.4	Groundwater Sampling Network and Systems	320
17.3.5	Collection of Groundwater Samples and Chemical Analysis of Tracers	320
17.3.6	Data Analysis—Spatial Moments Analysis for the Lithium Plume	322
17.4	Results and Discussion	324
17.4.1	Flow Pattern of the Lithium Plume	324
17.4.2	Breakthrough Curves of Bromide	328
17.4.3	Mass of Lithium Flowing through the Fe^0 Reactive Medium	329
17.4.4	Groundwater Velocity	331

17.5 Conclusions 331
17.6 References 332
Appendix **337**
Subject Index **339**

CHAPTER 1

Introduction

Irene M. C. Lo[1], Rao Surampalli[2], and Keith C. K. Lai[3]

1.1 Historical Development of Zero-valent Iron for Hazardous Waste Removal

To our best knowledge, the earliest report recording the use of zero-valent transition metal to remove chlorinated hydrocarbons was in the patent literature (Sweeney 1972; Sweeney 1973). The patent literature in 1972 reported the enhancement of the reductive degradation rate of aqueous chlorinated pesticides, such as dichloro-diphenyl-trichloroethane (DDT), by metallic zinc under acidic condition. Another patent literature in 1973 presented the higher reactivity of metallic couples, such as zinc-copper, iron-copper and aluminum-copper, than the zinc alone in the reductive degradation of aqueous DDT.

Probably due to the promising potential of zero-valent metals in the degradation of aqueous chlorinated hydrocarbons, Senzaki and Kumagai (1988) studied the possible degradation of aqueous 1,1,2,2-tetrachloroethane in contact with electrolytic-iron powder. It was found that the rate of degradation reaction increased with increasing temperature and the removal of dissolved oxygen from the water led several folds increase in the degradation rate. Methane, ethane and ethene, and large amount of hydrogen gas were also detected to generate during the degradation reaction between 1,1,2,2-tetrachloroethane and the electrolytic-iron powder.

In addition to 1,1,2,2-tetrachloroethane, Senzaki and Kumagai (1989) conducted additional experiments to investigate the removal of trichloroethylene (TCE) in the presence of electrolytic-iron powder. In analogy to 1,1,2,2-tetrachloroethane degradation, noticeable reductive dehalogenation of TCE was also observed and the TCE degradation rate increased when increasing the ratio of iron mass to the influent flux. Further study of the degradation of TCE by electrolytic-iron powder revealed that the TCE degradation rate was dependent upon the surface area of the iron, and TCE was found to be degraded to acetylene, ethane and ethene (Senzaki 1991).

Although the efficacy of metallic iron on the reductive degradation of chlorinated hydrocarbons was reported, only little attention was directed towards the application of zero-valent iron (Fe^0) for hazardous waste removal until the encouraging works published by Gillham and O'Hannesin (1992). Their laboratory

---

[1] Associate Professor, Dept. of Civil Engineering, The Hong Kong University of Science and Technology, Clear Water Bay, Kowloon, Hong Kong (corresponding author). E-mail: cemclo@ust.hk
[2]Engineer Director, United States Environmental Protection Agency, Kanasas City, Kansas 66117, United States.
[3] Research Associate, Dept. of Civil Engineering, The Hong Kong University of Science and Technology, Clear Water Bay, Kowloon, Hong Kong.

works indicated that Fe^0 could degrade a wide range of dissolved halogenated compounds in a reaction rate three to eight orders of magnitude higher than the naturally abiotic degradation process, thereby indicating the great potential of using Fe^0 for contaminant removal. The first field application of Fe^0 in permeable reactive barrier (PRB) technology for *in situ* remediation of groundwater contaminated by TCE and tetrachloroethylene (PCE) at Canadian Forces Base, Borden, Ontario in 1991 also showed about 95% and 91% removal of TCE and PCE, respectively, across the Fe^0 PRB (O'Hannesin and Gillham 1992). Because of the success of this field demonstration, a first commercial Fe^0 PRB was installed at the industrial site in Sunnyvale, California to remove the chlorinated hydrocarbons in groundwater in January 1995. In June 1996, another Fe^0 PRB was installed at the United States Coast Guard (USCG) Support Center near Elizabeth City, North Carolina for the removal of both chlorinated hydrocarbons and hexavalent chromium in groundwater (Powell et al. 1998). Nowadays, in addition to chlorinated hydrocarbons and hexavalent chromium, Fe^0 has been used to remove the hazardous wastes and inorganics, such as pentachlorophenol (Kim and Carraway 2000), nitroaromatic compounds (Klausen et al. 2001), nitrate, nitrite, bromate and chlorate (Alowitz and Scherer 2002; Westerhoff 2003), arsenic (Nikolaidis et al. 2003), selenium (McRae et al. 1997), technetium (Clausen et al. 1995) and uranium (Gu et al. 1998; Morrison et al. 2001), in groundwater and surface water.

1.2 Groundwater and Surface Water Standards

Heretofore, there is no universal agreement about cleanup standards of groundwater and surface water. In the United States, regulations under the Comprehensive Environmental Response, Compensation, and Liability Act (i.e., the Superfund law), and the Resource Conservation and Recovery Act (RCRA) require the contaminated groundwater to be cleaned up to meet the drinking water standards (National Research Council 1994). The last column of Table 1-1 illustrates the maximum contaminant levels (MCLs), provided by United States Environmental Protection Agency, for some of contaminants in drinking water (USEPA 2005). In addition to the drinking water standards, other countries including Netherlands, Sweden, Australia and France set their groundwater and surface water quality standards based upon the deviation of contaminant concentrations from the natural occurrence level of contaminants and the intended uses of water resources (NEPC 1999; Sanaterre Environmental 2000a; Sanaterre Environmental 2000b; Swedish EPA 2002). As shown in Table 1-1, Netherlands establishes target values and intervention values to assess groundwater quality. The former values are the baseline concentrations below which contaminants are known or assumed not to affect the natural properties of soils; whereas the intervention values are the maximum tolerable concentrations above which remedial action is required (Sanaterre Environmental 2000a). For instance, if TCE concentration in groundwater is below 0.01 μg/L, it is known that TCE does not adversely affect the natural properties of soils. However, immediate remedial action is needed if the TCE concentration is higher than 400 μg/L (Table 1-1).

Table 1-1. Groundwater and Surface Water Standards in Netherlands, Sweden, Australia, France and United States

[1]Rank	Compounds	[2]Dutch standards Groundwater [3]Target values	[4]Intervention values	[6]Swedish EPA Groundwater [7]Guideline values	[8]Surface water	[10,11]Australian groundwater investigation levels Fresh water aquatic ecosystems	Drinking water	Agricultural irrigation	[16]French guideline values [17]Fixed impact values [18]Sensitive use	[19]non-sensitive use	[20,21]USEPA [22]MCLs of drinking water
1	Trichloroethylene	0.01	400	--	20	--	--	--	10	50	5
2	Lead	1.7-15.0	75	10	1	1.0-5.0	10	200	50	250	15
3	Tetrachloroethylene	0.01	10	--	110	--	50	--	10	50	5
4	Benzene	0.2	30	10	300	300	1	--	1	5	5
5	Toluene	0.2	70	60	2	300	25-800	--	700	3500	1000
6	Chromium	1.0-2.5	30	50	[9] 20	[12] 10	[13] 50	[13] 100	[12] 50	[12] 250	[12] 100
7	Dichloromethane	6	400	1	--	--	4	--	20	100	5
8	Zinc	24-65	800	--	30	5.0-50	3000	2000	5000	10000	[23] 5000
9	1,1,1-Trichloroethane	0.01	130	--	--	--	--	--	2000	10000	200
10	Arsenic	7.2-10	60	50	50	50	7	100	50	250	10
11	Chloroform	0.01	300	--	2	--	--	--	100	500	--
12	1,1-Dichloroethane	7	900	--	--	--	0.003	--	--	--	--
13	1,2-Dichloroethene, trans	0.8	80	--	--	--	--	--	--	--	100
14	Cadmium	0.06-0.4	6.0	5	0.01	0.2-2.0	2	10	5	25	5
15	Manganese	--	--	--	--	--	100-500	2000	50	250	[23] 50
16	Copper	1.3-15	75	2000	4	2.0-5.0	1000-2000	200	1000	2000	1300
17	1,1-Dichloroethylene	0.01	10	--	--	--	--	--	30	150	7
18	Vinyl chloride	--	--	--	--	--	0.3	--	0.5	2.5	2
19	Barium	50-200	625	--	--	--	700	--	100	1000	2000
21	Ethylbenzene	4	150	20	90	--	3-300	--	300	1500	700
22	Nickel	2.1-15.0	75	50	150	15.0-150	20	20	50	250	--
23	Di(2-ethylhexy)phthalate	[5] 0.5	[5] 5	--	--	0.6	--	--	8	40	6
24	Xylenes	6.0	300	20	--	--	20-600	--	500	2500	10000
25	Phenol	0.2	600	--	1	50	--	--	--	--	--
--	Nitrate	--	--	--	--	--	[14] 50000	--	50000	100000	[14] 10000
--	Nitrite	--	--	--	--	--	[15] 3000	--	100	500	[15] 1000
--	1,2-Dichloroethene, cis	0.01	20	--	--	--	--	--	50	250	70

Note: all the units are in μg/L. [1]The number shown in the "Rank" column indicates the 25 most frequently detected groundwater contaminants at hazardous waste sites in United States; [2]Sanaterre Environmental (2000a); [3]The target values are the baseline concentration values below which compounds and/or elements are known or assumed not to affect the natural properties of the soil; [4]The intervention values are the maximum tolerable concentrations above which remediation is required; [5]Criteria for total phthalate; [6]Swedish EPA (2002); [7]Guideline values are the level which cannot be exceeded in order to avoid imposing risk to human health and/or the environment. The contamination levels of groundwater and surface water are classified as "not very serious" if contaminant concentrations are lower than the corresponding guideline values. However, the contamination levels are environmentally viewed as "moderately serious", "serious" and "very serious" if contaminant concentrations are 1-3, 3-10 and over 10 times, respectively higher the corresponding guideline values; [8]The guideline values are based upon the Canadian water quality criteria; [9]Trivalent chromium; [10]Groundwater investigation levels are defined as acceptable groundwater quality for various potential uses of groundwater; [11]NEPC (1999); [12]Total chromium; [13]Hexavalent chromium; [14]Nitrate-nitrogen; [15]Nitrite-nitrogen; [16]Sanaterre Environmental (2000b); [17]Fixed impact values for "water" environment are applicable to both groundwater and surface water; [18]Fixed impact values for sensitive use concern water resource which is destined for human consumption; [19]Fixed impact values for non-sensitive use concern water resource which is destined for industrial and commercial uses. They are derived from the fixed impact values for sensitive use by multiplying by 2 for the major elements, or by 5 for tracer elements; [20]USEPA (2005); [21]National primary drinking water regulations provided by United States Environmental Protection Agency; [22]Maximum contaminant levels; [23]National secondary drinking water regulations provided by United States Environmental Protection Agency.

Swedish Environmental Protection Agency assesses the contamination level in groundwater and surface water primarily in accordance with the deviation of contaminant concentrations from the acceptable (guideline values) or natural levels (reference levels) of the contaminants (Swedish EPA 2002). Guideline values refer to contaminant concentration levels which cannot be exceeded in order to avoid imposing risk to human health and/or the environment. On the other hand, reference values reflect the natural occurrence levels of contaminants. The Swedish guideline values for various contaminants in groundwater and surface water are summarized in Table 1-1. Based upon the Swedish criteria, the level of arsenic in groundwater and surface water should not exceed 50 μg/L; otherwise the arsenic in the groundwater and surface water will impose risk to human health or environment. Besides, the contamination level in groundwater and surface water is classified as "not very serious" if contaminant concentrations are lower than the corresponding guideline values. However, if contaminant concentrations are 1-3, 3-10 and more than 10 times higher than the relevant guideline values, the contamination levels are classified as "moderately serious", "serious" and "very serious", respectively.

National Environment Protection Council of Australia applies investigation levels and response levels for the assessment of groundwater quality (NEPC 1999). Investigation levels are the concentration of contaminants above which further appropriate investigation and evaluation of groundwater quality will be required. Response levels are the concentration of contaminants at a specific site based upon a site assessment for which some forms of response are required to provide an adequate margin of safety to protect public health and/or the environment. The magnitude of the investigation and response levels are highly dependent upon the intended uses of groundwater. Table 1-1 shows the Australian groundwater investigation levels at the point of extraction. As seen, if protection of fresh water aquatic ecosystems is the main point to be concerned, the investigation level of lead in groundwater is 1.0 to 5.0 μg/L. However, the investigation levels of lead in groundwater are 10 and 200 μg/L if the groundwater is used for drinking purpose and agricultural irrigation, respectively. As regards the response levels, the groundwater investigation levels listed in Table 1-1 can be applied as Australian groundwater response levels at the point of use. For instance, for drinking purpose, the response level of lead in groundwater at the point of use should not exceed 10 μg/L; otherwise some forms of remedial action are needed to take to protect the public health and/or the environment.

Fixed impact values are used as guideline values in France for contaminated water and groundwater management. The French fixed impact values, developed by the public health working group as part of national approach to the management and remediation of polluted sites, are established based upon studies of the toxicity of contaminants to human health and of the exposure of people to such contaminants. These values are only valid for preliminary site investigation and simplified risk assessment work, and do not represent threshold values for remediation works. There are two categories for the fixed impact values: sensitive use and non-sensitive use. The former category of the fixed impact values is used to assess the groundwater and surface water quality if the water resource is destined for human consumption (i.e., residential use); whereas the latter category of the fixed impact values is applied if the water resource is for industrial and commercial activity (Sanaterre Environmental 2000b). For example, as illustrated in Table 1-1, the concentration of chloroform should not exceed 100 μg/L if the water resource is destined for human consumption. For the industrial and commercial use, the fixed impact value of chloroform can up to 500 μg/L.

1.3 Comparison of the Fe^0-based Permeable Reactive Barriers and Pump-and-treat Systems in Hazardous Waste Removal

Conventional technology for removal of hazardous wastes in groundwater is pump-and-treat systems, which is an *in situ* remediation technology and has been used for more than 20 years in the remediation of groundwater contamination (Simon et al. 2002). Pump-and-treat systems involve extraction of contaminated groundwater from aquifers to ground surface through the extraction or pumping wells, treatment of the extracted groundwater aboveground, and discharge or re-injection of the treated groundwater into water bodies or aquifers (Fig. 1-1). Air stripping, carbon adsorption, biological treatment, chemical precipitation and oxidation, and ion exchange are the methods frequently used for aboveground treatment of contaminated groundwater (USEPA 2001).

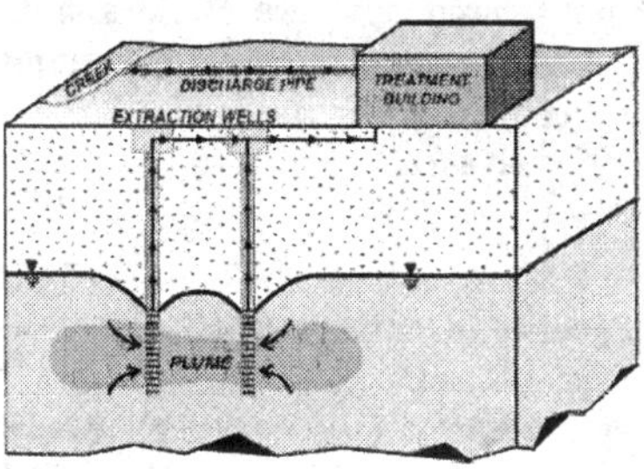

Fig. 1-1. Schematic diagram of pump-and-treat systems for groundwater remediation (Adapted from USEPA 2001)

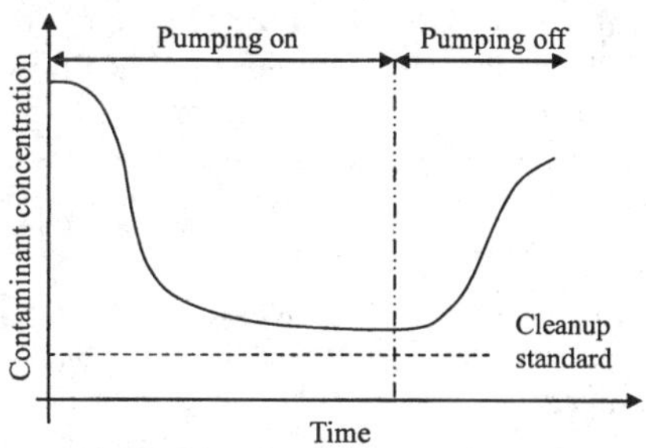

Fig. 1-2. Temporal change of contaminant concentration during groundwater remediation using pump-and-treat systems

The typical experience on pump-and-treat systems over the past decade, however, indicated that there had been little success in reducing contaminant concentrations to the cleanup goals (Travis and Doty 1990; National Research Council 1994). At many of the sites studied, there was an initial drop in contaminant concentrations by a factor of 2 to 10 but then contaminant concentrations leveled off and progressed slowly towards the cleanup standards, as illustrated in Fig. 1-2 (National Research Centre 1994). Once pumps were turned off, contaminant concentrations rose again (Travis and Doty 1990; Simon et al. 2002). Furthermore, because of the low solubility of most of groundwater contaminants such as chlorinated aliphatic hydrocarbons (CAHs), mass transfer limitation in aquifers, incomplete removal of the contaminant sources and slow desorption of contaminants from aquifer materials, long-term extraction of large volume of contaminated groundwater, which is usually within the range of decades at least, for aboveground treatment is generally required for pump-and-treat systems resulting in enormous operation and maintenance costs.

PRB technology using Fe^0 as reactive materials is an *in situ* approach for groundwater remediation in which Fe^0 is placed in underground trenches downgradient of the contaminant plume. Contaminated groundwater flows into the Fe^0 reactive medium of the PRBs by its natural gradient for *in situ* treatment. Unlike pump-and-treat systems, Fe^0 PRBs require no extraction or pumping wells and aboveground treatment systems, thereby resulting in a lower operating cost. Although cost comparison of groundwater remediation between pump-and-treat systems and PRBs shows that the latter requires higher capital cost because of the requirement of more detailed site characterization, emplacement of Fe^0 reactive materials into subsurface and purchasing of Fe^0, such cost can be compensated in a long run by a lower operating cost of Fe^0 PRBs (Simon et al. 2002). Reeter et al. (1999) reported that around 50 to 70% of cost saving can be achieved over the long term if Fe^0 PRBs rather than pump-and-treat systems are chosen as a remediation approach.

The other advantages of Fe^0 PRBs over conventional pump-and-treat systems are that the former requires no installation of invasive surface structures and equipment so that the affected property can be put to productive use. Besides, initial evidence indicates that Fe^0 used in PRBs is consumed very slowly and is therefore potentially suitable for a long-term passive treatment of hazardous wastes in groundwater (EnviroMetal Technologies Inc. 1998; Gavaskar et al. 2000).

1.4 References

Alowitz, M. J., and Scherer, M. M. (2002). "Kinetics of nitrate, nitrite, and Cr(VI) reduction by iron metal." *Environ. Sci. & Technol.*, 36(3), 299-306.

Clausen, J. L., Richards, W. L., Korte, N. E., and Liang, L. (1995). "ORNL/MMES research into remedial applications of zero-valence metals. 3: Removal of TCE, cis-1,2-DCE, vinyl chloride, and technetium." In *Proceedings of the 209th ACS National Meeting*, American Chemical Society, April 2-7, Anaheim, California.

EnviroMetal Technologies Inc. (1998). *Bench-scale treatability report of the EnviroMetal process at the Vapokon site, Fyn, Denmark*, ETI 31438.10, Guelph, Ontario, Canada N1K1S6.

Gavaskar, A., Gupta, N., Sass, B., Janosy, R., and Hicks, J. (2000). *Design guidance for application of permeable reactive barriers for groundwater remediation*, Strategic Environmental Research and Development Program (SERDP), Battelle, Columbus, Ohio.

Gillham, R. W., and O'Hannesin, S. F. (1992). "Metal-catalysed abiotic degradation of halogenated organic compounds." In *Proceedings of the 1992 IAH Conference Modern Trends in Hydrogeology*, May, Hamilton, Ontario, 10-13.

Gu, B., Liang, L., Dickey, M. J., Yin, X., and Dai, S. (1998). "Reductive precipitation of uranium(VI) by zero-valent iron." *Environ. Sci. & Technol.*, 32(21), 3366-3373.

Kim, Y. H., and Carraway, E. R. (2000). "Dechlorination of pentachlorophenol by zero-valent iron and modified zero valent irons." *Environ. Sci. & Technol.*, 34(10), 2014-2017.

Klausen, J., Ranke, J., and Schwarzenbach, R. P. (2001). "Influence of solution composition and column aging on the reduction of nitroaromatic compouneds by zero-valent iron." *Chemosphere*, 44, 511-517.

McRae, C. W., Blowes, D. W., and Ptacek, C. (1997). "Laboratory-scale investigation of remediation of As and Se using iron oxides." In *Proceedings of the 6th Symposium and Exhibition on Groundwater and Soil Remediation*, March 18-21, Montreal, Quebec, Canada.

Morrison, S. J., Metzler, D. R., and Carpenter, C. E. (2001). "Uranium precipitation in a permeable reactive barrier by progressive irreversible dissolution of zerovalent iron." *Environ. Sci. & Technol.*, 35(2), 385-390.

National Research Council (1994). *Alternatives for ground water cleanup*, Ground Water Cleanup Alternatives, Water Science and Technology Board, Board on Radioactive Waste Management and Commission on Geosciences, Environment and Resources, National Academy Press, Washington, DC, United States.

NEPC (1999). *Schedule B (1) guideline on the investigation levels for soil and groundwater*, Assessment of Site Contamination, National Environment Protection Council, Australia.

Nikolaidis, N. P., Dobbs, G. M., and Lackovic, J. A. (2003). "Arsenic removal by zero-valent iron: field, laboratory and modeling studies." *Wat. Res.*, 37(6), 1417-1425.

O'Hannesin, S. F., and Gillham, R. W. (1992). "Permeable reaction wall for in situ degradation of halogenated organic compounds." In *Proceedings of the 45th Canadian Geotechnical Conference*, Toronto, Ontario, Canada.

Powell, R. M., Blowes, D. W., Gillham, R. W., Schultz, D., Sivavec, T., Puls, R. W., Vogan, J. L., Powell, P. D., and Landis, R. (1998). *Permeable reactive barrier technologies for contaminant remediation*, EPA/600/R-98/125, Office of Research and Development, United States Environmental Protection Agency, Washington, DC 20460, United States.

Reeter, C., Chao, S., and Gavaskar, A. (1999). *Permeable reactive wall remediation of chlorinated hydrocarbons in groundwater, cost and performance report*, Environmental Security Technology Certification Program (ESTCP), United States Department of Defense, Washington DC, United States.

Sanaterre Environmental (2000a). "The Dutch target and intervention values." <http://www.sanaterre.com/guidelines/dutch.htm> (Dec. 3, 2005).

Sanaterre Environmental (2000b). "French guidelines: French guideline values used for contaminated land management." <http://www.sanaterre.com/guidelines/french.htm>, (Dec. 3, 2005).

Senzaki, T., and Kumagai, Y. (1988). "Removal of organochloro compounds in waste-water by reductive treatment – treatment of 1,1,2,2-tetrachloroethane with iron powder." *Kogyo Yosui*, 357, 2-7 (Japanese).

Senzaki, T., and Kumagai, Y. (1989). "Removal of organochloro compounds by reduction treatment (the second report) – treatment of trichloroethylene with iron powder." *Kogyo Yosui*, 369, 19-24 (Japanese).

Senzaki, T. (1991). "Removal of chlorinated organic compounds from waste water by reduction process (third report) – treatment of trichloroethylene with iron powder II." *Kogyo Yosui*, 391, 29-35.

Simon, T., Meggyes, C., and McDonald, C. (2002). *Advanced groundwater remediation: Active and passive technologies*, Telford, London.

Swedish EPA (2002). "Environmental Quality Criteria for Contaminated Sites." *Swedish EnviroNet*, Swedish Environmental Protection Agency, http://www.internat.environ.se/index.php3?main=/documents/legal/assess/assedoc/cont.htm> (Dec. 3, 2005).

Sweeney, K. H. (1972). *Reductive degradation of halogenated pesticides*, United States Patent No. 3,640,821.

Sweeney, K. H. (1973). *Decomposition of halogenated organic compounds using metallic couples*, United States Patent No. 3,737,384.

Travis, C. C., and Doty, C. B. (1990). "Can contaminated aquifer at superfund sites be remediated?." *Environ. Sci. & Technol.*, 24(10), 1464-1466.

USEPA (2001). *Groundwater pump and treat systems: Summary of selected cost and performance information at superfund-financed sites*, EPA/542/R-01/021a, Solid Waste and Emergency Response (5102G), United States Environmental Protection Agency, Washington DC 20460, United States.

USEPA (2005). "List of drinking water contaminants & MCLs." United States Environmental Protection Agency, < http://www.epa.gov/safewater/mcl.html#mcls>, (Dec. 3, 2005).

Westerhoff, P. (2003). "Reduction of nitrate, bromate, and chlorate by zero-valent iron (Fe^0)." *Wat. Res.*, 129(1), 10-16.

CHAPTER 2

Removals of Chlorinated Aliphatic Hydrocarbons by Fe^0: Full-scale PRB vs Column Study

Keith C. K. Lai[1], and Irene M. C. Lo[2]

Abstract: The thickness of the zero-valent iron (Fe^0) packed media in full-scale permeable reactive barriers (PRBs) for the remediation of chlorinated aliphatic hydrocarbons (CAHs) in groundwater is conventionally designed using the kinetic data from laboratory column experiments and first-order equation. To study the reliability of this conventional approach for the PRB thickness design, the performance of the full-scale Fe^0 PRB at Vapokon site, Denmark and the corresponding laboratory Fe^0 packed columns (LFPCs) on CAH dechlorination was studied and compared. Interestingly, the full-scale Fe^0 PRB was found having a better performance than the corresponding LFPCs on the dechlorination of tetrachloroethylene, trichloroethylene, 1,1-dichloroethylene, 1,1-dichloroethane and chloroform, most likely due to the enhancement from the sulfur-bearing minerals formed inside the Fe^0 PRB. On the contrary, the full-scale Fe^0 PRB showed a poorer performance on 1,1,1-trichloroethane, cis-dichloroethylene and vinyl chloride dechlorination. It is believed that the great difference in the longitudinal dispersivity (α_y) of the Fe^0 packed media between full-scale PRB (0.275 m) and the corresponding LFPCs (1.23×10^{-3} m) is one of the underlying reasons leading to this poorer dechlorination performance. This is because the dispersion of dissolved CAH inside the Fe^0 packed medium of the full-scale PRB could reduce the residence time for the dechlorination reactions. To achieve the overall effluent groundwater quality from the full-scale Fe^0 PRB at Vapokon site as good as that from the corresponding LFPCs, a factor of safety of about 3 should be applied to design the Fe^0 PRB thickness which is determined conventionally using the first-order equation.

CE Database subject headings: Barriers; Dispersion; Halogen organic compounds; Iron; Mineral deposits; Safety factors; Sulfur.

2.1 Introduction

Permeable reactive barrier (PRB) technology using zero-valent iron (Fe^0) as reactive media has received a great attention because of its cost-effective approach for *in situ* groundwater remediation (Scherer et al. 2000; USEPA 2002). Through the processes of chemical reduction, precipitation, sorption, and biologically mediated reduction, dissolved groundwater contaminants such as chlorinated aliphatic hydrocarbons (CAHs), chromate, phosphate, and sulfate (SO_4^{2-}) are either transformed into less

[1] Research Associate, Dept. of Civil Engineering, The Hong Kong University of Science and Technology, Clear Water Bay, Kowloon, Hong Kong.

[2] Associate Professor, Dept. of Civil Engineering, The Hong Kong University of Science and Technology, Clear Water Bay, Kowloon, Hong Kong (corresponding author). E-mail: cemclo@ust.hk

harmful products or immobilized inside the Fe^0 packed media of PRBs (Gillham and O'Hannesin 1994; Powell et al. 1995; Baker et al. 1997; Benner et al. 1997; Lo et al. 2005; Lai et al. 2006c). Therefore, the groundwater releasing from the downgradient side of Fe^0 PRBs can attain remediation concentration goals in downgradient aquifers.

Up until 2002, over 65% of full-scale Fe^0 PRBs were installed for the remediation of CAHs in contaminated groundwater because of their prevalence and recalcitrant in subsurface environment (USEPA 2002; Lee and Batchelor 2002a). Inside the Fe^0 packed media of PRBs, oxidation of Fe^0 can provide electron sources and thermodynamically favorable condition for the sequential dechlorination of CAHs, such as tetrachloroethylene (PCE) and trichloroethylene (TCE), into benign products including ethene and ethane. Currently, hydrogenolysis and reductive-β-elimination are found to be the main pathways involved in the reductive dechlorination of CAHs by Fe^0 (Powell et al. 1998; Arnold and Robert 2000).

To date, based upon the observed first-order dechlorination rate constant (k_{obs}) obtained from laboratory column experiments, the thickness of the Fe^0 packed media in full-scale PRBs for the remediation of CAH-contaminated groundwater is conventionally determined using a first-order equation, which is shown in Eq. (2.1) (Eykholt 1997; EnviroMetal Technologies Inc. 1998; Eykholt et al. 1999). However, the validity of the direct application of the laboratory k_{obs} for the thickness design is in doubt since the Fe^0 reactivity obtained from laboratory Fe^0 packed columns (LFPCs) may not be the same as that in full-scale Fe^0 PRBs. Besides, Eq. (2.1) may not be able to reliably describe longitudinal CAH concentration profiles along the Fe^0 packed media of full-scale PRBs since it does not consider the dispersion of dissolved CAHs.

$$\frac{C}{C_0} = \exp\left(\frac{-k_{obs}\,L_y}{v_y}\right) \tag{2.1}$$

where C is the CAH concentration at a longitudinal distance L_y (mg L^{-1}); C_0 is the initial CAH concentration (mg L^{-1}); k_{obs} is the observed first-order dechlorination rate constant (h^{-1}); L_y is the longitudinal distance along Fe^0 PRBs or LFPCs (m); v_y is the groundwater velocity (m h^{-1}).

The Fe^0 reactivity within full-scale PRBs is prone to be affected by the seasonal variation in groundwater temperature and velocity in aquifers since they can affect the rate of electron transfer between Fe^0 and CAHs, and mass transfer of CAHs from bulk solution to Fe^0 surface (Fogler 1992; Powell et al. 1998; Su and Puls 1999; Lai and Lo 2002). Besides, the differences in the type and amount of mineral precipitates coating on the Fe^0 surface between full-scale Fe^0 PRBs and the corresponding LFPCs can potentially lead to the disparity in the Fe^0 reactivity between these two systems. Such difference in the mineral precipitation is most likely due to the seasonal variation in the groundwater characteristics in aquifers, the difference in microbial activities, and the disparity in the created reducing conditions caused by the difference in the microbial activities between Fe^0 PRBs and the LFPCs (Liang et al. 2000; Furukawa et al. 2002; Gu et al. 2002). Some mineral precipitates such as calcite or aragonite ($CaCO_3$) and siderite ($FeCO_3$) can passivate Fe^0 for CAH dechlorination by masking the redox active sites so that exchange of electrons between Fe^0 and contaminants is hindered (Reardon 1995). Contrarily, precipitation of magnetite (Fe_3O_4), mackinawite (FeS), pyrite (FeS_2) and sulfate-containing green rust [$Fe_4^{II}Fe_2^{III}(OH)_{12}SO_4 \bullet 2H_2O$] can potentially exert beneficial effect on Fe^0 in CAH dechlorination since the ferrous and sulfide ions can act as electron donors and mediators for the dechlorination reactions (Butler and Hayes 1998; Butler and Hayes 2001; Lee and Batchelor 2002a; Lee and Batchelor 2002b).

In addition to the difference in Fe^0 reactivity, another reason leading to the difference in the performance on CAH dechlorination between Fe^0 PRBs and the LFPCs is the heterogeneity of aquifers and the Fe^0 packed PRBs causing the spatial variability of hydraulic conductivity (K) therein. Through the numerical modeling study, Elder et al. (2002) found that a higher variability in K of aquifers could increase the effluent contaminant concentration from PRBs. Furthermore, the possible heterogeneity of the Fe^0 packed media in full-scale PRBs inducing a spatial variability of K may also give a higher effluent CAH concentration than the LFPCs. This is because some of the mass of CAHs can move in advance of the advective front because of the high dispersion of dissolved CAHs (Bedient et al. 1999). Accordingly, the residence time of these CAH for the dechlorination reaction may greatly shorter than that within the LFPCs and thereby may result in a poorer performance of full-scale Fe^0 PRBs on CAH dechlorination (Eykholt et al. 1999).

Since a higher heterogeneity of the packed media can increase the dispersion of dissolved CAHs and therefore the effluent CAH concentration from full-scale Fe^0 PRBs, the assumption of plug-flow approximation (i.e., no dispersion of dissolved CAHs) inside Fe^0 packed media when using first-order equation for the PRB thickness design may overestimate the performance of full-scale Fe^0 PRBs on CAH dechlorination. It is generally accepted that this assumption is only valid for LFPCs but they may not true for full-scale PRBs (Gelhar et al. 1985). For the sake of correcting the possible difference in the performance on CAH dechlorination between full-scale Fe^0 PRBs and the corresponding LFPCs most likely caused by the disparity of the Fe^0 reactivity, the aquifer heterogeneity or the assumption of the plug-flow approximation, a factor of safety (FS) ranging from 1 to 10 was proposed for the designed thickness of full-scale Fe^0 PRBs by Elder et al. (2002). However, this range of the FS was determined from numerical modelings so that its application requires further experimental verification. In this chapter, the longitudinal dispersivities (α_y) of the packed media in a full-scale Fe^0 PRB at Vapokon, Denmark and its corresponding LFPC are first determined to evaluate whether the assumption of the plug-flow approximation or the difference in the α_y is capable of significantly deviating the CAH dechlorination performance of the full-scale Fe^0 PRB from that of the corresponding LFPCs. Afterwards, their experimental performances on CAH dechlorination and geochemical performances are compared so as to identify the underlying culprit(s) leading to the difference in their dechlorination performance. Finally, a FS for the designed Fe^0 PRB thickness is practically determined from the experimental data of the full-scale Fe^0 PRB at Vapokon site and the corresponding LFPCs for the correction of the difference in their CAH dechlorination performance.

2.2 Experimental Section

2.2.1 *Full-scale Fe^0 PRB Installed at Vapokon Site, Denmark*

The full-scale Fe^0 PRB to be studied was installed at Vapokon site, Denmark in September 1999 for the remediation of CAH-contaminated groundwater (Table 2-1). It is a type of funnel-and-gate PRB consisting of two 110 to 130 m long sheet pilings (funnel) and a permeable treatment zone (Fig. 2-1a). Inside the permeable treatment zone, there is a Fe^0 packed medium with dimensions of $14.5 \times 9 \times 0.8$ m (W × D × L) flanked by a 1 m thick layer of gravel pack on both upgradient and downgradient sides (Lai et al. 2006b). Totally, 270 tons of 100% Fe^0 (ETI-CC-1004, *Connelly-GPM Inc.*) were used to create the Fe^0 packed medium, thereby resulting in an initial porosity of 0.60 and ratio of Fe^0 surface area to solution volume (λ_{SA}) of about 7.79 m^2/mL (Lai et al. 2006c). Besides, a drainage system was also installed in the upgradient of the Fe^0 PRB to prevent any uncontaminated groundwater from flowing into the permeable treatment zone. As illustrated in Table 2-2, the Vapokon groundwater varying from a temperature of 8 to 14 ^{0}C contained considerable amount of total alkalinity (TAL), and high concentrations of calcium ion (Ca^{2+}) and SO_4^{2-} (Lo and Lai 2003; Lai et al. 2006c).

Between 1999 and 2002, 15 monitoring wells (3 piezometers in each well), 20 nylon and 6 copper multilevel samplers were installed over the permeable treatment zone (Fig. 2-1a) which resulted in 355 sampling screens vertically distributed between 1.9 and 9.7 meters below ground surface (mbgs). To determine the α_y of the Fe^0 packed medium, a natural gradient tracer experiment was implemented between 1 November 2002 and 8 August 2003 (Lai et al. 2006a; Lai et al. 2006b). Over the first 61 days of the tracer experiment, 127 L of tracer solution containing about 3000 mg/L of lithium (Li^+) as lithium bromide (LiBr) was continuously injected into the aquifer via the three injection wells (Fig. 2-1a) screened from 4.4 to 5.4 mbgs. In consequence, a Li^+ plume with a maximum concentration of about 50 mg/L was created in the downgradient of the injection wells. Groundwater samples were frequently collected from the monitoring wells and multilevel samplers for Li^+ analysis once the Li^+ injection was commenced so as to examine the movement of the Li^+ plume through the Fe^0 packed medium (Lai et al. 2006b). Spatial moments analysis of the Li^+ plume after the tracer experiment has given a groundwater velocity of 99.5 m/yr inside the Fe^0 packed medium (Lai et al. 2006a). To evaluate the CAH dechlorination performance and the geochemical performance of the Fe^0 PRB, another set of groundwater samples were collected from the sampling network in six different occasions between March 2000 and August 2003 for physical, inorganic and contaminant analyses.

2.2.2 *Laboratory Column Experiment*

The α_y of the Fe^0 packed medium in the LFPC, and its performance on the CAH dechlorination and geochemical performance were investigated by running laboratory column experiments with Vapokon groundwater. Three plexiglass^{TM} columns each has a length of 0.6 m and was packed by 100% Fe^0 (the same as the Fe^0 used in the PRB at Vapokon site) were set up resulting in a porosity of 0.56 and λ_{SA} of about 9.28 m^2/mL. Along each LFPC, there were 12 stainless steel sampling

Table 2-1. Concentration of CAHs in Contaminated Vapokon Groundwater and Normalized Dechlorination Rate Constant (k_{SA}) × 10^{-2} Obtained from the Full-scale Fe^0 PRB at Vapokon Site and the Corresponding LFPCs

Parameter		[1]Full-scale Fe^0 PRB at Vapokon site						Laboratory Fe^0 packed columns (LFPCs)		
		March 2000	September 2000	September 2001	September 2002	January 2003	August 2003	[2]Col 1-4	[2]Col 1-5	[3]ETI
PCE	C_0	12.00	20.00	19.00	4.81	4.86	7.30	17.34	9.66	
	k_{SA}	3.90	7.59	10.00	1.34	1.39	5.04	0.49	0.40	nac
	R^2	1.000	0.987	0.989	0.955	0.993	1.000	0.931-0.996	0.962-0.997	
TCE	C_0	5.00	34.00	19.00	23.66	17.47	22.00	29.63	13.25	
	k_{SA}	3.64	8.01	10.09	2.36	1.33	4.36	1.02	0.85	nac
	R^2	0.999	0.984	0.940	0.992	0.998	1.000	0.921-0.990	0.966-0.989	
cis-DCE	C_0	5.50	16.00	20.00	29.54	18.13	38.00	13.17	nd	0.45-0.98
	k_{SA}	0.18	0.17	0.21	0.17	0.14	0.28	nc	nc	0.63
	R^2	0.930	0.861	0.569	0.835	0.851	0.632			0.995
VC	C_0	0.55	0.10	0.71			2.40			0.02
	k_{SA}	0.95	0.60	0.34	na	na	0.64	na	na	1.90
	R^2	0.824	0.741	0.766			0.933			0.987
1,1-DCE	C_0	0.04	0.31	0.32			0.61	nd	nd	0.04-0.07
	k_{SA}	0.43	2.28	5.98	na	na	1.66	nc	nc	1.13
	R^2	0.850	0.965	0.915			0.975			0.976
1,1,1-TCA	C_0	4.40	37.00	12.00		15.32	31.00	33.20	19.95	
	k_{SA}	1.67	6.89	6.30	nac	1.12	7.43	13.40	18.50	nac
	R^2	1.000	0.996	0.987		0.998	0.999	0.993-1.000	0.992-1.000	
1,1-DCA	C_0	0.15	0.58	0.52	0.70	0.77	0.66			0.84
	k_{SA}	0.85	2.62	nc	nc	0.19	3.36	na	na	0.18
	R^2	0.975	0.951			0.524	0.786			0.996
TCM	C_0	0.64	0.77	0.69	nd	nd	0.26	nd	nd	0.08
	k_{SA}	0.91	2.55	4.67	nc	nc	1.71	nc	nc	0.99
	R^2	0.993	0.944	0.645			0.899			0.997

Note: na represents not available. nc refers to not calculable. nd represents not detectable and nac means not applicable. The unit of k_{SA} is in mL m^{-2} h^{-1}. C_0 refers to the initial concentration of CAHs with a unit of mg/L. R^2 means coefficient of determination.

[1]The k_{SA} for every CAH was calculated from their maximum concentration profiles along the Fe^0 PRB.

[2]The k_{SA} shown in the box is the average of the k_{SA} obtained from the steady-state concentration profiles.

[3]The laboratory Fe^0 packed column (LFPC) operated by EnviroMetal Technologies Inc. (1998) at a groundwater velocity of 343 m/yr and temperature of 23 °C possesses a porosity of 0.59 and ratio of Fe^0 surface area to solution volume (λ_{SA}) of 7.0 m^2/mL. The k_{SA} were corrected to the conditions of a groundwater velocity of 100 m/yr and temperature of 10 °C by dividing with a factor of 4 (Powell et al. 1998; Su et al. 1999; Lai and Lo 2002).

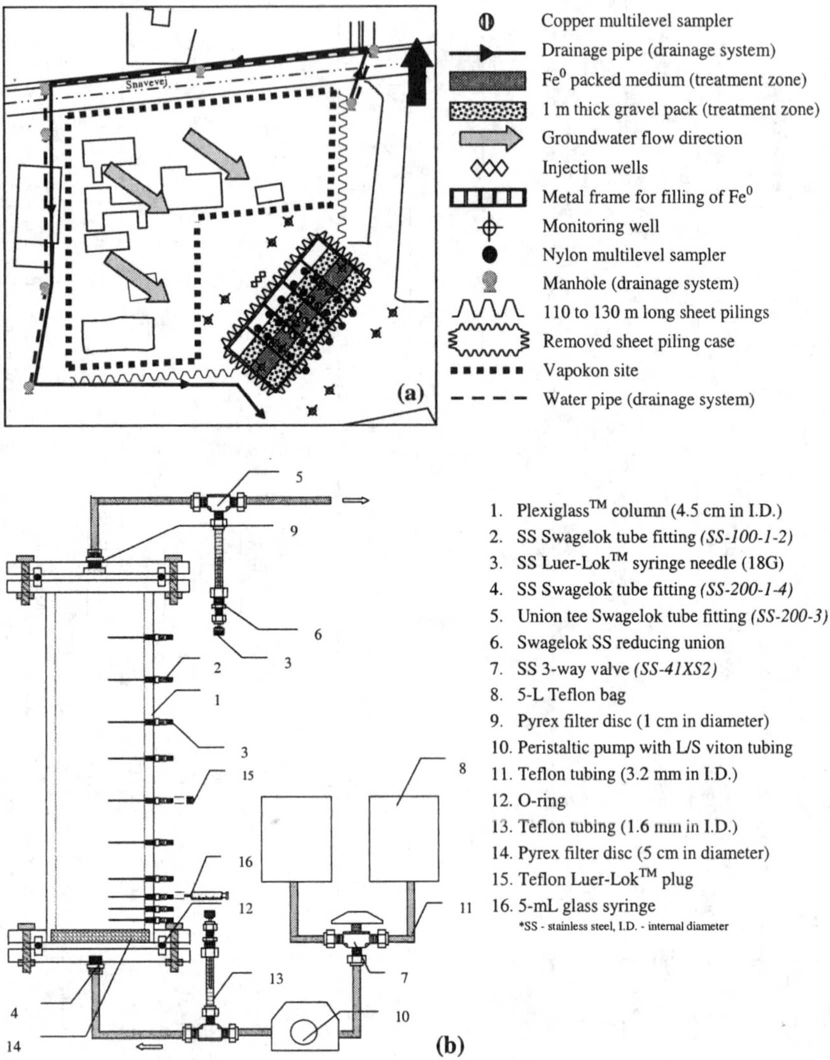

Fig. 2-1. (a) The schematic diagram of the funnel-and-gate Fe^0 PRB and drainage system at Vapokon site, and the distributions of the monitoring wells, injection wells, nylon and copper multilevel samplers over the Fe^0 PRB (adapted from Lai et al. 2006b), as well as (b) the schematic diagram of the experimental setup for laboratory column experiments (adapted from Lai 2004).

Table 2-2. Variation of the Physical and Inorganic Characteristics of Vapokon Groundwater along the Full-scale Fe^0 PRB and the Corresponding LFPCs

Parameter	[1]Full-scale Fe^0 PRB at Vapokon site					Laboratory Fe^0 packed columns (LFPCs)		
	March 2000	September 2000	September 2001	September 2002	January 2003	[2]Col 1-4	[2]Col 1-5	[2,3]ETI
Physical parameter								
	7.03	**7.38**	**7.33**		**7.16**	**8.06**	**8.03**	**7.30**
pH	8.98	9.69	9.53	na	9.44			
	9.50	10.56	10.62		10.21	9.54	8.40	7.80
Inorganic parameter								
	510	**646**		**338**	**299**	**299**	**151**	**350-383**
Ca^{2+} (as $CaCO_3$)	160	133	na	136	95.9			
	146	81.0		53.5	48.2	165	75.0	123-128
	64.9	**52.1**				**46.2**	**24.3**	**50.0-54.2**
Mg^{2+} (as $CaCO_3$)	37.2	41.2	na	na	na			
	21.4	27.7				42.7	17.6	41.7
	372	**323**			**248**	**270**	134	**316-324**
TAL (as $CaCO_3$)	77.7	99.2	na	na	125			
	19.3	30.7			35.4	nd	nd	13.0-14.0
	13.2	**13.2**		**0.18**	**0.15**	**1.05**	**0.51**	**0.26-4.3**
Fe_T	8.12	9.83	na	0.25	0.13			
	0.41	0.60		0.13	0.38	0.77	0.45	0.11-0.15
	189	**145**		**27.2**				**88.0-89.0**
SO_4^{2-}	72.8	47.6	na	3.70	na	na	na	
	108	39.6		2.20				85.0-86.0
	50.3	**43.6**		**46.1**	**43.9**	**44.5**	**22.92**	**60.0-65.0**
Cl^-	43.1	55.4	na	65.2	38.9			
	42.0	47.8		38.0	33.9	67.5	37.36	87.0

Note: The unit of the inorganic parameter is in mg/L. na represents not available. The bold figures represent the original characteristics of Vapokon groundwater.
[1]The figures shown in the box from the top to bottom represent the average value of the parameters collected from the upgradient location, middle and downgradient parts of the Fe^0 PRB, respectively.
[2]The upper and lower figures shown in the box were obtained from the groundwater samples collected from the influent and effluent ends of the LFPCs, respectively.
[3](EnviroMetal Technologies Inc. 1998)

ports each was normally capped by a Teflon Luer-Lok™ plug. However, during the sampling time, the sampling ports were connected by 5-mL glass syringes for the collection of about 3 mL of groundwater samples (Fig. 2-1b). In order to simulate the operating conditions of the full-scale Fe^0 PRB, all the column experiments were conducted at 10 ^{0}C in which the feed solutions stored into 5-L collapsible Teflon bags was fed into the LFPCs at an upward groundwater velocity between 99.7 and 101.3 m/yr to mimic the field conditions at Vapokon site.

Laboratory tracer experiment was implemented to evaluate the α_y of the LFPC in which the Vapokon groundwater spiked with about 212 mg/L of tritiated water (Tritium, *Amersham Pharmacia Biotech UK Ltd.*) and 78 mg/L of Li^+ as LiBr was fed into the LFPC fully saturated by Millipore water (Bitsch and Jensen 1990). Groundwater sampling was conducted hourly from the effluent sampling port for the assays of tritium activity and Li^+ concentration. Regarding the dechlorination performance and geochemical performance of the LFPC, contaminated groundwater collected from Vapokon site with (Col 1-5) and without (Col 1-4) half dilution by Millipore water were used as feed solutions. The purpose of running the LFPC with half dilution of the Vapokon groundwater is to take account of the influences from the variation of groundwater characteristics on the dechlorination performance of the LFPC (Tables 2-1 and 2-2). Before attaining steady-state CAH concentration profiles, groundwater samples were collected from the 12 sampling ports every 5 to 10 pore volumes for physical and chemical analyses.

2.2.3 Groundwater Sample Measurements

Measurement of the CAH concentration in the collected groundwater samples was implemented using the method of liquid-liquid extraction with pentane and gas chromatograph equipped with a Ni^{63} electron capture detector (standard method code 6232, APHA/AWWA/WEF 1998). Atomic absorption spectrometer (standard method code 3111, APHA/AWWA/WEF 1998) was applied for the measurement of Ca^{2+}, magnesium ion (Mg^{2+}), Li^+, and ferrous and ferric ions (Fe_T). However, the Li^+ concentration in the groundwater samples obtained from the natural gradient tracer experiment was measured by a flame photometer (PFP 7, *Jenway*) (standard method code 3500-Li, APHA/AWWA/WEF 1998). Determination of the concentrations of chloride ion (Cl^-) and SO_4^{2-} were performed using ion chromatograph (standard method code 4110, APHA/AWWA/WEF 1998). Titration method was utilized for TAL measurement (standard method code 2320, APHA/AWWA/WEF 1998). Concerning the concentration of tritiated water, aliquot of 1 mL of the groundwater sample was first added into 15 mL scintillation fluid ('HiSafe'3 Scintillation Cocktail, *Wallac OptiPhase*) and then assayed for tritium activity by a liquid scintillation analyzer (2000 TRI-CARB, *United Technologies Packard*) (standard method code 7500-^{3}H, APHA/AWWA/WEF 1998).

2.3 Data Analysis

2.3.1 Determination of Longitudinal Dispersivity

Tritiated water, which is an isotopic species of water molecule and viewed as an ideal tracer, was used as a main tracer in the laboratory tracer experiment. The addition of Li^+ during the laboratory tracer experiment, however, is to evaluate whether Li^+ is also conservative with respect to Fe^0. For the natural gradient tracer experiment, Li^+ was selected as a main tracer because tritiated water is radioactive in nature and its injection may disturb the natural tritium distribution in subsurface. After implementing both tracer experiments, the α_y of the Fe^0 packed medium in the LFPC and the Fe^0 PRB at Vapokon site were then determined from the obtained tritiated water and Li^+ breakthrough curves (BTCs), respectively. With using CXTFIT program, the experimental data of the tritiated water and Li^+ BTCs were fitted into the mathematical steady-state solutions of the deterministic equilibrium 1-D advection-dispersion equation (ADE), which is illustrated in Eq. (2.2), thereby resulting in the α_y of the Fe^0 packed media and the retardation factors of the tracers (van Genuchten 1981; Thorbjarnarson and Mackay 1994; Toride et al. 1999).

$$R\frac{\partial c}{\partial t} = D_y\frac{\partial^2 c}{\partial y^2} - v_y\frac{\partial c}{\partial y} \tag{2.2}$$

where c is the concentration of tritiated water or Li^+ (mg L^{-1}); t is the time (h^{-1}); D_y is the longitudinal dispersion coefficient (m^2 h^{-1}); R is the retardation factor; y is the distance in longitudinal direction (m).

2.3.2 Calculation of the Observed First-order Dechlorination Rate Constant

Gillham and O'Hannesin (1994) found that the dechlorination reaction of CAHs by Fe^0 was pseudo first-order with respect to CAH concentration. Therefore, the longitudinal CAH concentration profiles along laboratory Fe^0 packed columns and full-scale Fe^0 reactive barriers theoretically can be described by Eq. (2.3) (van Genuchten 1981; Eykholt et al. 1999). However, if the dispersion of dissolved CAHs within Fe^0 packed media is minimal, the flow through Fe^0 packed columns and Fe^0 reactive barriers occurs as a plug so that the outcomes from Eqs. (2.1) and (2.3) become similar (Eykholt 1997). As a result, the first-order equation [Eq. (2.1)] is conventionally used for the design and evaluation of the dechlorination performance of Fe^0 reactive barriers and laboratory Fe^0 packed columns, and in this study, it was also applied for determining the CAH k_{obs} from the full-scale Fe^0 PRB at Vapokon site and the corresponding LFPCs.

$$\frac{C}{C_0} = \exp\left(\frac{\left(v_y - v_y\sqrt{1+\frac{4k_{obs}D_y}{v_y^{\,2}}}\right)L_y}{2D_y}\right) \tag{2.3}$$

Through the pathway of hydrogenolysis, major chlorinated organics such as PCE, TCE, 1,1,1-trichloroethane (1,1,1-TCA) and chloroform (TCM) are sequentially dechlorinated into their corresponding chlorinated daughter products including cis-dichloroethylene (cis-DCE), 1,1-dichloroethylene (1,1-DCE), vinyl chloride (VC),

1,1-dichloroethane (1,1-DCA) and dichloromethane (DCM) (Vogel and McCarty 1987; Helland et al. 1995; Su and Puls 1999; Arnold and Roberts 2000). The following models (models 1, 2, and 3), which can take account of the concurrent generation and dechlorination of CAHs, were applied for the determination of the k_{obs} of different CAHs. The first-order equations for the models 1, 2 and 3 are shown in Eqs. (2.4) to (2.8), Eqs. (2.9) to (2.10) and Eq. (2.11), respectively. Since DCM has been proven to be not treatable by Fe^0, the first-order equation for the computation of DCM k_{obs} was absent (Powell et al. 1998). To calculate the CAH k_{obs} from the LFPCs, Eqs. (2.4) to (2.11) were directly used in Scientist®, which can perform integration and thereby calculated the k_{obs} by fitting the integrated solutions of the Eqs. (2.4) to (2.11) to the relevant experimental data (MicroMath 1995). Determination of the k_{obs} from the full-scale Fe^0 PRB was also in similar fashion but f_{PCE1} (0.983), f_{PCE2} (0.017), f_{TCE1} (0.014), f_{TCE2} (0.003), $f_{cis\text{-}DCE}$ (0.048) and $f_{1,1,1\text{-}TCA}$ (0.441) previously found by EnviroMetal Technologies Inc. (1998) were utilized for the fitting (Lai et al. 2006c).

Model 1

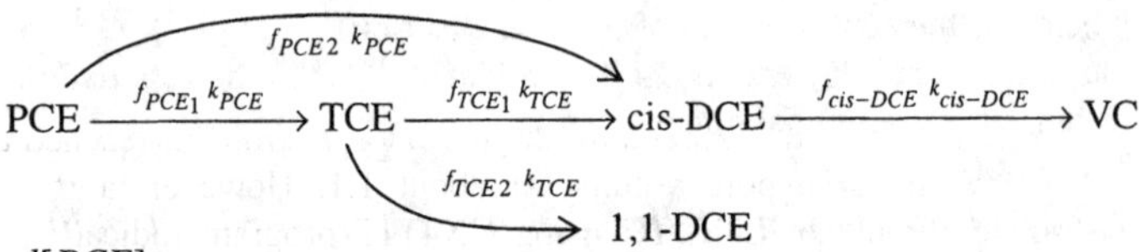

$$\frac{d[PCE]}{dt} = -k_{PCE}[PCE] \quad (2.4)$$

$$\frac{d[TCE]}{dt} = f_{PCE1}k_{PCE}[PCE] - k_{TCE}[TCE] \quad (2.5)$$

$$\frac{d[cis-DCE]}{dt} = f_{PCE2}k_{PCE}[PCE] + f_{TCE1}k_{TCE}[TCE] - k_{cis\text{-}DCE}[cis-DCE] \quad (2.6)$$

$$\frac{d[VC]}{dt} = f_{cis\text{-}DCE}k_{cis\text{-}DCE}[cis-DCE] - k_{VC}[VC] \quad (2.7)$$

$$\frac{d[1,1-DCE]}{dt} = f_{TCE2}k_{TCE}[TCE] - k_{1,1\text{-}DCE}[1,1-DCE] \quad (2.8)$$

Model 2

$$\text{1,1,1-TCA} \xrightarrow{f_{1,1,1\text{-}TCA}\,k_{1,1,1\text{-}TCA}} \text{1,1-DCA}$$

$$\frac{d[1,1,1-TCA]}{dt} = -k_{1,1,1\text{-}TCA}[1,1,1-TCA] \quad (2.9)$$

$$\frac{d[1,1-DCA]}{dt} = f_{1,1,1\text{-}TCA}k_{1,1,1\text{-}TCA}[1,1,1-TCA] - k_{1,1\text{-}DCA}[1,1-DCA] \quad (2.10)$$

Model 3

$$\text{TCM} \xrightarrow{f_{TCM}\,k_{TCM}} \text{DCM}$$

$$\frac{d[TCM]}{dt} = -k_{TCM}[TCM] \quad (2.11)$$

where f is the molar fraction; k is the CAH k_{obs} (h^{-1}); and [ ] is the CAH concentration (mg L^{-1}).

For comparison purpose, k_{obs} normalized to the λ_{SA} (i.e., k_{SA}), which can be obtained from Eq. (2.12), is reported in the latter section since k_{obs} was proven directly proportional to the λ_{SA} (Gillham and O'Hannesin 1994).

$$k_{SA} = \frac{k_{obs}}{\lambda_{SA}} \tag{2.12}$$

2.4 Results and Discussion

2.4.1 Longitudinal Dispersivity of the Fe^0 Packed Media

The tritiated water and Li^+ BTCs for the LFPC are illustrated in Fig. 2-2. As can be seen, tritiated water in the LFPC reached the effluent end of the column at a breakthrough time equivalent to a pore volume of one, thereby corroborating its non-reactivity with respect to Fe^0. Fitting of the tritiated water BTC with the mathematical steady-state solution of Eq. (2.2) gave a α_y of 1.23×10^{-3} m for the Fe^0 packed medium of the LFPC. In comparison to the tritiated water BTC, Li^+ was observed to be slightly retarded inside the LFPC in which it reached the effluent end of the column at a pore volume of about 1.1. However, a retardation of 1.12 determined from the Li^+ BTC using CXTFIT program indicated that Li^+ is also conservative with respect to Fe^0 so that it was appropriate to select Li^+ as a conservative tracer in the natural gradient tracer experiment (Lai et al. 2006b). Fig. 2-3 shows Li^+ BTCs obtained from the Fe^0 packed medium of the full-scale PRB at Vapokon site. In analogy to the results reported by Thorbjarnarson and Mackay (1994), spatial variability of the α_y was also observed for the full-scale Fe^0 PRB in which the α_y varied in a range of 0.061 m to 0.513 m with an average value of 0.275 m.

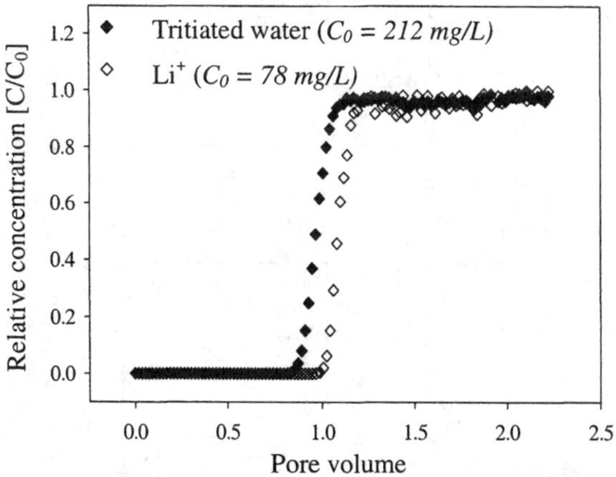

Fig. 2-2. Tritiated water and Li^+ breakthrough curves for the LFPC obtained from the laboratory tracer experiment (adapted from Lai et al. 2006b)

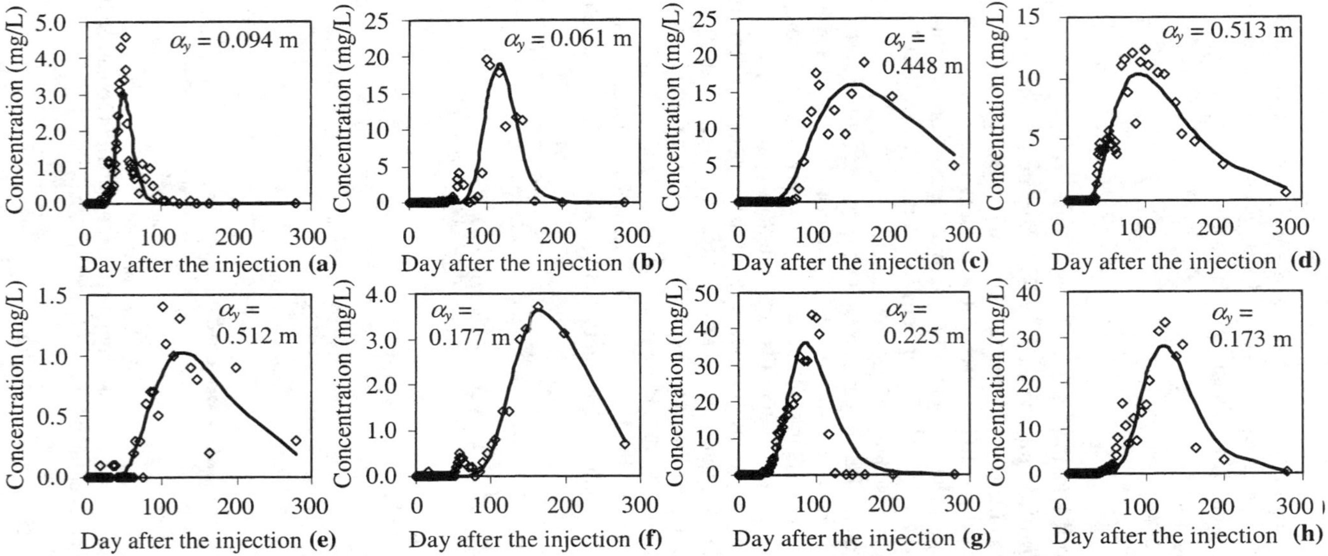

Fig. 2-3. Observed Li^+ breakthrough curves (open circles) obtained from the sampling points within the Fe^0 packed medium of the full-scale PRB at Vapokon site, and the corresponding fitted curves (solid curves) and longitudinal dispersivities (α_y) determined by CXTFIT program (adapted from Lai et al. 2006a)

Comparing the calculated α_y of the LFPC and the full-scale Fe^0 PRB, the solute dispersion in the full-scale Fe^0 PRB became more significant. Such an increase in the importance of the solute dispersion is believed to be mainly due to the scale dependency of the solute dispersion (Freyberg 1986; Garabedian et al. 1991). A review of the reported value of the field-scale α_y from 55 field sites by Gelhar et al. (1985) illustrates a noticeable increase in the α_y with transport scales. Therefore, it is generally agreed that field-measured α_y should be many orders of magnitude higher than the α_y determined from laboratory. As the scale of the measurement increases, the effect of the heterogeneity of porous media on the spread of solute becomes more significant, thereby resulting in the increase in a α_y with the transport scale (Moltyaner and Killey 1988; Garabedian et al. 1991; Bedient et al. 1999).

Another possible reason leading to a higher α_y in the full-scale Fe^0 PRB than the LFPC is likely ascribed to a higher spatial variability of the K of the full-scale Fe^0 PRB, which is due to a higher heterogeneity than the LFPC. In preparing the LFPC, the plexiglass™ column was first divided into 12 equal sections. Then each section was uniformly packed by the same amount of Fe^0 in order to ensure the homogeneity of the packed medium throughout the whole LFPC. However, for the full-scale Fe^0 PRB at Vapokon site, because of the compaction exerted by the weight of the upper portion of the Fe^0 packed medium, the middle and lower parts of the Fe^0 packed medium were initially less permeable than the upper part (Lai et al. 2006b). Continuous decrease in the K of the packed medium of the full-scale Fe^0 PRB probably caused by the formation of mineral precipitates on the Fe^0 surface further aggravated the spatial variability of the K therein. Measurement of the K of the Fe^0 packed medium at Vapokon site by slug test in March 2000 showed that there was two orders of magnitude variation (0.33 to 57.5 m/d) in the initial K with relative standard deviation (RSD) of 91.5%. In August 2003, there was three orders of magnitude variation (0.02 to 54.4 m/d) in the K with RSD of about 151.0%. Higher spatial variability of the K for the packed medium of the full-scale Fe^0 PRB led to a higher groundwater velocity variation inside the Fe^0 packed medium, thereby resulting in a higher α_y in the full-scale Fe^0 PRB than the LFPC.

2.4.2 Influence of the Longitudinal Dispersivity on the CAH Concentration along Fe^0 Packed Media

To evaluate whether the difference in the α_y between the full-scale Fe^0 PRB at Vapokon site and the corresponding LFPC can deviate the CAH dechlorination performance of the Fe^0 PRB from the LFPC, the influence of the α_y on the CAH concentration along Fe^0 packed media was investigated. Three situations were considered: (1) Fe^0 packed media without solute dispersion (i.e., $\alpha_y = 0$); (2) Fe^0 packed media with the α_y of 1.23×10^{-3} m; and (3) Fe^0 packed media with the α_y of 0.275 m. The longitudinal CAH concentration profile in the first situation was calculated using Eq. (2.1); whereas Eq. (2.3) was used to determine the concentration profiles at the α_y of 1.23×10^{-3} m and 0.275 m. As illustrated in Fig. 2-4, the longitudinal CAH concentration profiles along the Fe^0 packed medium without solute dispersion was nearly identical to the profiles obtained from the Fe^0 packed medium with the α_y of 1.23×10^{-3} m. This similarity in the CAH concentration

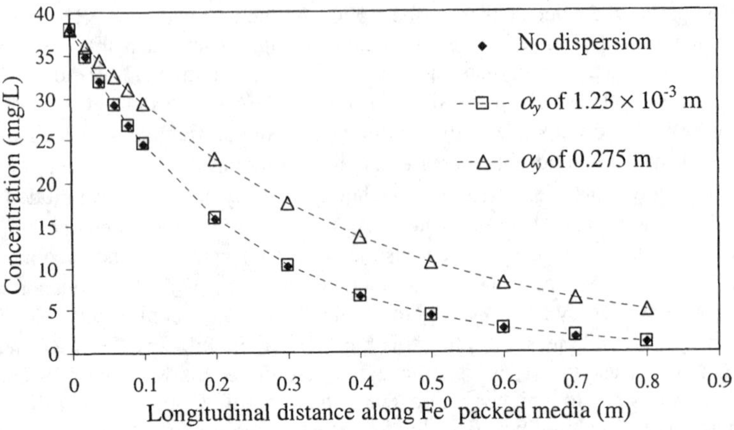

Fig. 2-4. Longitudinal CAH concentration profiles along the Fe^0 packed media with and without taking account of solute dispersion. All the CAH concentration profiles were determined based upon a k_{obs} of 0.05 hr^{-1}, initial CAH concentration of 38 mg/L and groundwater velocity of 99.5 m/yr.

profiles showed that the assumption of the plug-flow approximation is valid for the LFPC, and the first-order equation [Eq. (2.1)] can reliably describe longitudinal CAH concentration profiles along the LFPC.

On the contrary, there was a noticeable increase in the CAH concentration from the Fe^0 packed medium with the α_y of 1.23×10^{-3} m to that with the α_y of 0.275 m. In the former situation, the effluent CAH concentration (i.e., the concentration at a longitudinal distance of 0.8 m) was about 1.14 mg/L; while it was 4.84 mg/L in the situation with the α_y of 0.275 m. This increase in the CAH effluent concentration, as described, is mainly due to the increase in the mass flux of some CAHs with increasing α_y. The solute dispersion or groundwater velocity variation created by the heterogeneity of Fe^0 packed media inducing the spatial variability of the K can cause some of CAHs moving in advance of an advective front (Bedient et al. 1999). As a consequence, the residence time for the CAHs staying within the Fe^0 packed media for the dechlorination reaction decreases with increasing α_y, thereby resulting in a higher CAH effluent concentration at a higher α_y. Based upon the longitudinal CAH concentration profiles shown in Fig. 2-4, it is believed that the difference in the α_y of the Fe^0 packed media between the full-scale PRB at Vapokon site and the corresponding LFPC could adversely deviate the CAH dechlorination performance of the full-scale Fe^0 PRB from that of the LFPC. In other words, design of the thickness of the Fe^0 packed medium in the full-scale PRB at Vapokon site using first-order equation [Eq. (2.1)] based upon the CAH k_{obs} obtained from the corresponding LFPCs could overestimate the performance of the full-scale Fe^0 PRB on the CAH dechlorination. Thus a FS should be applied when designing the Fe^0 PRB thickness using Eq. (2.1) to correct the disparity in the dechlorination performance caused by the difference in the α_y.

2.4.3 The Performance on CAH Dechlorination

Figs. 2-5a to 2-5c illustrate the longitudinal concentration profiles of 1,1,1-TCA, PCE and TCE along the Fe^0 packed media of the full-scale PRB at Vapokon site and the corresponding LFPCs (i.e., Col 1-4 and Col 1-5). The concentration profiles along the Fe^0 PRB indicated the maximum contaminant concentrations being measured in each sampling occasion and they were considered to be the most representative of the concentration distribution through the Fe^0 packed medium of full-scale reactive barriers (O'Hannesin and Gillham 1998). On the other hand, the concentration profiles in Col 1-4 and Col 1-5 are the averages of the corresponding steady-state concentration profiles (i.e., the concentration profiles obtained after the column experiments reach a steady-state). As can be seen, the 1,1,1-TCA, PCE and TCE concentration profiles in Col 1-4 and Col 1-5 were nearly identical. Thus it is believed that the variation in the groundwater composition, such as the level of hardness and TAL, the concentration of Fe_T, SO_4^{2-}, Cl^-, and CAHs, did not exert significant influence to the LFPCs on CAH dechlorination.

Comparison of the performance on the CAH dechlorination between the full-scale Fe^0 PRB and the LFPCs indicated that the latter showed much higher performance on 1,1,1-TCA dechlorination. In Col 1-4 and Col 1-5, complete dechlorination of 1,1,1-TCA were achieved at a distance of about 0.05 m apart from the influent of the column, which is equivalent to a residence time of about 4.3 h. On the other hand, it was between a longitudinal distance of 0.4 and 0.58 m in the full-scale Fe^0 PRB corresponding to a residence time between 35.2 h and 51.1 h (Fig. 2-5a). In terms of reaction rate, the 1,1,1-TCA k_{SA} obtained from the Fe^0 PRB was in a range of 1.12×10^{-2} to 7.43×10^{-2} mL m^{-2} h^{-1}, while those determined from the corresponding LFPCs, on average, were about 3 times higher in which the 1,1,1-TCA k_{SA} obtained from Col 1-4 and Col 1-5 were 13.40×10^{-2} and 18.50×10^{-2} mL m^{-2} h^{-1}, respectively (Table 2-1). In stark contrast to 1,1,1-TCA, the Fe^0 PRB showed a better performance on PCE and TCE dechlorination. As shown in Figs. 2-5b and 2-5c, the PCE and TCE concentration profiles in both Col 1-4 and Col 1-5 generally possessed higher relative concentrations than those in the Fe^0 PRB. The PCE k_{SA} obtained from the Fe^0 PRB was in a range of 1.34×10^{-2} to 10.00×10^{-2} mL m^{-2} h^{-1}, which on average was about an order of magnitude higher than those determined from the LFPCs. In a similar manner, the TCE k_{SA} obtained from the Fe^0 PRB ranging from 1.33×10^{-2} to 10.09×10^{-2} mL m^{-2} h^{-1} were also approximately 5-fold higher than the TCE k_{SA} computed from Col 1-4 and Col 1-5.

Because of low concentrations in the feed solutions or the limitation from the analytical methods, the change of the concentrations of cis-DCE, VC, 1,1-DCE, 1,1-DCA and TCM along Col 1-4 and Col 1-5 was not detectable or available, thereby resulting in lacking of their k_{SA}. For comparison purpose, the k_{SA} results, and the relevant physical and inorganic information from the laboratory column experiment conducted by EnviroMetal Technologies Inc. (1998) were also included in Tables 2-1 and 2-2, respectively, since the Fe^0 and the feed solution being applied were also the same as those used in Col 1-4 and Col 1-5. Similar to the dechlorination of 1,1,1-TCA, the Fe^0 PRB also showed a poorer performance on cis-DCE and VC dechlorination in which the k_{SA} of cis-DCE and VC reported by EnviroMetal

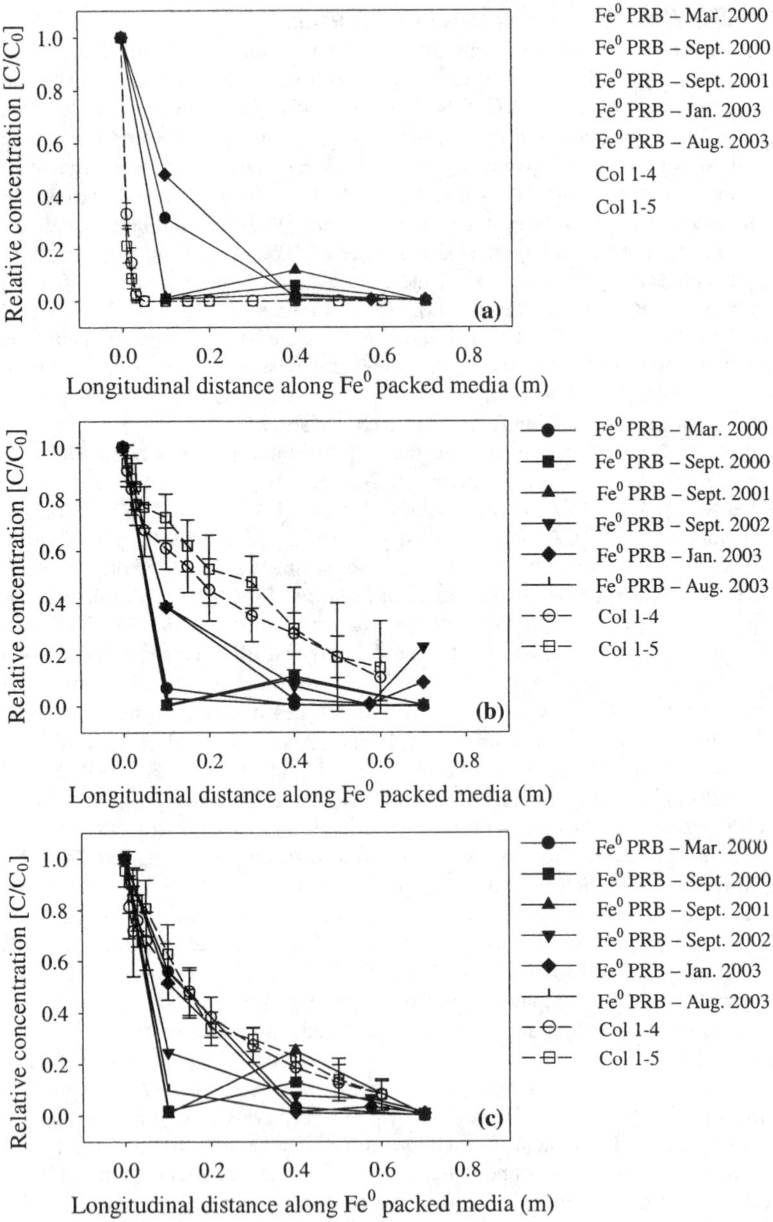

Fig. 2-5. The longitudinal concentration profiles of (a) 1,1,1-TCA, (b) PCE and (c) TCE along the Fe^0 packed media of the full-scale PRB, and the LFPCs (Col 1-4 and Col 1-5)

Technologies Inc. (1998) were about 3-fold higher than those determined from the Fe^0 PRB. Consistent with the dechlorination of PCE and TCE, generally, the Fe^0 PRB indicated a better performance on the dechlorination of 1,1-DCE, 1,1-DCA and TCM in which their k_{SA} were approximately a factor of 2 to 10 higher than those obtained from the laboratory column experiment.

Among various possible reasons, such as the seasonal variation in the groundwater temperature and velocity, the difference in the mineral precipitation and the microbial activities, etc. leading to a poorer performance of the Fe^0 PRB than the LFPCs on 1,1,1-TCA, cis-DCE and VC dechlorination, the influence of the α_y disparity caused by the spatial variability of the K of the Fe^0 packed full-scale PRB is believed to be one of the underlying culprits. To verify this hypothesis, the 1,1,1-TCA concentration profiles in Col 1-4 and Col 1-5 were recalculated for the Fe^0 packed media with a α_y of 0.275 m using Eq. (2.3). As can be seen in Fig. 2-6, with considering the α_y of the full-scale Fe^0 PRB at Vapokon site, the 1,1,1-TCA concentration profiles in Col 1-4 and Col 1-5 moved upward, thereby increasing the proximity to the 1,1,1-TCA concentration profiles in the Fe^0 PRB. Computation of the k_{SA} from the recalculated 1,1,1-TCA concentration profiles in Col 1-4 and Col 1-5 by Eq. (2.9) gave a value of 2.25×10^{-2} and 2.68×10^{-2} mL m^{-2} h^{-1}, respectively, which were in close proximity to the 1,1,1-TCA k_{SA} obtained from the Fe^0 PRB (i.e., 1.12×10^{-2} to 7.43×10^{-2} mL m^{-2} h^{-1}). The good agreement of the 1,1,1-TCA concentration profiles or k_{SA} between the Fe^0 PRB and the LFPCs after taking the α_y of 0.275 m into account suggested that the disparity of the α_y between the Fe^0 PRB and the LFPCs should be one of the main reasons for the poorer performance of the

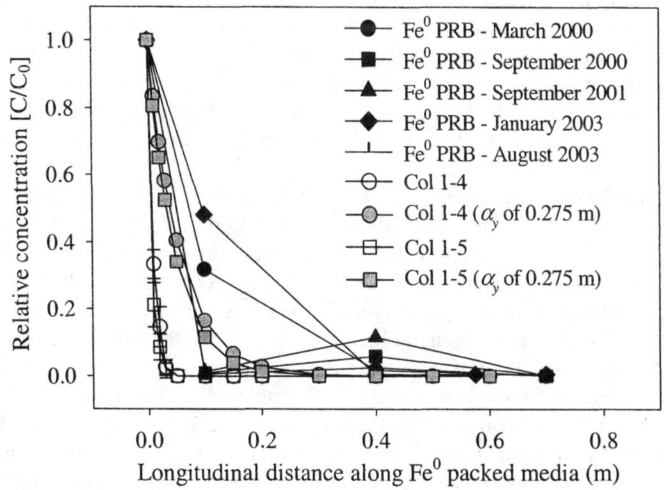

Fig. 2-6. Recalculation of the 1,1,1-TCA concentration profiles in Col 1-4 and Col 1-5 for the Fe^0 packed media with a α_y of 0.275 m and comparison of the recalculated 1,1,1-TCA concentration profiles with the concentration profiles in full-scale Fe^0 PRB at Vapokon site

Fe^0 PRB on 1,1,1-TCA dechlorination. Since the magnitude of the differences of cis-DCE and VC k_{SA} between the Fe^0 PRB and the LFPCs were similar to that of 1,1,1-TCA k_{SA}, it is expected that the difference in the performance on cis-DCE and VC dechlorination should also ascribe to the impact from the dispersion of dissolved cis-DCE and VC in the Fe^0 PRB.

In addition to the dispersion of the dissolved 1,1,1-TCA, cis-DCE and VC in the full-scale Fe^0 PRB, the higher α_y of the packed media in the Fe^0 PRB than the LFPC could also lead to a higher dispersion of the dissolved PCE, TCE, 1,1-DCE, 1,1-DCA and TCM inside the Fe^0 PRB. Therefore, the disparity in the α_y between the Fe^0 PRB and the LFPCs is believed to cause an adverse impact on the dechlorination of these CAHs by the Fe^0 PRB. The observation of a better performance of the Fe^0 PRB than the LFPCs on PCE, TCE, 1,1-DCE, 1,1-DCA and TCM dechlorination, however, indicated that there was a positive effect enhancing the performance of the Fe^0 PRB on their dechlorination, which may be pertained to the type of mineral precipitates coated on the Fe^0 surface. Within both Fe^0 PRB and LFPCs, $CaCO_3$ was the dominant mineral being precipitated on the Fe^0 surface since there was significant reduction in the amounts of Ca^{2+} and TAL in the groundwater. On the other hand, the precipitation of magnesite ($MgCO_3$) and dolomite [$CaMg(CO_3)_2$] were expected to be minimal because of the insignificant decrease in Mg^{2+} concentration in the groundwater across the Fe^0 PRB and LFPCs (Table 2-2). Within the Fe^0 PRB, however, the amount of Ca^{2+} being removed from the groundwater was generally higher than the quantity of TAL being precipitated out. Since $CaCO_3$ (13~14 mg/L) has a lower solubility than siderite (720 mg/L), it is highly expected that most of the removed TAL was precipitated as $CaCO_3$ and no or only trace amount of siderite was formed within the Fe^0 PRB (Dean 1985). In addition, there was a high likelihood of precipitation of pyrite, mackinawite, sulfate-containing green rust and magnetite within the Fe^0 PRB. This is because there were the reduction in the concentration of SO_4^{2-} and increase in pH (7.03 to 10.62) across the Fe^0 PRB, and also the decrease in or insignificant change of the Fe_T concentration inside the Fe^0 PRB even though there was a release of Fe_T from the Fe^0 oxidation (Yabusaki et al. 2001; Gu et al. 1999; Phillips et al. 2000).

Along the LFPCs, the amount of Ca^{2+} being removed from the groundwater was about 58 to 136 mg/L as $CaCO_3$ smaller than the quantity of TAL being precipitated out (Table 2-2). Therefore, it is believed that certain amount of the removed TAL was precipitated as $CaCO_3$; while some of it might precipitate as siderite since the insignificant change of the Fe_T concentration was also observed across the LFPCs. Magnetite might also be the mineral being formed within the LFPCs because of the alkaline condition and the availability of Fe_T but the dose might be lower than the amount of magnetite formed within the Fe^0 PRB since the pH within the LFPCs were comparatively low and also some of the Fe_T within the LFPCs, as described, might precipitate as siderite rather than magnetite (Ritter et al. 2002). Another important point is that precipitation of sulfur-bearing minerals including pyrite, mackinawite and sulfate-containing green rust within the LFPCs was expected to be minimal. Probably owing to the lack of the biologically mediated reduction of SO_4^{2-} to sulfide, there was only insignificant change of SO_4^{2-} concentration along the LFPC (Table 2-

2) and thereby limited the occurrence of the reaction between sulfide ion and the available iron species (Butler and Hayes 1998).

It has been proven from various batch experiments that magnetite, pyrite, mackinawite and sulfate-containing green rust can transform chlorinated alkenes, such as PCE, TCE and 1,1-DCE, and chlorinated alkanes into less chlorinated or non-chlorinated products (Kriegman-King and Reinhard 1994; Butler and Hayes 1998; Butler and Hayes 1999; Butler and Hayes 2001; Lee and Batchelor 2002a; Ferrey et al. 2004). This is because the ferrous and sulfide ions in these minerals can act as electron donors and mediators for the reductive dechlorination reactions (Butler and Hayes 1998; Lee and Batchelor 2002a). Surprisingly, treatment of granular iron with sulfide ion was proven to be able to increase the rate of CAH dechlorination since iron sulfide was found to be capable of providing more reactive sites per unit surface area than the iron metal for the dechlorination reaction (Lipczynska-Kochany et al. 1994; Sivavec and Horney 1997; Hassan 2000). As a result, the high possibility of the precipitation of comparatively high amount of magnetite, and the formation of pyrite, mackinawite and sulfate-containing green rust inside the Fe^0 packed medium of the full-scale PRB not only nullified the impact from the dispersion inside the Fe^0 PRB on PCE, TCE, 1,1-DCE, 1,1-DCA and TCM dechlorination but also leads to the higher performance than the corresponding LFPCs on the dechlorination of these CAHs.

2.4.4 A Factor of Safety for the Designed Fe^0 PRB Thickness

As aforesaid, the conventional approach for the design of the thickness of the Fe^0 packed media in full-scale reactive barriers involves the direct application of the kinetic data from laboratory column experiments and the first-order equation. Observation of the poorer performance of the full-scale Fe^0 PRB at Vapokon site than the corresponding LFPCs on 1,1,1-TCA, cis-DCE and VC dechlorination, however, indicated the overestimation of the dechlorination performance of the full-scale Fe^0 PRB from the corresponding LFPCs by the conventional approach of the PRB thickness design. Therefore, a FS to ensure the effluent groundwater quality from the full-scale Fe^0 PRB at Vapokon site to be the same as or even better than the effluent groundwater quality from the corresponding LFPCs is required for the design of Fe^0 PRB thickness. To practically determine the FS from the experimental data obtained from the full-scale Fe^0 PRB at Vapokon site and the corresponding LFPCs, the longitudinal concentration profiles of different CAHs along the Fe^0 PRB and the LFPC were calculated based upon the k_{SA} shown in Table 2-1 using Eqs. (2.4) to (2.11) and then compared. As can be seen in Figs. 2-7a and 2-7b, the effluent concentration (i.e., the concentration at a longitudinal distance of 0.8 m) of PCE, TCE, VC, 1,1-DCE, 1,1,1-TCA, 1,1-DCA and TCM from the Fe^0 PRB were either the same as or lower than those from the LFPC. However, the effluent concentration of cis-DCE from the Fe^0 PRB (7.69 mg/L) was observed to be higher than that from the LFPC (1.04 mg/L). For the sake of reducing the cis-DCE effluent concentration from the Fe^0 PRB from 7.69 mg/L to 1.04 mg/L or below, the thickness of the Fe^0 packed medium in the full-scale PRB at Vapokon site should be increased from 0.8 m to approximately 2.35 m, which corresponds to the FS of about 3. This FS actually

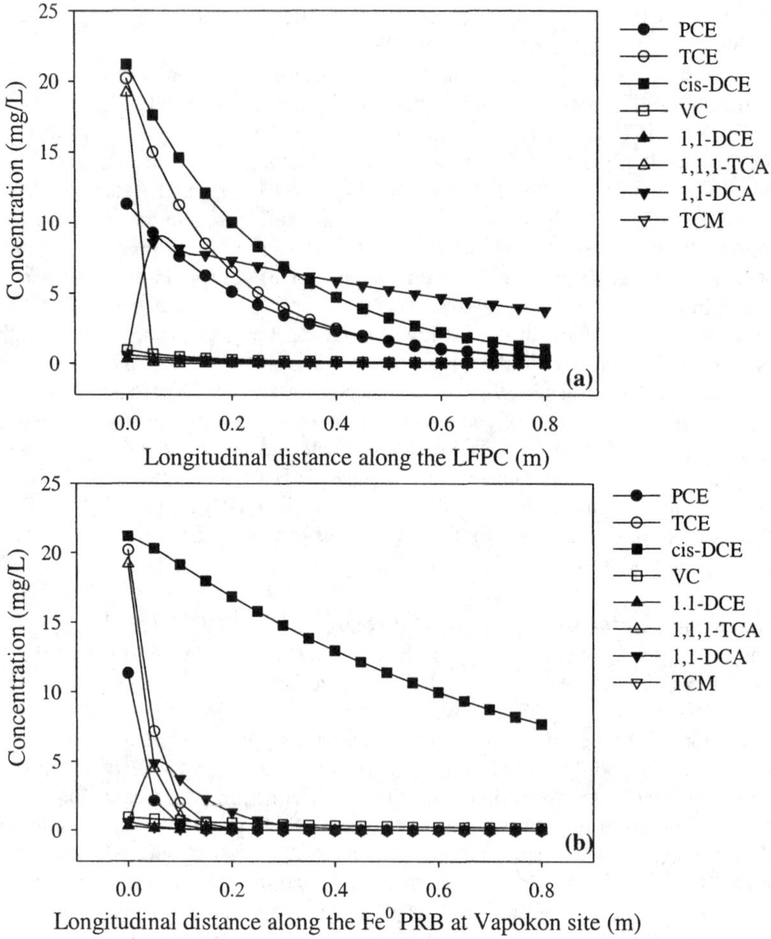

Fig. 2-7. The calculated CAH concentration profiles along (a) the LFPC and (b) the full-scale Fe^0 PRB at Vapokon site. The calculated concentration profiles were determined based upon the k_{SA} from Col 1-4 and k_{SA} provided by EnviroMetal Technologies Inc. (1998), and the average k_{SA} from the Fe^0 PRB, which were shown in Table 2-1.

corrects the influence mainly from the difference in the α_y between the Fe^0 PRB at Vapokon site and the LFPC, as well as also the possible impacts from the seasonal variation in the groundwater temperature and velocity in the aquifer, and the spatial variability of the aquifer's K on the difference of the performance in CAH dechlorination between the Fe^0 PRB at Vapokon site and the LFPCs. In comparison to the range of FS proposed by Elder et al. (2002) using groundwater flow program

MODFLOW and particle tracking program Path3D, the FS of 3 determined from the experimental data was observed falling into the proposed range of the FS (1 to 10).

2.5 Conclusions

Investigation of the performance of the full-scale Fe^0 PRB at Vapokon site and the corresponding LFPCs on the CAH dechlorination indicated that there was a difference in the dechlorination performance between these two systems. The full-scale Fe^0 PRB showed a poorer performance on 1,1,1-TCA, cis-DCE and VC dechlorination; whereas it had a better performance than the corresponding LFPCs on the dechlorination of PCE, TCE, 1,1-DCE, 1,1-DCA and TCM. The observation of the difference in the dechlorination performance between the Fe^0 PRB and the LFPCs indicated that the conventional approach for the design of the thickness of full-scale Fe^0 reactive barriers based upon the kinetic data obtained from corresponding laboratory column experiments and the assumption of plug-flow approximation may overestimate the dechlorination performance of full-scale Fe^0 reactive barriers. Therefore, a FS is required for the designed Fe^0 PRB thickness during the design stage to compensate the overestimation.

Most likely due to the spatial variability of the K of the Fe^0 packed medium in the full-scale PRB at Vapokon site, the α_y of its Fe^0 packed medium (0.275 m) was much higher than the α_y of the corresponding LFPCs (1.23×10^{-3} m). Study of the influence of the α_y on the CAH concentration along Fe^0 packed media revealed that increasing the α_y of Fe^0 packed media from 1.23×10^{-3} m to 0.275 m could noticeably increase the effluent CAH concentration from PRBs. Therefore, the significant difference in the α_y between the full-scale Fe^0 PRB at Vapokon site and the corresponding LFPCs is believed to be one of the main underlying reasons leading to the poorer performance of the full-scale Fe^0 PRB on the 1,1,1-TCA, cis-DCE and VC dechlorination. The better performance of the full-scale Fe^0 PRB than the LFPCs on the dechlorination of PCE, TCE, 1,1-DCE, 1,1-DCA and TCM, however, is believed to be due to the precipitation of sulfur-bearing minerals inside the full-scale Fe^0 PRB. This is because the sulfur-bearing minerals, such as pyrite, mackinawite and sulfate-containing green rust, were found to be able to act as electron donors and mediators, and provide more reactive sites than the iron metal for the dechlorination reactions. As a result, these sulfur-bearing minerals not only nullified the impact from the dispersion of dissolved PCE, TCE, 1,1-DCE, 1,1-DCA and TCM inside the Fe^0 PRB at Vapokon site but also enhanced the performance of the Fe^0 PRB on the dechlorination of these CAHs.

To ensure that the effluent CAH concentrations from full-scale Fe^0 PRBs at Vapokon site are the same as or even lower than those from the corresponding LFPCs, a FS of about 3 was practically determined for the designed Fe^0 PRB thickness based upon the experimental data from the full-scale Fe^0 PRB and the LFPCs. This FS primarily takes account of the influence from the disparity in the α_y, as well as also considers the possible impacts from the seasonal variation in the groundwater temperature and velocity in the aquifer, and the aquifer heterogeneity on the difference in the dechlorination performance between the full-scale Fe^0 PRB at Vapokon site and the corresponding LFPCs. In addition, a FS of 3 was observed falling into the proposed range of the FS (1 to 10) determined by numerical modeling, thereby indicating the validity of the proposed range of the FS for the correction of the difference in dechlorination performance between full-scale Fe^0 reactive barriers and the corresponding laboratory column experiments.

Acknowledgements

We extend our deepest appreciation to the Hong Kong RGC Research Grants Council (HKUST 6086/00E) for funding this research. The authors would like to thank *Connelly-GPM Inc.* for the provision of the granular iron.

2.6 References

Arnold, W. A., and Roberts, A. L. (2000). "Pathways and kinetics of chlorinated ethylene and chlorinated acetylene reaction with Fe(0) particles." *Environ. Sci. & Technol.*, 34(9), 1794-1805.

APHA, AWWA and WEF (1998). *Standard methods for the examination of water and wastewater*, 20th Ed., American Public Health Association, American Water Works Association and Water Environment Federation, Washington DC, United States.

Baker, M. J., Blowes, D. W., and Ptacek, C. J. (1997). "Phosphorous adsorption and precipitation in a permeable reactive wall: Applications for wastewater disposal systems, In *Proceedings of International Containment Technology Conference and Exhibition*, February 9-12, St. Petersburg, Florida, United States.

Bedient, P. B., Rifai, H. S., and Newell, C. J. (1999). *Ground water contamination transport and remediation*, 2nd Ed., Prentice Hall PTR, Upper Saddle River, NJ, United States.

Benner, S. G., Blowes, D. W., and Ptacek, C. J. (1997). "Porous reactive wall for prevention of acid mine drainage: Results of full-scale field demonstration." In *Proceedings of International Containment Technology Conference and Exhibition*; February 9-12, St. Petersburg, Florida, United States.

Bitsch, K., and Jensen, K. H. (1990). "Natural gradient dispersion test in a sandy aquifer using tritium as tracer." In *Proceedings of the International Conference and Workshop on Transport and Mass Exchange Processes in Sand and Gravel Aquifers*, Ontario, Canada.

Butler, E. C., and Hayes, K. F. (1998). "Effects of solution composition and pH on the reductive dechlorination of hexachloroethane by iron sulfide." *Environ. Sci. & Technol.*, 32(9), 1276-1284.

Butler, E. C., and Hayes, K. F. (1999). "Kinetics of the transformation of trichloroethylene and tetrachloroethylene by iron sulfide." *Environ. Sci. & Technol.*, 33(12), 2021-2027.

Butler, E. C., and Hayes, K. F. (2001). "Factors influencing rates and products in the transformation of trichloroethylene by iron sulfide and iron metal." *Environ. Sci. & Technol.*, 35(19), 3884-3891.

Dean, J. A. (1985). *Lange's handbook of chemistry*, 13th Ed., McGraw-Hill Inc., New York.

Elder, C. R., Benson, C. H., and Eykholt, G. R. (2002). "Effects of heterogeneity on influent and effluent concentrations from horizontal permeable reactive barriers." *Water Resour. Res.*, 38(8), 10.1029/2001WR001259.

EnviroMetal Technologies Inc. (1998). *Bench-scale treatability report of the EnviroMetal process at the Vapokon site, Fyn, Denmark*, ETI 31438.10, Guelph, Ontario, Canada N1K1S6.

Eykholt, G. R. (1997). "Uncertainty-based scaling of iron reactive barriers, In *In situ remediation of the geoenvironment geotechnical special publication ASCE*, Vol. 71, pp 41-55, American Society of Civil Engineers, New York, United States.

Eykholt, G. R., Elder, C. R., and Benson, C. H. (1999). "Effects of aquifer heterogeneity and reaction mechanism uncertainty on a reactive barrier." *J. Hazard. Mater.*, 68(1), 73-96.

Ferrey, M. L., Wilkin, R. T., Ford, R. G., and Wilson, J. T. (2004). "Nonbiological removal of cis-dichloroethylene and 1,1-dichloroethylene in aquifer sediment containing magnetite." *Environ. Sci. & Technol.*, 38(6), 1746-1752.

Fogler, H. S. (1992). *Elements of chemical reaction engineering*, 2nd Ed., Prentice-Hall Inc., Englewood Cliffs, New Jersey.

Freyberg, D. L. (1986). "A natural gradient experiment on solute transport in a sand aquifer 2. Spatial moments and the advection and dispersion of nonreactive tracers." *Water Resour. Res.*, 22(13), 2031-2046.

Furukawa, Y., Kim, J. W., Watkins, J., and Wilkin, R. T. (2002). "Formation of ferrihydrite and associated iron corrosion products in permeable reactive barriers of zero-valent iron." *Environ. Sci. & Technol.*, 36(24), 5469-5475.

Garabedian, S. P., LeBlanc, D. R., Gelhar, L. W., and Celia, M. A. (1991). "Large-scale natural gradient tracer test in sand and gravel, Cape Cod, Massachusetts 2. Analysis of spatial moments for a nonreactive tracer." *Water Resour. Res.*, 27(5), 911-924.

Gelhar, L. W., Mantoglou, A., Welty, C., and Rehfeldt, K. R. (1985). "A review of field-scale
physical solute transport processes in saturated and unsaturated porous media." In *Final project report EPRI EA-4190*, Electric Power Research Institute, Palo Alto, California, United States.

Gillham, R. W., and O'Hannesin, S. F. (1994). "Enhanced degradation of halogenated aliphatics by zero-valent iron." *Ground Water*, 32(6), 958-967.

Gu, B., Phelps, T. J., Liang, L., Dickey, M. J., Roh, Y., Kinsall, B. L., Palumbo, A. V., and Jacobs, G. K. (1999). "Biogeochemical dynamics in zero-valent iron columns: Implications for permeable reactive barriers." *Environ. Sci. & Technol.*, 33(13), 2170-2177.

Gu, B., Watson, D. B., Wu, L. Y., Phillips, D. H., White, D. C., and Zhou, J. Z. (2002). "Microbiological characteristics in a zero-valent iron reactive barrier." *Environ. Monit. Assess.*, 77, 293-309.

Hassan, S. M. (2000). "Reduction of halogenated hydrocarbons in aqueous media: I. Involvement of sulfur in iron catalysis." *Chemosphere*, 40(12), 1357-1363.

Helland, B. R., Alvarez, P. J. J., and Schnoor, J. L. (1995). "Reductive dechlorination of carbon tetrachloride with elemental iron." *J. Hazard. Mater.*, 41(2-3), 205-216.

Kriegman-King, M. R., and Reinhard, M. (1994). "Transformation of carbon tetrachloride by pyrite in aqueous solution." *Environ. Sci. & Technol.*, 28(4), 692-700.

Lai, K. C.-K., and Lo, I. M.-C. (2002). "Bench-scale study of the effects of seepage velocity on the dechlorination of TCE and PCE by zero-valent iron." In *Proceedings of the 6th International Symposium on Environmental Geotechnology*, July 2-5, Seoul, Korea.

Lai, K. C.-K. (2004). *Laboratory and full-scale studies of a permeable reactive barrier on the dechlorination of chlorinated aliphatic hydrocarbons*, Ph.D. Thesis, Civil Engineering Department, The Hong Kong University of Science and Technology, Hong Kong, China.

Lai, K. C.-K., Lo, I. M.-C., and Kjeldsen, P. (2006a). "Natural gradient tracer test for the permeable reactive barrier in Denmark 2. Spatial Moments Analysis and Dispersion of Conservative Tracer" *J. of Hazardous, Toxic and Radioactive Waste Management, ASCE*, in press.

Lai, K. C.-K., Lo, I. M.-C., and Kjeldsen, P. (2006b). "Natural gradient tracer test for the permeable reactive barrier in Denmark 1. Field study of tracer movement." *J. of Hazardous, Toxic and Radioactive Waste Management, ASCE*, in press.

Lai, K. C.-K., Lo, I. M.-C., Birkelund, V., and Kjeldsen, P. (2006c). "Field monitoring of a permeable reactive barrier for removal of chlorinated organics." *J. Environ. Eng. ASCE*, 132(2), 199-210.

Lee, W. J., and Batchelor, B. (2002a). "Abiotic reductive dechlorination of chlorinated ethylenes by iron-bearing soil minerals. 2. Green rust." *Environ. Sci. & Technol.*, 36(24), 5348-5354.

Lee, W. J., and Batchelor, B. (2002b). "Abiotic reductive dechlorination of chlorinated ethylenes by iron-bearing soil minerals. 1. Pyrite and magnetite." *Environ. Sci. & Technol.*, 36(23), 5147-5154.

Liang, L. Y., Korte, N., Gu, B., Puls, R., and Reeter, C. (2000). "Geochemical and microbial reactions affecting the long-term performance of in situ iron barriers." *Adv. Environ. Res.*, 4(4), 273-286.

Lo, I. M.-C., and Lai, K. C.-K. (2003). "Field monitoring of the performance of PRB at Vapokon site in Denmark." In *Proceedings of the RTDF Permeable Reactive Barriers (PRBs) Action Team Meeting*, Remediation Technologies Development Forum, Oct. 15-16, pp 98-112, Niagara Falls, New York, United States.

Lo, I. M.-C., Lam, C. S.-C., and Lai, K. C.-K. (2005). "Competitive effects of TCE on Cr(VI) removal by zero-valent iron." *J. Environ. Eng., ASCE*, 131(11), 1598-1606.

Lipczynska-Kochany, E., Harms, S., Milburn, R., Sprah, G., and Nadarajah, N. (1994). "Degradation of carbon tetrachloride in the presence of iron and sulfur containing compounds." *Chemosphere*, 29(7), 1477-1489.

MicroMath Scientist® (1995). *Experimental data fitting/Microsoft Windows version 2.0*, Salt Lake City, Utah 84121, United States.

Moltyaner, G. L., and Killey, R. W. D. (1988). "Twin lake tracer tests: Longitudinal dispersion." *Water Resour. Res.*, 24(10), 1613-1627.

O'Hannesin, S. F., and Gillham, R. W. (1998). "Long-term performance of an in-situ 'iron wall' for remediation of VOCs." *Ground Water*, 36(1), 164-170.

Phillips, D. H., Gu, B., Watson, D. B., Roh, Y., Liang, L., and Lee, S. Y. (2000). "Performance evaluation of a zerovalent iron reactive barrier: Mineralogical characteristics." *Environ. Sci. & Technol.*, 34(19), 4169-4176.

Powell, R. M., Puls, R. W., Hightower, S. K., and Sabatini, D. A. (1995). "Coupled iron corrosion and chromate reduction: Mechanisms for subsurface remediation." *Environ. Sci. & Technol.*, 29(8), 1913-1922.

Powell, R. M., Blowes, D. W., Gillham, R. W., Schultz, D., Sivavec, T., Puls, R. W., Vogan, J. L., Powell, P. D., and Landis, R. (1998). *Permeable reactive barrier technologies for contaminant remediation*, EPA/600/R-98/125, Office of Research and Development, United States Environmental Protection Agency, Washington, DC 20460, United States.

Reardon, E. J. (1995). "Anaerobic corrosion of granular iron: Measurement and interpretation of hydrogen evolution rates." *Environ. Sci. & Technol.*, 29(12), 2936-2945.

Ritter, K., Odziemkowski, M. S., and Gillham, R. W. (2002). "An in situ study of the role of surface films on granular iron in the permeable iron wall technology." *J. Contam. Hydrol.*, 55(1-2), 87-111.

Scherer, M. M., Richter, S., Valentine, R. L., and Alvarez, P. J. J. (2000). "Chemistry and microbiology of permeable reactive barriers for *in situ* groundwater clean up." *Crit. Rev. Environ. Sci. & Technol.*, 30(3), 363-411.

Sivavec, T. M., and Horney, D. P. (1997). "Reduction of chlorinated solvents by Fe(II) minerals." In *Proceedings of the 213th ACS National Meeting*, Vol. 37(1), pp 115-117, San Francisco, CA, American Chemical Society, Washington, DC, United States.

Su, C. M., and Puls, R. W. (1999). "Kinetics of trichloroethene reduction by zerovalent iron and tin: Pretreatment effect, apparent activation energy, and intermediate products." *Environ. Sci. & Technol.*, 33(1), 163-168.

Thorbjarnarson, K. W., and Mackay, D. M. (1994). "A forced-gradient experiment on solute transport in the Borden aquifer 2. Transport and dispersion of the conservative tracer." *Water Resour. Res.*, 30(2), 385-399.

Toride, N., Leij, F. J., and van Genuchten, M. Th. (1999). *The CXTFIT code for estimating transport parameters from laboratory or field tracer experiments version 2.1*, research report no. 137. United States Salinity Laboratory Agricultural Research Service, United States Department of Agriculture, Riverside, California, United States.

USEPA (2002). *Field applications of in situ remediation technologies: Permeable reactive barriers*, EPA/68/W-00/084, Technology Innovation Office, Office of Solid Waste and Emergency Response, United States Environmental Protection Agency, Washington, DC 20460, United States.

van Genuchten, M. Th. (1981). "Analytical solutions for chemical transport with simultaneous adsorption, zero-order production and first-order decay." *J. of Hydrology*, 49, 213-233.

Vogel, T. M., and McCarty, P. L. (1987). "Abiotic and biotic transformations of 1,1,1-trichloroethane under methanogenic conditions." *Environ. Sci. & Technol.*, 21(12), 1208-1213.

Yabusaki, S., Cantrell, K., Sass, B., and Steefel, C. (2001). "Multicomponent reactive transport in an in situ zero-valent iron cell." *Environ. Sci. & Technol.*, 35(7), 1493-1503.

CHAPTER 3

Zero-valent Iron and Organo-clay for Chromate Removal in the Presence of Trichloroethylene

Hyun-Hee Cho[1], and Jae-Woo Park[2]

Abstract: Hybrid barriers of zero-valent iron (Fe^0) filings and hexadecyltrimethylammonium (HDTMA)-bentonite were simulated in columns to remove chromate and trichloroethylene (TCE) via reduction and sorption. The normalized reduction rate constant (k_{SA}) of chromate with the mixture of Fe^0 filings and HDTMA-bentonite was about twice higher than that with Fe^0 filings only, suggesting that chromate reduction was enhanced when HDTMA-bentonite was mixed with Fe^0 filings. For the column of two separate layers of Fe^0 and HDTMA-bentonite, the reduction rate of chromate slightly increased, but not as much as in the column with mixtures of Fe^0 filings and HDTMA-bentonite. In addition, the sorption rates of both TCE and chromate in the column with mixtures of Fe^0 filings and HDTMA-bentonite were lower than in the columns of both HDTMA-bentonite only and two separate layers of Fe^0 and HDTMA-bentonite. Based on the observation from this study, hybrid barriers can be more effective for mixed contaminants.

CE Database subject headings: Hybrid methods; Barriers; Cast iron; Bentonite; Chromium; TCE.

3.1 Introduction

Chromium is one of the most common groundwater pollutants at industrial sites and military facilities due to its widespread use as a metal corrosion inhibitor (Melitas et al. 2001). In groundwater environments, chromium occurs in two stable oxidation states, hexavalent chromium [Cr(VI)] and trivalent chromium [Cr(III)]. Many Cr(VI) species are known as toxic, while Cr(III) species are generally benign (Powell et al. 1995; Blowes et al. 1997; Pratt et al. 1997).

Groundwater remediation using permeable reactive barriers (PRBs) can offer significant economic advantage of *in situ* treatment (Gillham and O'Hannesin 1994; Orth and Gillham 1996). Zero-valent iron (Fe^0) has been a focus of investigation in recent years among the reactive media that are used in the remediation of sites contaminated with inorganic compounds (e.g., chromate) as well as chlorinated organic compounds (Powell et al. 1995; Blowes et al. 1997; Johnson et al. 1998; Farrell et al. 2000; Melitas et al. 2001; Cho and Park 2005). The removal of chromate from aqueous solutions by Fe^0 is mainly due to reduction to Cr(III) with

---

[1]Postdoctoral Fellow, Dept. of Geography and Environmental Engineering, Johns Hopkins University, 313 Ames Hall, 3400 N, Charles St., Baltimore, MD 21218-2686, United States.

[2] Associate Professor, Dept. of Civil Engineering, Hanyang University, 17 Haengdang-dong, Sungdong-gu, Seoul 133-791, South Korea (corresponding author). E-mail: jaewoopark@hanyang.ac.kr

which Fe(III) will form chromium-iron hydroxide [$Cr_x Fe_{1-x}(OH)_3$] precipitations (Li et al. 1999).

Immobilization can be used in the liners for landfills and storage tanks of hazardous materials (Neupane and Park 1999; Neupane and Park 2000). Organo-bentonite, the bentonite modified with cationic surfactants by ion exchange, can effectively sorb organic contaminants due to its organic sorption capacity (Xu and Boyd 1995; Lo 2001). Recent researches reported that organo-bentonite can be effectively applied for the immobilization of both organic contaminants and heavy metals (Lee et al. 2002; Yoo et al. 2004).

Combined use of both reduction and immobilization can be desirable for some sites. Li et al. (1999) reported that surfactant-modified zeolite (SMZ)/Fe^0 pellets were observed to enhance the reduction of tetrachloroethylene (PCE) and chromate. Zhang et al. (2002) reported that the reduction rate of PCE with SMZ/Fe^0 in column system was three times higher than that with unmodified zeolite/Fe^0. Even though SMZ/Fe^0 pellets exhibit high hydraulic conductivity and reaction capacity, the process of pellet preparation can be complicated and expensive. This may limit the practical field-applicability of the technology. The combination of Fe^0 filings and HDTMA-bentonite was introduced and their efficiency was proven for TCE removal (Cho et al. 2005). Fe^0 filings and HDTMA-bentonite are simply mixed rather than pelletized, so it can be rather easily applied in the field.

However, in order to enhance the field applicability, more studies are needed to better understand Fe^0 filings and HDTMA-bentonite systems. Previous studies reported that the removal of chromate and TCE as multi-contaminants using Fe^0 could be affected by co-existing compound (Mayer et al. 2001; Cho and Park 2005). Therefore, the objectives of this study are: 1) to study the reduction of chromate in Fe^0 filings and HDTMA-bentonite column systems and 2) to investigate the effect of co-existing reducible contaminants on the reduction and sorption of chromate and TCE in the proposed hybrid barrier.

3.2 Experimental Section

3.2.1 Materials and Their Characterization Methods

Fe^0 filings were purchased from Fisher Scientific (USA) and its specific surface area was reported to be 1.06 m^2/g based on BET N_2-method. Total organic carbon (TOC) content, determined using an elemental analyzer (CHNS-932, LECO, USA), was 1.94%. Total Fe^0 content, determined with sequential X-ray Fluorescence Spectrophotometer (XRF-1700, *Shimadzu*, Japan), was 92.5%. Ottawa sand (20-40 mesh) was obtained from *Fisher Scientific* (USA). TCE (99.9%, *Fisher Scientific*) was used as received. Chromate solution was prepared by dissolving K_2CrO_4 (*Fisher Scientific*) into deionized water. The clays used in this study were commercial bentonites (*Donghae Chemical Co.*, Korea), which are primarily composed of low-charge Na-montmorillonite with cation exchange capacity of 63.9 meq/kg. HDTMA bromide ($HDTMA^+Br^-$) as obtained from *Aldrich Chemical*.

3.2.2 Preparation of Organo-bentonite

HDTMA-bentonite was prepared using the procedure in Lee et al. (2002). Twenty grams of pre-dried bentonites were wetted with 100 mL of distilled water, and then mixed with 200 mL of HDTMA solution. The mixture was agitated for 24 h at room temperature, and then centrifuged at 1554 *g* for 10 min. Subsequently, the mixture was washed three times with distilled water. The mixture was then dried at 80°C, activated at 105°C in an oven for 1 h, and mechanically ground. The properties of untreated bentonites and organo-bentonites were also characterized with X-ray powder diffraction (XRD, *Mac Science* MXP-1500) and with an elemental analyzer (*Elementar Vario EL*).

3.2.3 Column Experiments

A laboratory-scale barrier system was simulated using glass columns of 40 cm long and 5.0 cm internal diameter, as shown in Fig. 3-1. The columns contained three sampling ports at 10 cm intervals. Column end caps, fittings, and tubings were made of Teflon. The columns were connected with viton tubings to a peristaltic pump with eight channels that were adjusted to a flow rate of 480 mL/d. Experiments were performed at 20±2°C in a constant temperature room.

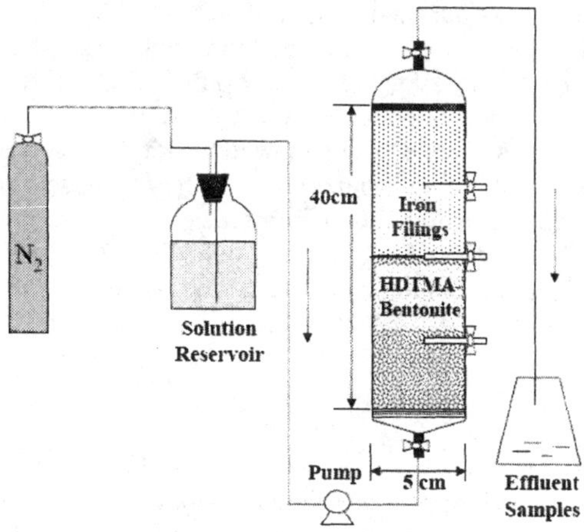

Fig. 3-1. Design of experimental column set-up (C4)

Four columns were packed with reactive materials and fed with degassed synthetic groundwater (8×10^{-5} M $NaHCO_3$). This synthetic groundwater was used as the aqueous solutions for all subsequent experiments. Column 1 (C1) was packed with 80% sand and 20% Fe^0 filings, and column 2 (C2) was packed with 80% sand and 20% HDTMA-bentonite (Table 3-1). Columns 3 and 4 (C3 and C4) contained hybrid barriers of both reactive and sorptive materials. A mixture of HDTMA-bentonite and Fe^0 filings was placed in column 3, while column 4 has two distinct layers of each material. The immobilizing barrier contained HDTMA-bentonite and sand with a ratio of 20:80. The reactive barrier contained Fe^0 filings and sand with a ratio of 20:80. All injections were conducted in an upward flow mode. The influent solutions were prepared with 50 mg/L of chromate and 100 mg/L of TCE in de-ionized water purged with nitrogen gas to remove oxygen gas.

Table 3-1. Compositions of Reactive Materials Used in Dynamic Column Experiments

Column	Reactive mixture composition (mass %)
C1	80 % sand, 20 % Fe^0 filings (from 0 to 40 cm)
C2	80 % sand, 20 % HDTMA-bentonite (from 0 to 40 cm)
C3	60 % sand, 20 % Fe^0 filings, 20 % HDTMA-bentonite (from 0 to 40 cm)
C4	80 % sand, 20 % HDTMA-bentonite (from 0 to 20 cm) 80 % sand, 20 % Fe^0 filings (from 20 to 40 cm)

For each column, the chromate and TCE concentrations in the effluent were monitored. Effluent samples were collected from effluent sampling ports using 5-mL *Hamilton* gastight syringes. The samples were immediately transferred to 2-mL glass vials and capped with Teflon-lined septa. Influent samples were periodically collected to monitor input concentrations. Flow rates were determined from the cumulative volume of effluent during the experiments.

Chromate and TCE concentration were measured with high performance liquid chromatography (*Waters* 515, St. Milford, MT, USA). For the chromate concentration, the analyses were conducted at a wavelength of 365 nm and with a flow rate of 0.6 mL/min, a mobile phase of water with 20% acetonitrile (*Fisher*, optima grade), and a 3.9 × 300 mm μ-bondapak C18 reverse phase column (*Waters*, St. Milford, MT, USA). Retention time of chromate was 2.5 min and the linear response was 0.1 to 240 mg/L based on peak area. For TCE, the analyses were conducted at a wavelength of 220 nm and with a flow rate of 1.0 mL/min, a mobile phase of 100% acetonitrile, and a 3.9 × 300 mm μ-bondapak C18 reverse phase column.

3.2.4 Data Analysis

One-dimensional transport of a solute undergoing reversible sorption and first-order decay in porous media can be expressed (Schnoor 1996; Zhang et al. 2002; Cho et al. 2005):

$$\frac{\partial C}{\partial t}(1+K_d\frac{\rho}{\theta}) = D\frac{\partial^2 C}{\partial x^2} - v\frac{\partial C}{\partial x} - \mu C \tag{3.1}$$

where C is the solute concentration in aqueous phase, D is the dispersion coefficient, ρ is the porous medium bulk density, θ is the porosity, t is time, x is distance, v is the seepage velocity, K_d is the linear distribution coefficient, and μ is the overall first-order reaction rate constant. We hypothesized that the sorption of chromate and TCE to HDTMA-bentonite and Fe^0 filings followed linear sorption at low solute concentrations.

The analytical solution for Eq. (3.1) with the initial concentration of zero and boundary condition of constant influent concentration (at x = 0) and zero concentration gradient (at x = ∞) is (Schnoor 1996):

$$\frac{C(x,t)}{C_0}=\frac{1}{2}exp\left[\frac{(v-u)x}{2D}\right]erfc\left[\frac{Rx-ut}{2(DRt)^{1/2}}\right]+\frac{1}{2}exp\left[\frac{(v+u)x}{2D}\right]erfc\left[\frac{Rx+ut}{2(DRt)^{1/2}}\right] \quad (3.2)$$

where *C(x,t)* is the concentration at a given distance and time, C_0 is the influent concentration. The retardation factor *R* and *u* are defined as:

$$R=1+\frac{\rho}{\theta}K_d \quad (3.3)$$

$$u=v\left[1+\frac{4\mu D}{v^2}\right]^{1/2} \quad (3.4)$$

Tracer test was conducted with bromide tracers to determine the porosity, the seepage velocity, and the dispersion coefficient of the reactive media. A tracer solution of KBr at 100 mg/L was flowed into four columns at of 480 mL/d (Table 3-2). Parameter values were estimated using Eq. (3.2) with least-squares optimization.

Table 3-2. Surface Characteristics and Parameters Used in Column Experiments

	C1	C2	C3	C4
Dry density (mg/m^3)	1.88	1.41	1.72	1.63
Porosity	0.34	0.25	0.33	0.30
Seepage velocity (×10^{-4} cm/sec)	4.83	4.87	4.07	6.84
Dispersion coefficient (×10^{-3} cm^2/sec)	2.04	2.14	2.42	4.92

3.3 Results and Discussion

Properties of untreated bentonite and HDTMA-bentonite, including the d-spacing, mineralogy, and TOC, are listed in Table 3-3. The mineralogy of the bentonite in this research was characterized as smectite from XRD analysis. TOC value and interlayer spacing of HDTMA-bentonite were quite different from those of the untreated bentonite. TOC value increased linearly as the mass of HDTMA sorbed on the bentonite increased. Increased TOC value and interlayer spacing will facilitate more organic contaminants to be sorbed to the interlayer of the HDTMA-bentonites (Lee et al. 2002; Yoo et al. 2004). Lee et al. (2002) reported that the desorbed concentration increased as the initial sorbed concentration of HDTMA was higher, but it was less than 2% of the total adsorbed mass. Therefore, the effect from the desorbed HDTMA would be minimal in this research.

Table 3-3. Interlayer Spacings and Organic Carbon Contents of the Bentonites in this Study

	Interlayer spacing (Å)	Organic carbon contents (%)
bentonite	12.27	0.54
HDTMA-bentonite	19.98	15.08

The removal of chromate and TCE in four columns was studied in single- and dual-solute systems (Figs. 3-2, 3-4 to 3-6). The breakthrough of chromate and TCE in C1 with Fe^0 filings is shown in Fig. 3-2. The transport model was fitted to the experimental results using least-squares algorithm and used to obtain the value of K_d and the normalized reaction rate constant (k_{SA}) with respect to the surface area of the Fe^0 (Table 3-4). In the presence of TCE, the reduction rate of chromate slightly increased (t value = 3.45, two-tailed *t* test) (Fig. 3-2a). TCE as well as chromate were

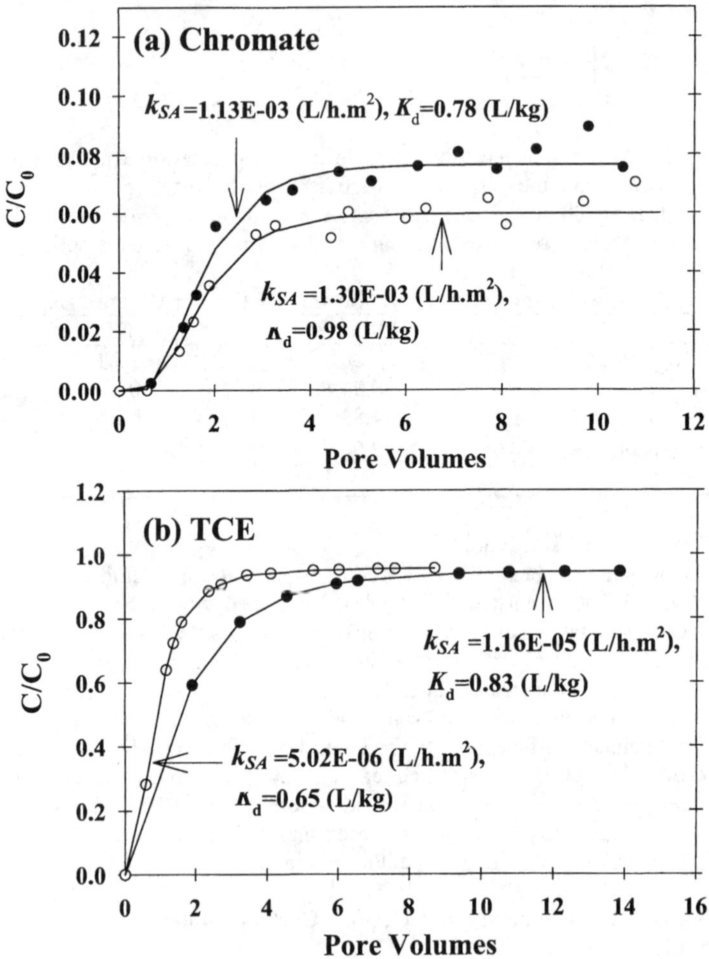

Fig. 3-2. Effluent concentrations of (a) chromate and (b) TCE in C1 (Fe^0 filings + sand). Closed and opened symbols are the concentrations in single- and in dual-solute systems, respectively.

Table 3-4. Distribution Coefficients and Reduction Rates from Column Experiments

	K_d (L/kg)				k_{SA} (× 10^{-3} L/h.m^2)			
	C1	C2	C3	C4	C1	C2	C3	C4
Chromate Only	0.78	3.21	2.67	3.45	1.13	-	2.01	1.42
Chromate with TCE	0.98	2.87	2.05	3.03	1.30	-	2.28	1.24
TCE only	0.83	3.24	2.38	3.70	1.16	-	8.45	1.05
TCE with chromate	0.65	3.51	2.63	3.91	0.50	-	6.53	1.11

reduced by the Fe^0 filings, but the reduction of chromate was even faster than dechlorination of TCE with Fe^0 filings (Blowes et al. 1997). Therefore, chromate reduction in C1 was not hindered by the presence of TCE as the competition between chromate and TCE for reactive sites scarcely occurred.

Fig. 3-2b shows the effect of chromate on TCE removal in C1. The reduction rate of TCE with chromate was two times lower than that without it. Chromate competed with TCE for reactive sites, as the reaction of chromate by Fe^0 was faster than that of TCE. The presence of chromate could result in the formation of more passive surface film on the Fe^0, which interfered with TCE reduction (Cho and Park 2005). With chromate, TCE could not access to the reactive sites where it could if alone, and then those Fe^0 surface became passive so that less TCE could be reduced.

Usually, the corrosion of Fe^0 and the reduction of chromate are enhanced at lower pH. The effluent pH at equilibrium (after 10 pore volumes) apparently increased to 9.1 while the influent pH was 6.8. Therefore, the reduction rate of chromate significantly decreased after 10 pore volumes (Fig. 3-2a). Fe(III) from Fe^0 oxidation and Cr(III) from Cr(VI) reduction formed Cr(III)- Fe(III) precipitation and this could inhibit the reaction of chromate by Fe^0. The SEM image of Fe^0 surface reacted with chromate in C1 is shown in Fig. 3-3. More of Cr(III)- Fe(III) as $Cr_xFe_{1-x}(OH)_3$ could be found on the surface reacted with chromate, compared to the non-reacted surface.

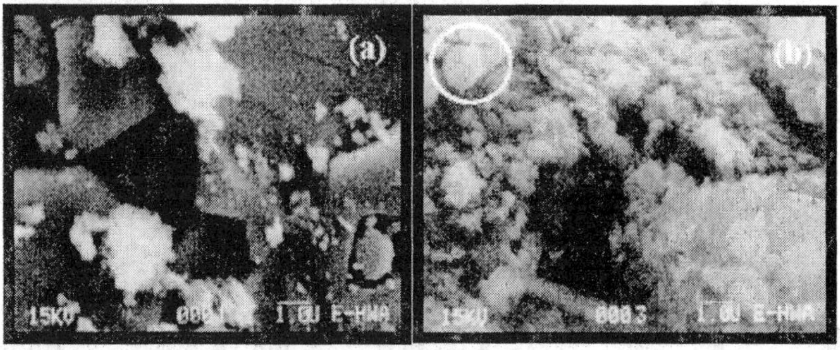

Fig. 3-3. SEM images of mixed Cr(III)-Fe(III) solid precipitations in C1: (a) non-reacted Fe^0 filings and (b) reacted Fe^0 filings.

Removal of chromate and TCE in C2 with HDTMA-bentonite is shown in Fig. 3-4a. Previous studies reported that significantly more organic contaminants were sorbed onto HDTMA-bentonite than heavy metals (Lee et al. 2002; Yoo et al. 2004). Chromate, however, is negatively-charged and HDTMA has positively-charged head group that can react with chromate, so chromate was sorbed as much as TCE in C2. There was no big difference between the sorption of chromate without TCE (3.21 L/kg) and with TCE (2.87 L/Kg) as different sorptive sites were involved with TCE

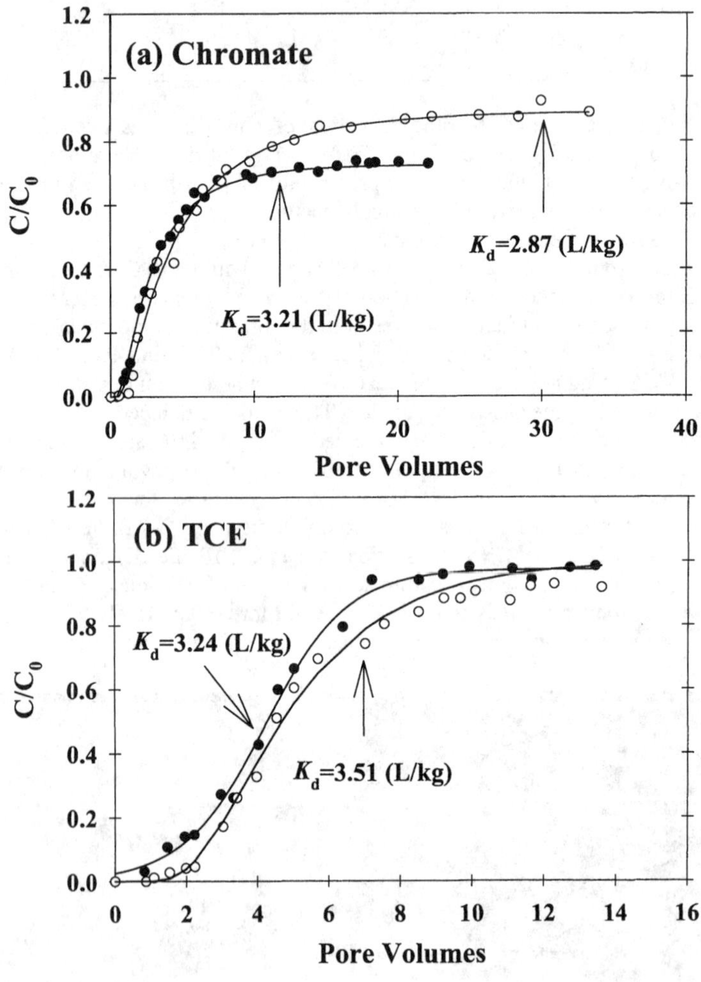

Fig. 3-4. Effluent concentrations of (a) chromate and (b) TCE in C2 (HDTMA-bentonite + sand). Closed and opened symbols are the concentrations in single- and dual-solute systems.

and chromate. Likewise, the sorption rate of ` TCE without chromate was similar to that with chromate (Fig. 3-4b).

Fig. 3-5 shows the concentration change of chromate and TCE in C3. Both the reduction and sorption of chromate in C3, which contains the Fe^0 filings and HDTMA-bentonite, were higher than those in C1 due to the presence of HDTMA-bentonite. Since the inlet TCE concentration, effluent pH (8.9±0.2), mass of reactive material (1500±50 g), porosity, and dry density for the four columns were very similar (Table 3-2), the enhanced chromate reduction in C3 was definitely due to the

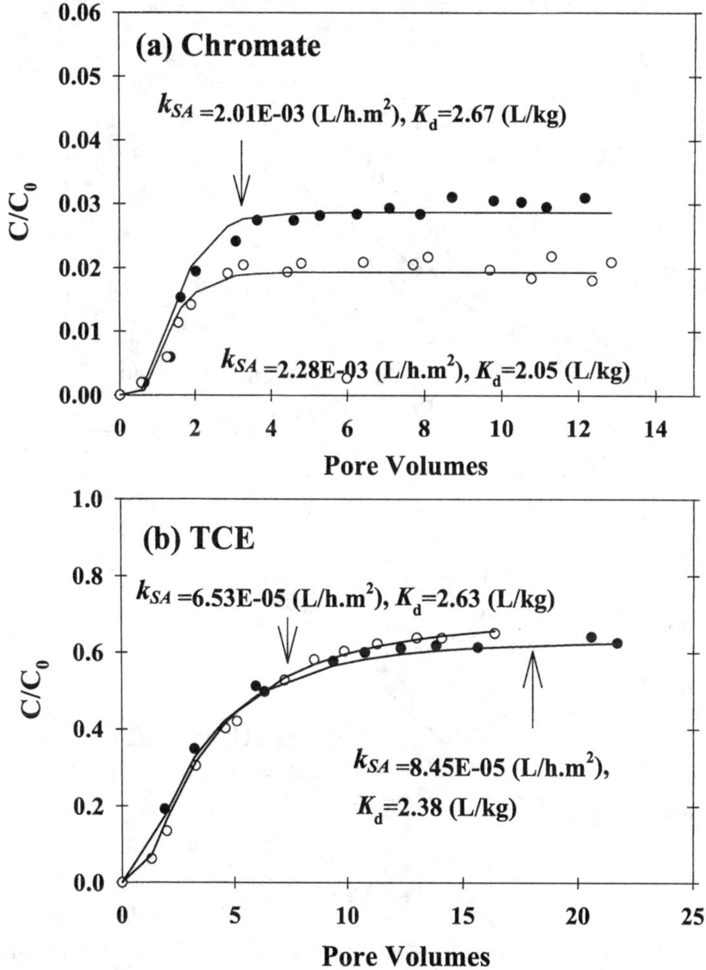

Fig. 3-5. Effluent concentrations of (a) chromate and (b) TCE in C3 (Fe^0 filings + HDTMA-bentonite + sand). Closed and opened symbols are the concentrations in single- and dual-solute systems, respectively.

presence of HDTMA-bentonite. Cho et al. (2005) showed that the reduction rate of TCE by Fe^0 with the coexisting HDTMA-bentonite significantly increased because TCE could easily approach to the vicinity of the Fe^0 surface due to the partitioning of TCE into sorbed organo-bentonite, which promoted facilitated mass transfer of TCE to the Fe^0 surface. For chromate, it was more likely to be paired with the ionic group of the sorbed HDTMA-bentonite and still readily available for reduction by Fe^0.

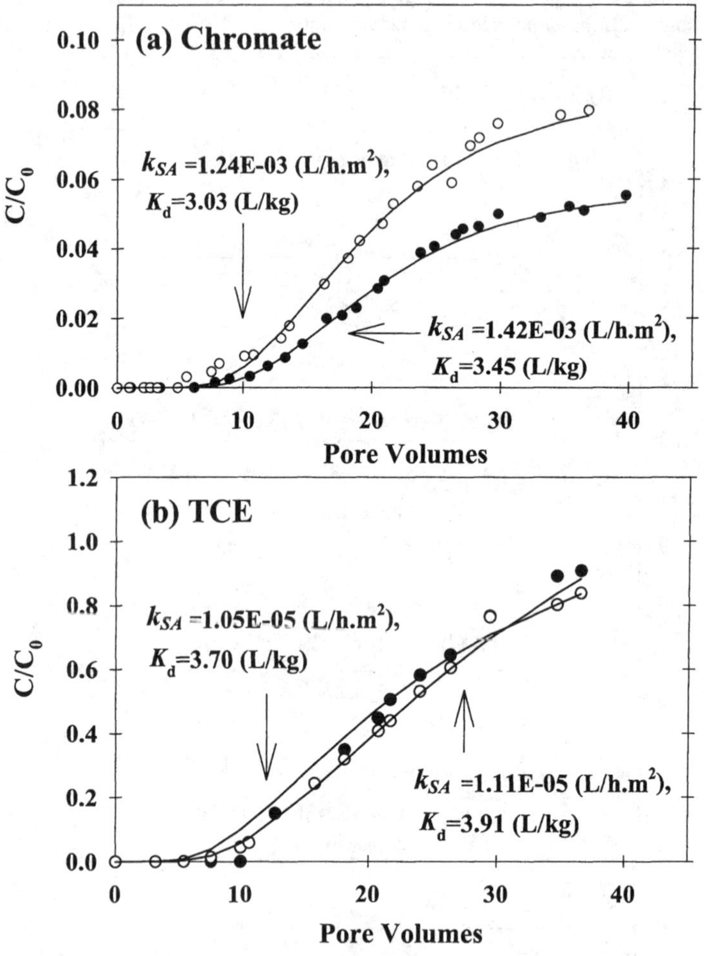

Fig. 3-6. Effluent concentrations of (a) chromate and (b) TCE in C4 (two layers: HDTMA-bentonite + sand and Fe^0 filings + sand). Closed and opened symbols are the concentrations in single- and dual-solute systems, respectively.

In addition, the observed enhancement in TCE reduction by the Fe^0 filings and HDTMA-bentonite mixture may be attributed to increased sorption of TCE directly to the Fe^0 surfaces coated with HDTMA, which was released from HDTMA-bentonite. Cho and Park (2003) reported that significant increases in TCE accumulation during PCE reduction were observed in batch studies when Fe^0 was modified with cationic, anionic, and nonionic surfactants.

Fig. 3-6 shows the effluent concentration of chromate and TCE in C4, which was composed of two layers of Fe^0 filings and HDTMA-bentonite. The k_{SA} values of both chromate and TCE were higher than those in C1, but lower than those in C3. However, the sorption rates in C4 were similar to those in C2 and much higher than those in both C1 and C3. The increase in K_d value, however, did not cause a delay in the breakthrough. In the presence of TCE, the reduction rate of chromate by Fe^0 filings was similar to that without TCE since TCE existed at low concentration because TCE was immobilized in the first layer with HDTMA-bentonite.

3.4 Conclusions

The operation of Fe^0 in PRB is limited because the reactivity of Fe^0 decreases over time and the accumulation of secondary mineral increases, although the efficiencies of the reductive Fe^0 were verified to remove the chlorinated hydrocarbons of wide range. Incorporation of HDTMA-bentonite into Fe^0 filings, suggested in this research, can effectively retard the transport of inorganic contaminants as well as organic contaminants from landfill leachate as well as oil spills in subsurface environment. For the hybrid barriers (C3 and C4), the competition between chromate and TCE could hardly be observed in dual-solute systems. Therefore, if subsurface environments are contaminated with multi-contaminants, the proposed hybrid system can be applied successfully while the conventional reactive barriers with Fe^0 alone may not be successful.

3.5 References

Blowes, D. W., Ptacek, C. J., and Jambor, J. L. (1997). "In-situ remediation of Cr(VI)-contaminated ground water using permeable reactive wall: Laboratory studies." *Environ. Sci. & Technol.*, 31(12), 3348-3357.

Cho, H. H., and Park, J.-W. (2003). "Reductive dechlorination of PCE using zero valent iron plus surfactants." In *Chlorinated solvent and DNAPL remediation: Innovative strategies for subsurface cleanup*, ACS symposium series 837, 141-153.

Cho, H. H., and Park, J.-W. (2005). "Effect of coexisting compounds on the sorption and reduction of trichloroethylene with iron." *Environ. Toxicol. Chem.*, 24(1), 11-16.

Cho, H. H., Lee, T., Hwang, S. J., and Park, J.-W. (2005). "Iron and organo-bentonite for the reduction and sorption of trichloroethylene." *Chemosphere*, 58, 103-108.

Farrell, J., Kason, M., Melitas, N., and Li, T. (2000). "Investigation of the long-term performance of zero-valent iron for reductive-dechlorination of trichloroethylene." *Environ. Sci. & Technol.*, 34(3), 514-521.

Gillham, R. W., and O'Hannesin, S. F. (1994). "Enhanced degradation of halogenated aliphatics by zero-valent iron." *Ground Water*, 32, 958-967.

Johnson, T. L., Fish, W., Gorby, Y. A., and Tranyek, P. G. (1998). "Degradation of carbon tetrachloride by iron metal: Complexation effects on the oxide surface." *J. Contam. Hydrol.*, 29, 377-396.

Lee, J. J., Choi, J., and Park, J.-W. (2002). "Simultaneous sorption of lead and chlorobenzene by organobentonite." *Chemosphere*, 49, 1309-1315.

Li, Z., Jones, H. K., Bowman, R. S., and Helferich, R. (1999). "Enhanced reduction of chromate and PCE by palletized surfactant-modified zeolite/zerovalent iron." *Environ. Sci. & Technol.*, 33(23), 4326-4330.

Lo, I. M.-C. (2001). "Organoclay with soil-bentonite admixture as waste contaminant barriers." *J. Environ. Eng. ASCE*, 127, 756-759.

Mayer, K. U., Blower, D. W., and Frind, O. E. (2001). " Reactive transport modeling of an in situ reactive barrier for the treatment of hexavalent chromium and trichloroethylene in ground water." *Water Resour. Res.*, 37, 3091-3103.

Melitas, N., Chuffe-Moscoso, O., and Farrell, J. (2001). "Kinetics of soluble chromium removal from contaminated water by zerovalent iron media: Corrosion inhibition and passive oxide effects." *Environ. Sci. & Technol.*, 35(19), 3948-3953.

Neupane, D., and Park, J.-W. (1999). "Binding of dialkylated disulfonated diphenyl oxide surfactant onto alumina in the aqueous phase." *Chemosphere*, 38, 1-12.

Neupane, D., and Park, J.-W. (2000). "Partitioning of naphthalene to gemini surfactant-treated alumina." *Chemosphere*, 41, 787-1792.

Orth, W. S., and Gillham, R. W. (1996). "Dechlorination of trichloroethene in aqueous solution using Fe^0." *Environ. Sci. & Technol.*, 30(1), 66-71.

Powell, R. M., Puls, R. W., Hightower, S. K., and Sabatini, D. A. (1995). "Coupled iron corrosion and chromate reduction: Mechanism for subsurface remediation." *Environ. Sci. & Technol.*, 29(8), 1913-1922.

Pratt, A. R., Blowes, D. W., and Ptacek, C. J. (1997). "Products of chromate reduction on proposed remediation material." *Environ. Sci. & Technol.*, 31(9), 2492-2498.

Schnoor, J. L. (1996). *Environmental modeling: Fate and transport of pollutants in water, air, and soil*, John Wiley & Sons, Inc., New York.

Xu, S., and Boyd, S. A. (1995). "Cationic surfactant adsorption by swelling and nonswelling layer silicates." *Langmuir*, 11, 2508-2514.

Yoo, J. Y., Choi, J., Lee, T., and Park, J.-W. (2004) "Organobentonite for sorption and degradation of phenol in the presence of heavy metals." *Water, Air, and soil Pollution*, 154(1-4), 225-237.

Zhang, P., Tao, X., Li, Z., and Bowman, R. S. (2002). "Enhanced perchloroethylene reduction in column systems using surfactant-modified zeolite/zero-valent iron pellets." *Environ. Sci. & Technol.*, 36(16), 3597-3603.

CHAPTER 4

Competitive Effects on the Dechlorination of Chlorinated Aliphatic Hydrocarbons by Zero-valent Iron

Irene M. C. Lo[1], Keith C. K. Lai[2], Chester C. S. Lam[3], and Hilary I. Inyang[4]

Abstract: Dechlorination of chlorinated aliphatic hydrocarbons (CAHs) by zero-valent iron (Fe^0) was found to be influenced by the competitive effects exerted by other groundwater contaminants. Laboratory column study of the competitive effects on CAH dechlorination by Fe^0 indicated that the presence of 1,1,1-trichloroethane (1,1,1-TCA) in the trichloroethylene (TCE)-contaminated groundwater could decrease the normalized dechlorination rate constant (k_{SA}) of TCE from 3.04×10^{-2} to 2.74×10^{-2} mL m^{-2} hr^{-1}. In a similar fashion, introduction of chloroform (TCM) into the synthetic groundwater containing TCE and 1,1,1-TCA led to a 40 to 54% drop in TCE and 1,1,1-TCA k_{SA}, thus indicating competition among TCE, 1,1,1-TCA and TCM during dechlorination reactions induced by Fe^0. Activation energy ranging from 34.3 to 53.7 kJ/mol for the simultaneous dechlorination of TCE, 1,1,1-TCA and TCM by Fe^0 showed that the process of the electron transfer from Fe^0 to the CAHs is the dominant step limiting the rate of the dechlorination reactions so that the electron released from Fe^0 is most likely in competition with TCE, 1,1,1-TCA and TCM during the dechlorination reactions. In addition to CAHs, abiotic reduction of hexavalent chromium [Cr(VI)] by Fe^0 also exerted effects on TCE dechlorination leading to a 31% drop in TCE k_{SA} after the addition of Cr(VI) into the TCE-contaminated groundwater. Groundwater geochemical factors such as alkalinity and contaminant concentration could potentially influence competition among TCE, 1,1,1-TCA, TCM and Cr(VI) during the abiotic reduction of chemical substances by Fe^0.

CE Database subject headings: Chromium; TCE; Trihalomethanes; Competition; Iron.

---

[1]Associate Professor, Dept. of Civil Engineering, The Hong Kong University of Science and Technology, Clear Water Bay, Kowloon, Hong Kong (corresponding author). E-mail: cemclo@ust.hk
[2]Research Associate, Dept. of Civil Engineering, The Hong Kong University of Science and Technology, Clear Water Bay, Kowloon, Hong Kong.
[3]Research Assistant, Dept. of Civil Engineering, The Hong Kong University of Science and Technology, Clear Water Bay, Kowloon, Hong Kong.
[4]Duke Energy Distinguished Professor and Director, Global Institute for Energy and Environmental Systems, University of North Carolina, Charlotte, North Carolina, United States.

4.1 Introduction

In the United States, there are about 350,000 sites contaminated with a wide range of toxic chemicals (Scherer et al. 2000). Among various groundwater contaminants, chlorinated aliphatic hydrocarbons (CAHs) such as trichloroethylene (TCE), 1,1,1-trichloroethane (1,1,1-TCA) and chloroform (TCM) are the most prevalent because of their wide spread use as cleansing and degreasing agents (National Research Council 1994; O'Hannesin and Gillham 1998). Since CAHs are carcinogenic and recalcitrant in the natural environment, cost-effective methods of cleaning CAH-contaminated groundwater are highly desirable.

In recent years, permeable reactive barrier (PRB) technology using zero-valent iron (Fe^0) as reactive media has received much interest as an alternative to pump-and-treat technology for removing CAHs from contaminated groundwater (McMahon et al. 1999). Dechlorination of CAHs by Fe^0 is a surface reaction that involves a series of redox reactions (Weber 1996; Scherer et al. 2000). Under both aerobic and anaerobic conditions, the electrons released through Fe^0 oxidation enable sequential reduction of CAHs into less chlorinated or non-chlorinated forms with concomitant release of hydroxide ions and hydrogen gas (Powell et al. 1998). As described by Arnold and Roberts (2000), hydrogenolysis [Eq. (4.1)] and reductive-β-elimination [Eq. (4.2)] are the main pathways involved in the dechlorination reactions of CAHs by Fe^0. Besides, catalytic hydrogenolysis of CAHs can also occur through the use of the hydrogen gas released from the Fe^0 oxidation [Eq. (4.3)] (Matheson and Tratnyek 1994).

$$RHCl = RHCl + 2e^- + H^+ \rightarrow RH_2 = RHCl + Cl^- \tag{4.1}$$

$$RHCl = RHCl + 2e^- \rightarrow RH \equiv RH + 2Cl^- \tag{4.2}$$

$$RHCl = RHCl + H_2 \xrightarrow{Catalyst} RH_2 = RHCl + H^+ + Cl^- \tag{4.3}$$

Fe^0 reactivity in the remediation of CAH-contaminated groundwater has been widely studied. Investigations include dechlorination mechanisms and kinetics, as well as the influence of groundwater geochemistry, seepage velocity, temperature and microbiology (Arnold and Roberts 2000; Gu et al. 2002; Lai and Lo 2002; Lo et al. 2003; Andrea et al. 2005). However, only a few studies have addressed the competitive effects induced by the presence of other CAHs or other groundwater contaminants such as heavy metals and nutrients on the dechlorination reaction with Fe^0 (Schlicker et al. 2000; Gandhi et al. 2002; Lo et al. 2004). In this paper, competitive effects among 1,1,1-TCA, TCM and hexavalent chromium [Cr(VI)] on the dechlorination of TCE, which is ranked as the most frequently detected groundwater contaminant in the United States, was primarily studied (National Research Council 1994). Furthermore, the competitive effects of TCE on the Fe^0 reactivity in the remediation of 1,1,1-TCA and Cr(VI) are also discussed. Simultaneous dechlorination of TCE, 1,1,1-TCA and TCM by Fe^0 was studied at various temperatures so as to investigate whether the released electron from Fe^0 oxidation is being competed by varied groundwater contaminants during the abiotic reduction process. Competition between TCE and Cr(VI) in groundwater that contained high alkalinity was also investigated in order to explore the influence of groundwater geochemistry on the competition.

4.2 Materials and Methods

4.2.1 Materials

The Fe^0 filings (ETI-CC-1004) used in this study were obtained from *Connelly-GPM Inc*. The grain size, surface area and particle density of the granular Fe^0 were 0.25 to 2.00 mm, 1.80 m^2/g and 6.43 g/cm^3, respectively. Columns were all packed with 100% granular Fe^0 with a porosity of approximately 0.60 and bulk density of about 2.57 g/cm^3. Synthetic groundwater containing TCE, 1,1,1-TCA or TCM was prepared by first dissolving corresponding amounts of TCE (ACS Reagent, *Acros Organics*), 1,1,1-TCA (UN2831, *Fisher Scientific*) or TCM (A3505E, *Labscan*) into methanol (ChromAR®HPLC, *Mallinckrodt*). The methanolic standards were then spiked into Millipore water in a ratio of 20 μL to 100 mL of water to make the desired concentration (standard method code 6232B). Groundwater containing Cr(VI) or bicarbonate ion (HCO_3^-) was prepared by dissolving potassium dichromate (12255, *Riedel-deHaën)* or sodium hydrogen carbonate (S5761, *Sigma*) into Millipore water (APHA/AWWA/WEF 1998). After the preparation, the synthetic groundwater was transferred into 5-L collapsible Teflon bag for further storage or connecting to the Fe^0 packed column.

Columns used in this study were made of rigid PVC with 3.6 cm in internal diameter and 30 cm long (Fig. 4-1). The synthetic groundwater was fed into the columns in an up-flow manner by a multi-channel peristaltic pump at a groundwater velocity of 400 ± 10 m/yr. Except for viton tubing used in the pump, all the tubing used in the influent and effluent lines of the columns were made of Teflon. One pore volume (PV) of the Fe^0 packed column is equivalent to 183 mL and the required time of running one PV is about 6.6 hrs.

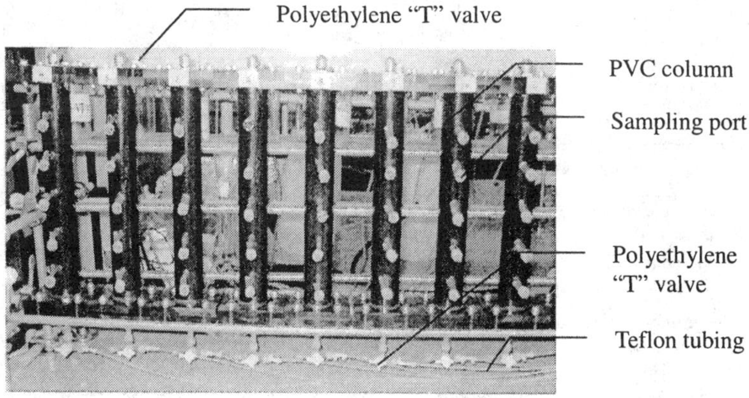

Fig. 4-1. Experimental setup of the column experiments

4.2.2 Experimental Methods

For the purpose of investigating competition among TCE, 1,1,1-TCA, TCM and Cr(VI) during their abiotic reduction by Fe^0, eleven columns were set up as summarized in Table 4-1. Col-A fed by TCE-contaminated groundwater, was the control column for the investigation of the competitive effects of 1,1,1-TCA (Col-C) and Cr(VI) (Col-I) on TCE dechlorination. Col-B fed by the synthetic groundwater merely containing 1,1,1-TCA, was the control column for studying the competitive effects induced by TCE on the 1,1,1-TCA dechlorination (Col-C). Col-C, Col-D and Col-E were designed to remediate the synthetic groundwater containing 160, 100 and 60 mg/L of TCE and 1,1,1-TCA, respectively, so as to explore the relationship between the competitive effects and contaminant concentration. Col-F, Col-G and Col-H were conducted at 4, 10 ad 22 ^{0}C, respectively to treat the synthetic groundwater spiked with 100 mg/L of TCE, 1,1,1-TCA and TCM. Col-J fed by the synthetic groundwater only containing Cr(VI) was the control column for investigating the competitive effects of TCE on the Cr(VI) reduction by Fe^0 (Col-I). In addition to TCE and Cr(VI), 200 mg/L as $CaCO_3$ of HCO_3^- was also spiked into the synthetic groundwater for Col-K to study the competition between TCE and Cr(VI) in high alkalinity groundwater.

Table 4-1. Operating Conditions and Compositions of the Synthetic Groundwater for the Column Experiments

Laboratory Columns	[a]Col-A	[b]Col-B	Col-C	Col-D	Col-E
Operating Conditions					
Seepage Velocity (m/yr)	400	400	400	400	400
Temperature (^{0}C)	22	22	22	22	22
Porosity	0.6	0.6	0.6	0.6	0.6
Composition of the Synthetic Groundwater					
[TCE] (mg/L)	160	na	160	100	60
[1,1,1-TCA] (mg/L)	na	160	160	100	60
[TCM] (mg/L)	na	na	na	na	na
[Cr(VI)] (mg/L)	na	na	na	na	na
[HCO_3^-] (mg/L as $CaCO_3$)	na	na	na	na	na

Laboratory Columns	Col-F	Col-G	Col-H	Col-I	[c]Col-J	Col-K
Operating Conditions						
Seepage Velocity (m/yr)	400	400	400	400	400	400
Temperature (^{0}C)	4	10	22	22	22	22
Porosity	0.6	0.6	0.6	0.6	0.6	0.6
Composition of the Synthetic Groundwater						
[TCE] (mg/L)	100	100	100	160	na	160
[1,1,1-TCA] (mg/L)	100	100	100	na	na	na
[TCM] (mg/L)	100	100	100	na	na	na
[Cr(VI)] (mg/L)	na	na	na	25	25	25
[HCO_3^-] (mg/L as $CaCO_3$)	na	na	na	na	na	200

Note: na refers to not applicable. a indicates the control column for Col-C and Col-I. b refers to the control column for Col-C and c represents the control column for Col-I.

Groundwater samples were collected every 5 or 10 pore volumes from the sampling ports along the column as well as from the polyethylene "T" valves connected to both ends of the column. In total, five sampling ports equipped with Teflon-faced septum were positioned along the column every 2.5 or 5 cm from the inlet end (see Fig. 4-1). Before collecting the groundwater samples, about 0.5 mL of the solution was first withdrawn so as to remove the stagnant water from the tubing of the sampling ports. To minimize the disturbance from the groundwater sampling to the Fe^0 packed column, a maximum 3 mL of sample solution was collected from a sampling port in each time of groundwater sampling and then it was transferred to a 8 mL-glass vial for chemical analyses.

The concentrations of TCE, 1,1,1-TCA and TCM in the groundwater samples were immediately measured after sampling using the method of liquid-liquid extraction with pentane and a gas chromatograph (HP 5890 Series II GC, *Hewlett-Packard*) equipped with an electron capture detector (ECD). Total chromium concentration was determined using a flame atomic absorption spectrophotometer (Z-8200, *Hitachi*) and 1,5-diphenylcarbohydrazide spectrophotometric method (DR/2000, *Hach*) was used in measuring Cr(VI) concentration. Groundwater alkalinity was measured by titration method using 0.02 N sulfuric acid as the titrant and a bromcresol green as the indicator solution (APHA/AWWA/WEF 1998).

4.2.3 Data Analysis

The results of research by Gillham and O'Hannesin (1994) have supported the notion that dechlorination reaction of CAHs by Fe^0 is in pseudo first-order kinetics with respect to CAH concentration. As a consequence, the rate of dechlorination of CAHs by Fe^0 can be expressed by Eq. (4.4).

$$\frac{dC}{dt} = -k_{obs}C \tag{4.4}$$

where C is the CAH concentration (mg L^{-1}), k_{obs} is the observed first-order dechlorination rate constant (hr^{-1}) and t is time (hr).

In a laboratory column packed with Fe^0 for CAH dechlorination, the k_{obs} of CAHs can be simply determined from their concentration profiles along the column in which they are equal to the slope of the graph of $-\ln(C/C_0)$ against L/v_s [i.e., Eq. (4.5)].

$$-\ln\left(\frac{C}{C_0}\right) = k_{obs}\frac{L}{v_s} \tag{4.5}$$

where L is the longitudinal distance along a Fe^0 packed column (m), C_0 is the initial CAH concentration (mg L^{-1}), and v_s is the longitudinal seepage velocity (m hr^{-1}). In case the ratio of the Fe^0 surface area to solution volume (i.e., λ_s, m^2 mL^{-1}) for a Fe^0 packed medium is known, the normalized dechlorination rate constant (k_{SA}, mL m^{-2} hr^{-1}) of CAHs can be directly determined using Eq. (4.6).

$$k_{SA} = \frac{k_{obs}}{\lambda_s} \tag{4.6}$$

4.3 Results and Discussion

4.3.1 *Dechlorination of CAHs by Fe^0*

Analysis of the CAH concentration in the collected groundwater samples indicated a marked decrease in CAH concentration along the Fe^0 packed column. Fig. 4-2 illustrates the typical CAH concentration profile along the Fe^0 packed column. As seen, there was a sharp reduction in TCE concentration in the first 5 cm of the column and afterwards, the TCE concentration decreased gradually along the rest of the column. Reactivity of Fe^0 in CAH dechlorination was usually high at an initial stage. As shown in Fig. 4-2, complete dechlorination of TCE could be achieved inside Col-I at 4.4 PV. Afterwards, the TCE concentration profiles moved upward gradually and tended to converge with each other after reaching 80.6 PV. Generally, the CAH concentration profile along the column tended to become steady after running for 20 to 90 PV. In this paper, TCE, 1,1,1-TCA and TCM k_{SA} determined from their steady concentration profiles were only considered and discussed to investigate competition among TCE, 1,1,1-TCA and TCM during dechlorination by Fe^0. For instance, the TCE k_{SA} for Col-I computed from the steady concentration profiles using Eqs. (4.5) and (4.6) ranged from 1.23 to 3.01 × 10^{-2} mL m^{-2} hr^{-1}. Table 4-2 summarizes the k_{SA} of TCE, 1,1,1-TCA and TCM obtained at different PV for Col A to Col-I and Col-K.

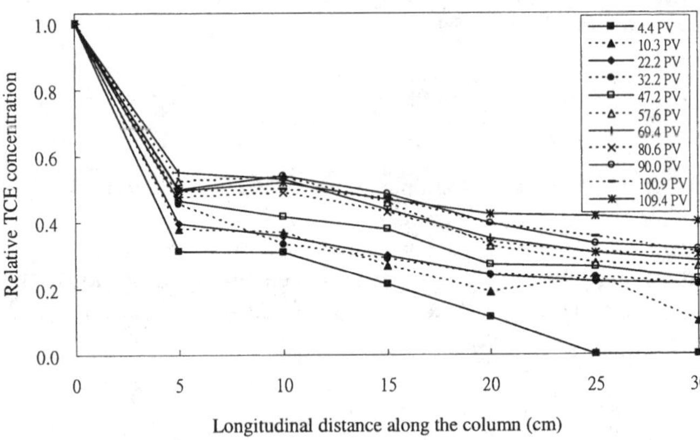

Fig. 4-2. TCE concentration profiles along Col-I

4.3.2 *Competition between TCE and 1,1,1-TCA*

Competitive effect of 1,1,1-TCA on the TCE dechlorination was investigated by comparing the TCE k_{SA} obtained from Col-A and Col-C. As shown in Table 4-3, TCE k_{SA} was about 3.04 × 10^{-2} mL m^{-2} hr^{-1} when there was only 160 mg/L of TCE in the

Table 4-2. Normalized Dechlorination Rate Constant (k_{SA}) of TCE, 1,1,1-TCA and TCM for Col-A to Col-I and Col-K Obtained at Varied PV

Col-A PV	TCE $k_{SA} \times 10^{-2}$	Col-B PV	1,1,1-TCA $k_{SA} \times 10^{-2}$	Col-C PV	TCE $k_{SA} \times 10^{-2}$	1,1,1-TCA $k_{SA} \times 10^{-2}$	Col-D PV	TCE $k_{SA} \times 10^{-2}$	1,1,1-TCA $k_{SA} \times 10^{-2}$
10.0	3.81	10.0	58.74	10.0	na	34.31	8.8	3.97	40.77
18.9	3.61	18.6	53.73	18.6	3.70	29.88	19.8	3.78	29.25
31.4	**3.14**	31.4	55.45	31.2	3.67	24.24	29.8	3.64	28.76
39.3	**2.95**	39.3	41.01	39.1	3.77	22.58	40.6	**3.23**	24.93
48.3	**3.02**	48.3	**34.15**	48.2	**2.86**	**17.91**	50.5	**3.18**	**24.41**
58.5	**3.06**	58.4	**38.09**	57.3	**2.61**	**17.55**	58.8	**2.97**	**22.95**

Col-E PV	TCE $k_{SA} \times 10^{-2}$	1,1,1-TCA $k_{SA} \times 10^{-2}$	Col-F PV	TCE $k_{SA} \times 10^{-2}$	1,1,1-TCA $k_{SA} \times 10^{-2}$	TCM $k_{SA} \times 10^{-2}$	Col-G PV	TCE $k_{SA} \times 10^{-2}$	1,1,1-TCA $k_{SA} \times 10^{-2}$
10.0	4.37	36.87	5.0	1.24	8.36	1.67	5.0	1.31	18.51
17.9	4.17	36.48	13.7	1.05	7.03	1.48	13.7	1.12	15.09
31.1	4.12	34.31	19.2	0.82	5.72	1.24	19.2	0.87	11.12
39.1	4.07	**33.65**	26.4	0.74	5.45	1.15	26.4	0.80	10.24
48.2	**4.02**	**33.49**	30.0	**0.57**	4.95	**0.99**	30.0	0.85	12.94
58.2	**3.93**	**33.32**	37.6	**0.55**	**5.85**	**0.87**	37.6	**0.71**	9.75
			41.2	**0.54**	**5.53**	**1.07**	41.2	**0.73**	**8.89**
			57.5	**0.51**	**5.37**	**0.99**	46.6	**0.76**	**9.32**
							57.5	**0.75**	**9.08**

Col-G PV	TCM $k_{SA} \times 10^{-2}$	Col-H PV	TCE $k_{SA} \times 10^{-2}$	1,1,1-TCA $k_{SA} \times 10^{-2}$	TCM $k_{SA} \times 10^{-2}$	Col-I PV	TCE $k_{SA} \times 10^{-2}$	Col-K PV	TCE $k_{SA} \times 10^{-2}$
5.0	2.38	3.6	2.21	22.08	7.26	4.4	7.75	4.4	13.96
13.7	1.94	9.1	1.90	15.88	5.16	10.3	6.54	10.4	3.18
19.2	1.51	16.3	1.88	17.86	4.86	22.2	3.10	22.0	2.43
26.4	3.74	19.7	**1.54**	12.32	4.79	32.2	3.33	32.0	2.43
30.0	**1.42**	23.5	**1.46**	19.36	5.08	47.2	3.42	47.0	2.09
37.6	**1.42**	34.3	**1.40**	**14.30**	**3.96**	57.6	3.70	57.3	2.68
41.2	**1.59**	51.5	**1.47**	**14.27**	**4.01**	69.4	3.44	69.0	2.8
46.6	na					80.6	**2.40**	80.2	1.51
57.5	**1.56**					90.0	**3.01**	89.5	**1.15**
						100.9	**2.10**	100.4	**1.16**
						109.4	**1.23**	108.7	**0.80**

Note: Bold figure is steady k_{SA} computed from steady concentration profile. na refers to not available.

Table 4-3. Experimental results of Col-A to Col-K

Laboratory Columns	Col-A	Col-B	Col-C	Col-D	Col-E
TCE k_{SA}	3.04×10^{-2}	na	2.74×10^{-2}	3.13×10^{-2}	3.98×10^{-2}
1,1,1-TCA k_{SA}	na	36.12×10^{-2}	17.73×10^{-2}	23.68×10^{-2}	33.49×10^{-2}
TCM k_{SA}	na	na	na	na	na
Cr(VI) Removal Capacity	na	na	na	na	na

Laboratory Columns	Col-F	Col-G	Col-H	Col-I	Col-J	Col-K
TCE k_{SA}	0.54×10^{-2}	0.74×10^{-2}	1.44×10^{-2}	2.10×10^{-2}	na	1.00×10^{-2}
1,1,1-TCA k_{SA}	5.59×10^{-2}	9.10×10^{-2}	14.30×10^{-2}	na	na	na
TCM k_{SA}	0.98×10^{-2}	1.50×10^{-2}	3.99×10^{-2}	na	na	na
Cr(VI) Removal Capacity	na	na	na	2.2	4.1	2.5

Note: na refers to not applicable. The k_{SA} listed in this table is obtained by averaging the steady k_{SA} shown in Table 4-2 and its unit is mL m^{-2} hr^{-1}. The unit of Cr(VI) removal capacity is mg Cr/ g Fe^0. Cr(VI) removal capacity is determined based upon the normalized migration rate of the Cr(VI) front obtained at a relative Cr(VI) concentration of 0.3.

synthetic groundwater (i.e., Col-A). However, TCE k_{SA} decreased to 2.74×10^{-2} mL m^{-2} hr^{-1} in the presence of the same concentration of 1,1,1-TCA in the synthetic groundwater (i.e., Col-C). In a similar fashion, the presence of TCE in the groundwater also exerted detrimental effect on the 1,1,1-TCA dechlorination by Fe^0.

The 1,1,1-TCA k_{SA} dropped from 36.12 × 10^{-2} (Col-B) to 17.73 × 10^{-2} mL m^{-2} hr^{-1} (Col-C) with the addition of TCE, thereby indicating the competition between TCE and 1,1,1-TCA during the dechlorination by Fe^0. In comparison to the TCE and 1,1,1-TCA dechlorination reactions, it was found that the detrimental impact exerted by TCE dechlorination on 1,1,1-TCA dechlorination was greater than that exerted by 1,1,1-TCA dechlorination on TCE dechlorination. This is because there was only about 10% decrease in TCE k_{SA} in the presence of 1,1,1-TCA, whereas there was more than 50% drop in 1,1,1-TCA k_{SA} with the addition of TCE.

To explore the relationship between the competition and contaminant concentration, competition between TCE and 1,1,1-TCA during the dechlorination was studied at varied levels of contaminant concentration. Col-C was repeated but the concentration of TCE and 1,1,1-TCA was at 100 mg/L for Col-D and 60 mg/L for Col-E. As shown in Table 4-3, both TCE and 1,1,1-TCA k_{SA} values increased with decreasing the contaminant concentration. TCE and 1,1,1-TCA k_{SA} increased by about 45 and 89%, respectively, when the contaminant concentration decreased from 160 to 60 mg/L. The possible reasons for this phenomenon may be due to a decrease in the competitive effects induced by 1,1,1-TCA on TCE dechlorination or vice versa at low levels of contaminant concentration.

4.3.3 Competition among TCE, 1,1,1-TCA and TCM at Various Temperatures

Information on the competitive effect of TCM on TCE and 1,1,1-TCA dechlorination can be obtained by comparing their k_{SA} values obtained from Col-D and Col-H. According to Table 4-3, the TCE and 1,1,1-TCA k_{SA} for Col-D fed by the synthetic groundwater containing 100 mg/L of TCE and 1,1,1-TCA were 3.13 × 10^{-2} and 23.68 × 10^{-2} mL m^{-2} hr^{-1}, respectively. However, addition of another 100 mg/L of TCM into the groundwater noticeably reduced the TCE k_{SA} to 1.44 × 10^{-2} mL m^{-2} hr^{-1} and decreased the 1,1,1-TCA k_{SA} to 14.30 × 10^{-2} mL m^{-2} hr^{-1} (Col-H). Approximately, 40 to 54% drop in TCE and 1,1,1-TCA k_{SA} after the addition of TCM conclusively proves that there is competitive effect of TCM on TCE and 1,1,1-TCA dechlorination.

As stated above, CAH dechlorination reaction induced by Fe^0 is a surface reaction involving a series of redox reactions. Therefore, close contact between CAHs and the reactive sites on the Fe^0 surface as well as the transfer of electron from Fe^0 to the CAHs are the key steps for the occurrence of the dechlorination reaction (Fogler 1992; Weber 1996). The presence of more than one type of CAHs in the synthetic groundwater inside the Fe^0 packed column may result in competition for reactive sites or the released electron or both during the dechlorination reactions.

To investigate whether the process of the electron transfer from Fe^0 to the CAHs is a slow step or rate-limiting step of the reactions leading to the competition of the released electron among TCE, 1,1,1-TCA and TCM during the dechlorination reactions, Col-H was repeated at 4 (Col-F) and 10 ^{0}C (Col-G). The aim was to compute the corresponding activation energies (E_a) for the simultaneous dechlorination of TCE, 1,1,1-TCA and TCM by Fe^0. As shown in Table 4-3, by decreasing the temperature from 22 to 4 ^{0}C, there was approximately 60 to 75% drop in TCE, 1,1,1-TCA and TCM k_{SA}. In analogy to the system that only contained TCE in a feed solution (Su and Puls 1999), the relationship between k_{SA} and temperature in the system containing TCE, 1,1,1-TCA and TCM can also be described by Arrhenius equation [Eq. (4.7)], which is illustrated in Fig. 4-3. The E_a for the simultaneous dechlorination of TCE, 1,1,1-TCA and TCM by Fe^0 could be obtained from the slope of a plot of ln k_{SA} against $1/T$, which were 37.1, 34.3 and 53.7 kJ/mol, respectively. It is generally accepted that an E_a of greater than 15 kJ/mol indicates the domination of the electron transfer process in limiting the rate of the dechlorination reaction because of the slow rate of the electron transfer from Fe^0 to CAHs (Su and Puls 1999). In the circumstance in which there is limited supply of the released electrons, the possibility was high for TCE, 1,1,1-TCA and TCM to compete for the released electron during the dechlorination reactions, thereby leading to competition among TCE, 1,1,1-TCA and TCM.

$$\ln k_{SA} = \ln A - \frac{E_a}{RT} \tag{4.7}$$

where A is the pre-exponential factor (mL m^{-2} hr^{-1}), E_a is the activation energy (kJ mol^{-1}), R is the ideal gas constant (kJ mol^{-1} K^{-1}) and T is temperature (K).

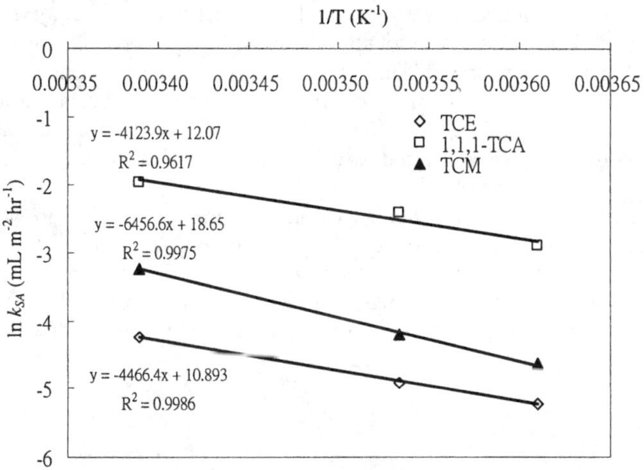

Fig. 4-3. Relationship between the TCE, 1,1,1-TCA and TCM k_{SA}, and temperature obtained from Col-F, Col-G and Col-H

4.3.4 Competition between TCE and Cr(VI)

In addition to CAHs, Cr(VI) is another groundwater contaminant that can be effectively remediated by a Fe^0 PRB (Wilkin and Puls 2003; RTDF 2001). Theoretically, Fe^0 donates electrons to first reduce Cr(VI) to trivalent chromium [Cr(III)] and it is simultaneously oxidized to ferric iron [Eq. (4.8)]. The reduced

Cr(III) is then removed from groundwater through the precipitation of chromium hydroxide [Eq. (4.9)], co-precipitation of mixed chromium-iron hydroxide solid as delineated in [Eq. (4.10)] (Powell et al. 1995; Blowes et al. 1997) or mixed chromium-iron oxyhydroxide solid as shown in [Eq. (4.11)] (Eary and Rai 1988; Schwertmann et al. 1989). Since abiotic reduction of Cr(VI) to Cr(III) also requires the electron released from Fe^0, TCE may compete with Cr(VI) during the abiotic reduction by Fe^0. As illustrated in Table 4-3, introduction of 25 mg/L of Cr(VI) into the synthetic groundwater containing 160 mg/L of TCE (Col-I) resulted in about 31% drop in TCE k_{SA} (from 3.04×10^{-2} in Col-A to 2.10×10^{-2} mL m^{-2} hr^{-1} in Col-I), thereby showing the competitive effect of Cr(VI) on TCE dechlorination.

$$Cr_2O_7^{2-} + 2Fe^0_{(s)} + 7H_2O_{(l)} \Leftrightarrow 2Cr^{3+} + 14OH^- + 2Fe^{3+} \tag{4.8}$$
$$2Cr^{3+} + 6OH^- \Leftrightarrow 2Cr(OH)_{3(s)} \tag{4.9}$$
$$(1\text{-}x)Fe^{3+} + xCr^{3+} + 3H_2O_{(l)} \Leftrightarrow Cr_xFe_{1-x}(OH)_{3(s)} + 3H^+ \tag{4.10}$$
$$(1\text{-}x)Fe^{3+} + xCr^{3+} + 2H_2O_{(l)} \Leftrightarrow Cr_xFe_{1-x}OOH_{(s)} + 3H^+ \tag{4.11}$$

Competition of TCE on Cr(VI) reduction by Fe^0 was studied by comparing the capacity of Fe^0 to remove Cr(VI) from the groundwater (i.e., Cr(VI) removal capacity) with and without the introduction of TCE. When Cr(VI)-contaminated groundwater was continuously fed into the Fe^0 packed column, chromium or chromium-iron precipitates were formed which exhausted Fe^0, thereby leading to the migration of Cr(VI) fronts through the column. The breakthrough curves, as illustrated in Fig. 4-4a, showed the Cr(VI) migration pattern in Col-K in which the propagation of the Cr(VI) migration front was almost constant throughout the experiment as shown in Fig. 4-4b (Lo et al. 2005). The normalized migration rate of the Cr(VI) front, which is defined as the distance traveled by the Cr(VI) front at a desired relative concentration per number of PV of Cr(VI)-contaminated groundwater passing through the column divided by the volume of one PV, provides a good parameter to evaluate Cr(VI) removal capacity of Fe^0. The slower the normalized migration rate of the Cr(VI) front, the higher the Cr(VI) removal capacity of Fe^0. After knowing the normalized migration rate, the Cr(VI) removal capacity can be estimated using Eq. (4.12). Table 4-3 summarizes the Cr(VI) removal capacity of Fe^0 obtained from Col-I, Col-J and Col-K.

$$RC = \frac{[Cr(VI)]}{M \cdot A \cdot \rho_b} \tag{4.12}$$

where A is the cross section area of the column (cm^2), $[Cr(VI)]$ is the initial concentration of Cr(VI) (mg L^{-1}), M is the normalized migration rate of Cr(VI) front (cm cm^{-3}) at a desired relative concentration, RC is the Cr(VI) removal capacity of Fe^0 (mg Cr/g Fe^0) and ρ_b is the bulk density of Fe^0 filled in the column (g L^{-1}).

From Table 4-3, it is clear that the addition of TCE into the Cr(VI)-contaminated groundwater affected the Cr(VI) removal capacity of Fe^0. The latter decreased from

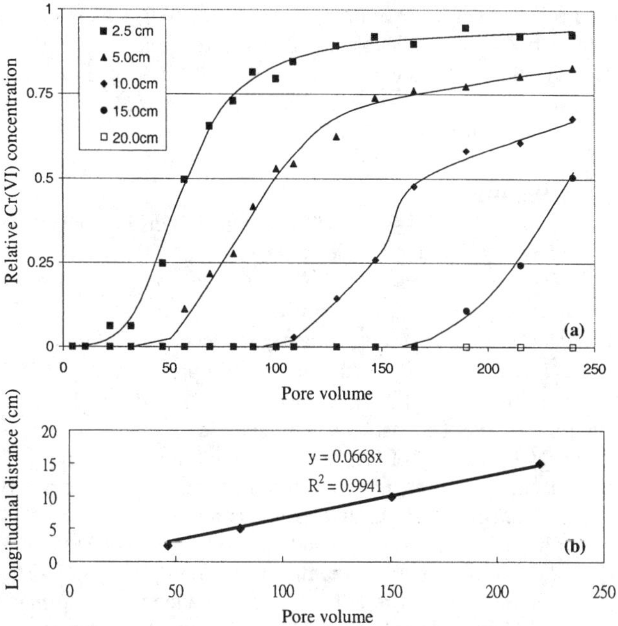

Fig. 4-4. (a) Cr(VI) breakthrough curves of Col-K (this Cr(VI) breakthrough pattern can be applied to Col-I and Col-J) and (b) migrated distance of the Cr(VI) moving front at relative Cr(VI) concentration of 0.3 in Col-K (the slope of the line indicates the migration rate of the Cr(VI) front along the column).

4.1 in Col-J to 2.2 mg Cr/g Fe^0 in Col-I. Similar to the competition among TCE, 1,1,1-TCA and TCM, TCE and Cr(VI) may also compete for the released electron from Fe^0 or the reactive site on the Fe^0 surface or both during the abiotic reduction by Fe^0. It is observed that TCE and Cr(VI) were reduced simultaneously in Col-I and the dechlorination of TCE still occurred even the Fe^0 was exhausted for the Cr(VI) reduction. This possibly indicates that there are some difference in the reactive sites of TCE and Cr(VI) reduction by Fe^0. Thus, competition between TCE and Cr(VI) during the abiotic reduction is less likely attributed to the competition for reactive sites on the Fe^0 particle surface.

Comparison of the TCE k_{SA} and Cr(VI) removal capacity of Fe^0 between Col-I and Col-K showed that there was another 52% drop in TCE k_{SA}, whereas the Cr(VI) removal capacity slightly increased from 2.2 to 2.5 mg Cr/g Fe^0 after introducing 200 mg/L as $CaCO_3$ of alkalinity into the synthetic groundwater. The decrease in the TCE k_{SA} after the addition of alkalinity was probably due to the formation of carbonate precipitates on the Fe^0 surface. Carbonate precipitates can block reactive sites on Fe^0 surface, thereby significantly reducing the rate of TCE dechlorination. The decrease in the rate of the TCE dechlorination consequently reduced the amount of the released electron used for the TCE dechlorination. Therefore, more released electrons could be used by Cr(VI) for the abiotic reduction to Cr(III), thereby resulting in the increase in the Cr(VI) removal capacity. The difference in the TCE k_{SA} and Cr(VI) removal capacity of Fe^0 after the addition of alkalinity indicated that the competition between TCE and Cr(VI) during the abiotic reduction by Fe^0 could be influenced by groundwater geochemistry.

4.4 Conclusions

Competition among TCE, 1,1,1-TCA, TCM and Cr(VI) during their abiotic reduction by Fe^0 occurred with a consequent decrease in the dechlorination rate constant of CAHs and Cr(VI) removal capacity of Fe^0. As regards the competition between TCE and 1,1,1-TCA during dechlorination reactions, the presence of 1,1,1-TCA in the TCE-contaminated groundwater consequently reduced the TCE k_{SA} from 3.04×10^{-2} to 2.74×10^{-2} mL m^{-2} hr^{-1}. Furthermore, the presence of TCE in synthetic groundwater also decreased 1,1,1-TCA k_{SA} from 36.12×10^{-2} to 17.73×10^{-2} mL m^{-2} hr^{-1}. In addition, in the Fe^0 packed columns fed by the synthetic groundwater containing both TCE and 1,1,1-TCA, their k_{SA} were found increasing with decreasing the contaminant concentration, most likely due to the small competitive effects at low contaminant concentration. In a similar manner, the addition of TCM into the synthetic groundwater containing TCE and 1,1,1-TCA led to 40 to 54% drop in TCE and 1,1,1-TCA k_{SA} and thus indicates competition among TCE, 1,1,1-TCA and TCM. E_a of 37.1, 34.3 and 53.7 kJ/mol for the simultaneous dechlorination of TCE, 1,1,1-TCA and TCM, respectively showed that the process of electron transfer from Fe^0 to CAHs was the dominant step limiting the rate of the dechlorination reactions. Therefore, electron released from Fe^0 through oxidation was most likely in competition with TCE, 1,1,1-TCA and TCM during dechlorination by Fe^0.

Competition between TCE and Cr(VI) during their abiotic reduction by Fe^0 was also observed. Abiotic reduction of Cr(VI) by Fe^0 could lead to a 31% drop in TCE k_{SA}. Similarly, the presence of TCE in the Cr(VI)-contaminated groundwater decreased Cr(VI) removal capacity of Fe^0 from 4.1 to 2.2 mg Cr/ g Fe^0. Since the dechlorination of TCE still occurred even though the Fe^0 was exhausted for the Cr(VI) reduction, competition between TCE and Cr(VI) during the abiotic reduction is less likely caused by competition for reactive sites on the Fe^0 surface. Another important point is that groundwater geochemistry might influence the competition between TCE and Cr(VI). Introduction of alkalinity at 200 mg/L as $CaCO_3$ into the groundwater containing TCE and Cr(VI) caused about 52% reduction in TCE k_{SA}, probably due to the detrimental impacts of carbonate precipitates formed on the Fe^0 surface. The decrease in the rate of the TCE dechlorination led to an increase in Cr(VI) removal capacity of Fe^0 since more of the released electrons could be used for the abiotic reduction of Cr(VI).

4.5 References

Andrea, P. D., Lai, K. C.-K., Kjeldsen, P., and Lo, I. M.-C. (2005). "Effect of groundwater inorganics on the reductive dechlorination of TCE by zero-valent iron." *J. of Water, Air and Soil Pollution*, 162, 401-420.

APHA, AWWA and WEF (1998). *Standard methods for the examination of water and wastewater*, 20th Ed., American Public Health Association, American Water Works Association and Water Environment Federation, Washington DC, United States.

Arnold, W. A., and Roberts, A. L. (2000). "Pathways and kinetics of chlorinated ethylene and chlorinated acetylene reaction with Fe(0) particles." *Environ. Sci. & Technol.*, 34(9), 1794-1805.

Blowes, D. W., Ptacek, C. J., and Jambor, J. L. (1997). "In-situ remediation of Cr(VI)-contaminated groundwater using permeable reactive walls: Laboratory studies." *Environ. Sci. & Technol.*, 31(12), 3348-3357.

Eary, L. E., and Rai, D. (1988). "Chromate removal from aqueous wastes by reduction with ferrous ion." *Environ. Sci. & Technol.*, 22(8), 972-977.

Fogler, H. S. (1992). *Elements of chemical reaction engineering*, 2nd Ed., Prentice-Hall Inc., Englewood Cliffs, New Jersey.

Gandhi, S., Oh, B. T., Schnoor, J. L., and Alvarez, P. J. J. (2002). "Degradation of TCE, Cr(VI), sulfate, and nitrate mixtures by granular iron in flow-through columns under different microbial conditions." *Water Res.*, 36(8), 1973-1982.

Gillham, R. W., and O'Hannesin, S. F. (1994). "Enhanced degradation of halogenated aliphatics by zero-valent iron." *Ground Water*, 32, 958-967.

Gu, B., Watson, D. B., Wu, L. Y., Phillips, D. H., White, D. C., and Zhou, J. Z. (2002). "Microbiological characteristics in a zero-valent iron reactive barrier." *Environ. Monit. Assess.*, 77, 293-309.

Lai, K .C.-K., and Lo, I. M.-C. (2002). "Bench-scale study of the effects of seepage velocity on the dechlorination of TCE and PCE by zero-valent iron." In *Proceeding of the 6th International Symposium on Environmental Geotechnology*, Seoul, Korea, 181-186.

Lo, I. M.-C., Lam, C. S.-C., and Chan, K. K. (2003). "Effects of groundwater characteristics on chromate removal by permeable reactive barriers." In *Proceeding of the International Conference on Contaminated Land*, Gent, Belgium.

Lo, I. M.-C., Lai, K. C.-K., and Kjeldsen, P. (2004). "Field monitoring of Fe^0 PRB for removal of chlorinated hydrocarbons." In *Proceeding of the 4th International Conference of Remediation of Chlorinated and Recalcitrant Compounds*, Monterey, California.

Lo, I. M.-C., Lam, C. S.-C., and Lai, K. C.-K. (2005). "Competitive effects of TCE on Cr(VI) removal by zero-valent iron." *J. Environ. Eng. ASCE*, 131(11), 1598-1606.

Matheson, L. J., and Tratnyek, P. G. (1994). "Reductive dehalogenation of chlorinated methanes by iron metal." *Environ. Sci. & Technol.*, 28(12), 2045-2053.

McMahon, P. B., Dennehy, K. F., Sandstrom, M. W. (1999). "Hydraulic and geochemical performance of a permeable reactive barrier containing zero-valent Iron, Denver Federal Centre." *Ground Water*, 37(3), 396-404.

National Research Council (1994). *Alternatives for ground water cleanup*, National Academy Press, Washington.

O'Hannesin, S. F., and Gillham, G. W. (1998). "Long-term performance of an in situ "iron wall" for remediation of VOCs." *Ground Water*, 36(1), 164-170.

Powell, R. M., Puls, R. W., Hightower, S. K., and Sabatini, D. A. (1995). "Coupled iron corrosion and chromate reduction: Mechanisms for subsurface remediation." *Environ. Sci. & Technol.*, 29(8), 1913-1922.

Powell, R. M., Blowes, D. W., Gillham, R. W., Schultz, D., Sivavec, T., Puls, R. W., Vogan, J. L., Powell, P. D., and Landis, R. (1998). *Permeable reactive barrier technologies for contaminant remediation*, EPA/600/R-98/125, United States Environmental Protection Agency, Washington, DC.

RTDF (2001). *Summary of the remediation technologies development forum permeable reactive barriers action team meeting, Florida*, Remediation Technologies Development Forum.

Scherer, M. M., Richter. S., Valentine, R. L., and Alvarez, P. J. (2000). "Chemistry and microbiology of permeable reactive barriers for *in situ* groundwater clean up." *Crit. Rev. Environ. Sci. & Technol.*, 30(3), 363-411.

Schlicker, O., Eber, M., Fruth, M., Weidner, M., Wüst, W., and Dahmke, A. (2000). "Degradation of TCE with iron: The role of competing chromate and nitrate reduction." *Ground Water*, 38(3), 403-409.

Schwertmann, U., Gasser, U., and Sticher, H. (1989). "Chromium-for-iron substitution in synthetic goethites." *Geochim. Cosmochim. Acta*, 53(6), 1293-1297.

Su, C., and Puls, R. W. (1999). "Kinetics of trichloroethene reduction by zerovalent iron and tin: Pretreatment effect, apparent activation energy, and intermediate products." *Environ. Sci. & Technol.*, 33(1), 163-168.

Weber, E. J. (1996). "Iron-mediated reductive transformations: Investigation of reaction mechanism." *Environ. Sci. & Technol.*, 30(2), 716-719.

Wilkin, R. T., and Puls, R. W. (2003). *Capstone report on the application, monitoring, and performance of permeable reactive barriers for ground-water remediation: Volume 1 – performance evaluations at two sites*, EPA/600/R-03/045a, United States Environmental Protection Agency, Cincinnati.

CHAPTER 5

Removal of Hexavalent Chromium from Groundwater using Zero-valent Iron Media

James Farrell[1]

Abstract: Groundwater contaminated by hexavalent chromium [Cr(VI)] is a worldwide problem resulting from the widespread industrial use of chromium. Zero-valent iron (Fe^0) and Fe(II)-containing minerals have been found to be effective at removing dissolved Cr(VI) from water via reductive precipitation of trivalent chromium [Cr(III)] compounds. Permeable reactive barriers (PRBs) containing Fe^0 filings can be effective for removing Cr(VI) from groundwater for concentrations up to ~10,000 μg/L. However, both field and laboratory studies indicate that higher Cr(VI) concentrations result in passivation of the Fe^0 surfaces and make Fe^0 ineffective for long-term Cr(VI) removal. Results from a field test in Elizabeth City, North Carolina indicate that Fe^0 barriers can remove Cr(VI) to levels below 1 μg/L for more than 8 years with no loss in performance, and only minor amounts of pore clogging by precipitated minerals. Spectroscopic evidence indicates that Cr(III) forms solid solutions with a variety of Fe(II) and Fe(III) containing oxides, such as carbonate green rust, siderite, and ferric hydroxide. Microbiological activity that results in precipitation of sulfide minerals, such as mackinawite, also appears to contribute to Cr(VI) removal.

CE Database subject headings: Barriers; Chromium; Field investigations; Iron; Kinetics; Mineral deposits.

5.1 Introduction

Chromium is one of the most common groundwater contaminants due to its widespread use in electroplating, wood preserving, leather tanning, and textile and pigment manufacturing. The United States Environmental Protection Agency (USEPA) lists chromium as the second most common metal contaminant at sites on the National Priorities List (USEPA 2000), and estimates that there are more than 1300 sites with Cr(VI) contaminated groundwater in the United States. Although chromium is an essential element in human metabolism, it is toxic to humans at high concentrations and has also been implicated as a carcinogen, mutagen and teratogen (Baruthio and Sferete 1992). The USEPA has set a maximum contaminant level for chromium in drinking water of 100 μg/L, and an ambient water quality standard for freshwater aquatic life protection of 11 μg-Cr(VI)/L (USEPA 2000). The World Health Organization's guideline for total Cr in drinking water is 50 μg/L.

[1] Associate Professor, Dept. of Chemical and Environmental Engineering, University of Arizona, Tucson, AZ 85721. E-mail: farrellj@engr.arizona.edu

Under most environmental conditions chromium exists in either its hexavalent [Cr(VI)] or its trivalent oxidation state [Cr(III)] (Pourbaix 1966). At circumneutral pH values, the predominant Cr(VI) species in solution are chromate (CrO_4^{2-}) and bichromate ($HCrO_4^-$) for concentrations below ~125 mM, while dichromate ($Cr_2O_7^{2-}$) predominates at higher concentrations. The predominant Cr(III) species in solution under most environmental conditions are $CrOH^{2+}$, $Cr(OH)_2^+$ and $Cr(OH)_3$ (Palmer and Wittbrodt 1991). The aqueous solubility of Cr(VI) species are generally more than 3 orders of magnitude greater than those of Cr(III). However, the solubility of Cr(III) species may be increased via formation of complexes with chloride, fluoride, sulfate, ammonium, cyanide, and organic acids (Pourbaix 1966). In the absence of complex ion formation, Cr(III) precipitates from pure water as the metastable solid $Cr(OH)_3{\bullet}nH_2O$ at Cr(III) concentrations above 60 μg/L. This solid then ages to form a more stable Cr_2O_3 solid phase which has a Cr(III) solubility of 0.024 ng/L at neutral pH in pure water (Pourbaix 1966). The dramatically lower aqueous solubilities of Cr(III) versus Cr(VI) species can be exploited for removing Cr(VI) from contaminated waters.

5.2 Removal Mechanisms

Laboratory studies have shown that Cr(VI) can be reduced to Cr(III) by metallic iron or zero-valent iron (Fe^0) according to (Gould 1982; Powell et al. 1995):

$$Fe^0_{(s)} + CrO_4^{2-}{}_{(aq)} + 4H_2O_{(aq)} \rightarrow Cr(OH)_{3(s)} + Fe(OH)_{3(s)} + 2OH^-_{(aq)} \quad (5.1)$$

In addition to precipitation of $Cr(OH)_{3(s)}$, Cr(III) may also form $Cr_2O_{3(s)}$, or solid solutions with Fe(III), according to (Powell et al. 1995):

$$x\ Fe(OH)_{3(s)} + (1-x)\ Cr(OH)_{3(s)} \rightarrow (Fe_xCr_{1-x})(OH)_{3(s)} \quad (5.2)$$

where x can range from 0 to 1. In this solid solution, Cr(III) is expected to have an aqueous solubility below that in pure $Cr(OH)_3{\bullet}nH_2O$. Direct evidence that the removal mechanism involves reduction at the iron surface has been shown for chromium removal in both laboratory and field studies (Blowes et al. 1997; Pratt et al. 1997; Wilkin et al. 2005). Investigators have also reported that Cr(III) was associated with goethite (α-FeOOH) coated regions on iron surfaces, but not with those coated by maghemite (γ-Fe_2O_3) (Blowes et al. 1997). A Cr(III)-goethite solid solution was observed for up to 10% Cr(III) substitution for Fe(III) (Pratt et al. 1997), while at higher levels of Cr(III) incorporation, the goethite transformed to hematite (α-Fe_2O_3) resulting in a mixed Cr_2O_3/Fe_2O_3 solid solution.

In addition to reduction by Fe^0, Cr(VI) may also be reduced by atomic hydrogen produced from reduction of water (Gould 1982), by aqueous Fe(II) (Fendorf and Li 1996; Lee and Hering 2003), or by Fe(II) containing oxides (Peterson et al. 1997) and green rust phases (Williams and Scherer 2001) produced from iron corrosion. Corroding iron may also remove Cr(VI) from solution without reduction to Cr(III). Several investigations have reported that chromate forms both monodentate and bidentate inner-sphere surface complexes with ferric oxides (Grossl et al. 1997; Fendorf et al. 1997; Zachara et al. 1987).

Precipitated Cr(III) in ferric oxide solid solutions appears to be resistant to reoxidation upon exposure to oxygen. Blowes et al. (1997) did not observe Cr(III) oxidation on Fe^0 filings after three months of exposure to the atmosphere. This is consistent with the irreversible nature of the Cr(III)/Cr(VI) redox couple (Miller and Zittel 1964), and chromium speciation often differs from that expected based on the solution redox potential. Oxidation of Cr(III) by O_2 is very slow at pH values less than 9 (Powell et al. 1995; Lingane et al. 1966), and thus dissolved oxygen plays only a minor role in Cr(III) oxidation in many groundwaters. However, other species present in groundwater, such as MnO_2, may rapidly oxidize Cr(III) compounds (Loyaux-Lawinczak et al. 2001).

5.3 Reaction Kinetics

The kinetics of Cr(VI) reduction by corroding iron are very dependent on the type of Fe^0, the solution chemistry, the aqueous Cr(VI) concentration, and the condition of the Fe^0 surfaces at the time of their exposure to chromate containing solutions. The type of Fe^0 appears to be the most important factor affecting chromium removal kinetics. Rates of Cr(VI) removal in batch reactors have been found to vary by more than two orders of magnitude between different types of Fe^0 (Powell et al. 1995; Powell 1994; Miehr et al. 2004). With respect to solution chemistry parameters, Cr(VI) removal rates are most affected by the solution pH value. Fig. 5-1 shows surface area normalized removal rates as a function of pH in 10,000 μg/L Cr(VI) solutions determined from data taken in bath reactors containing iron wires or Fe^0 filings. The data for the iron wires indicates a reaction order with respect to the H^+ activity of 0.5 (Gould 1982), while the Fe^0 filings data indicates a reaction order of 0.7 (Alowitz and Scherer 2002).

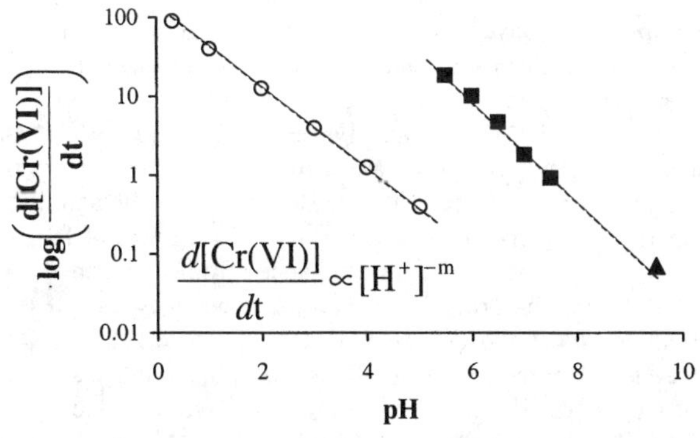

Fig. 5-1. Surface area normalized removal rates at a Cr(VI) concentration of 10,000 μg/L based on rate constants for iron wires in Gould (1982) (circles), and for Fe^0 filings in Alowitz and Scherer (2002) (squares) and Powell et al. (1995) (triangle). The slope of the dashed lines (m) indicates that the reaction order with respect to the H^+ activity ($[H^+]$) is 0.5 for the iron wire data and 0.7 for the Fe^0 filings.

The background electrolyte composition and concentration have also been found to affect Cr(VI) removal kinetics. Gould (1982) reported that increasing the ionic strength of a sodium sulfate electrolyte between 30 and 90 mM decreased rates of Cr(VI) removal by more than a factor of 2, but further increases in ionic strength did not affect removal rates. Other investigators have reported that silica released by dissolution of aluminosilicate aquifer materials can strongly affect Cr(VI) removal rates. In short-term batch testing, the addition of kaolinite [$Al_2Si_2O_5(OH)_4$] to Fe^0 filings increased the rate of Cr(VI) removal by up to a factor of four, as compared to Fe^0 systems without such minerals (Powell et al. 1995). The mechanism for the increased reaction rate was attributed to pH buffering arising from kaolinite dissolution according to:

$$Al_2Si_2O_5(OH)_4 + 7\,H_2O \rightarrow 2\,Al(OH)_4^- + 2\,H_4SiO_4 + 2\,H^+ \quad (5.3)$$

The investigators proposed that the silicic acid (H_4SiO_4) produced from kaolinite dissolution reacts with iron oxyhydroxides produced by the corroding iron, or with mineral surfaces, such as goethite. Coordination of silicic acid with α-FeOOH releases protons under alkaline conditions and removes silicic acid from solution, thereby promoting further mineral dissolution. Other investigators have also found that the buffer composition may impact Cr(VI) removal rates by more than a factor of 2 (Alowitz and Scherer 2002).

The mechanisms responsible for water chemistry effects on Cr(VI) removal rates are not well-understood, but most can likely be attributed to their impact on the formation and maintenance of a ferric oxide passive film on the Fe^0 surfaces. Prior studies have shown that Fe^0 filings exposed to the atmosphere are coated with an inner layer of magnetite (Fe_3O_4) and an outer layer of Fe_2O_3 (Ritter et al. 2003). When placed in solution, the outer ferric oxide becomes partially reduced by the underlying Fe^0 (Ritter et al. 2002). In waters of low oxidizing potential, this autoreduction process converts the Fe_2O_3 into Fe_3O_4, which is a non-passivating oxide due to its high electrical conductivity. This autoreduction process does not occur in waters of high oxidizing potential, such as those containing high concentrations of Cr(VI) (Melitas and Farrell 2002). Thus, except at very low pH values, oxidation of Fe^0 by chromate will lead to Fe^0 surface passivation for sufficiently high chromate concentrations. Depending on the solution chemistry, the critical concentration for chromate passivation is in the range of 10^{-4} to 10^{-3} M (5.2 to 52 mg/L) (Uhlig and Revie 1985). The presence of other oxidants, such as dissolved oxygen, will decrease the critical concentration, whereas the presence of chloride ions increases the critical concentration (Uhlig and Revie 1985). After passivation is achieved, there is still some reduction of chromate, but at much slower rates.

Rates of Cr(VI) reduction by corroding iron have been measured in short-term batch tests by several investigators (Gould 1982; Alowitz and Scherer 2002). However, due to passivation effects and multiple reduction mechanisms, rates of Cr(VI) removal in short-term batch tests are not likely to be representative of steady-state removal rates under field conditions. For example, in studies using high iron surface area concentrations with respect to the amount of aqueous Cr(VI), chromate reduction may occur primarily by reaction with Fe(II) in the oxides coating the Fe^0

filings at the time of their immersion into the chromate solutions. Thus, these tests may not probe turnover of the reactive sites on the Fe^0 surfaces, and thereby neglect the effects of chromium induced surface passivation. Short-term batch tests, therefore, exploit only a small fraction of the iron's reactive capacity and may not be representative of long-term performance where continuous Fe^0 corrosion is required for sustained Cr(VI) removal. This flaw in many short-term batch tests may account for the different kinetic models for Cr(VI) removal that have been observed in different studies.

For a fixed Fe^0 surface area and water chemistry, rates of chromium reduction by corroding iron have been described by kinetic models of the form (Gould 1982; Powell et al. 1995; Melitas and Farrell 2002):

$$\frac{d[Cr(VI)]}{dt} = -k[Cr(VI)]^n \tag{5.4}$$

where k is a rate constant and n is the reaction order with respect to the aqueous chromate concentration. Short-term batch tests using high surface area concentrations with respect to the amount of soluble Cr(VI) have reported n values ranging from 0 to 1 (Gould 1982; Powell et al. 1995; Alowitz and Scherer 2002). Part of this disparity in best-fit n values can be attributed to differences in initial Cr(VI) concentrations ($\leq$10,000 to 213,000 μg/L) among investigations which affected the amount of surface passivation. Additionally, because the duration of these tests ranged from 1 minute to 9 hours, some tests were measuring initial reaction rates while others were measuring reactivity over a more prolonged period. Initial reaction rates may be dominated by Cr(VI) reduction by Fe(II) oxides initially coating the iron surfaces, while reaction rates after several hours elapsed may be representative of reduction by the underlying Fe^0 phase.

The concentration dependence of steady-state Cr(VI) removal rates and iron surface passivation was investigated by Melitas et al. (2001). Chromate removal rates by iron wires exposed to solutions with initial Cr(VI) concentrations ranging from 100 to 10,000 μg/L were investigated over a 13 day period in batch reactors with large solution volumes with respect to the amount of reactive surface area (225 mL/cm^2). Measurable amounts of Cr(VI) removal in these reactors required multiple turnover of reactive sites on the Fe^0 surfaces, and steady state removal rates were observed in all reactors. The steady state removal rates in reactors with different initial concentrations were zeroth order ($n = 0$) in the Cr(VI) concentration. However, the steady state zeroth order removal rates were found to decrease with increasing initial Cr(VI) concentration, as illustrated in Fig. 5-2. This behavior is consistent with Cr(VI) removal rates that were controlled by the anodic reaction releasing Fe^{2+} into the solution, and not by the cathodic reaction involving Cr(VI) reduction.

Anodic control of iron corrosion is illustrated in Fig. 5-3 showing Tafel polarization diagrams for iron wires immersed in 5000 and 10,000 μg/L chromate solutions. Comparison of cathodic currents in the linear Tafel region shows that increasing the Cr(VI) concentration by a factor of 2 increased the cathodic current by a factor of ~2, and had little effect on the cathodic Tafel slope. However, comparison of anodic currents in the linear Tafel region shows that increasing the concentration

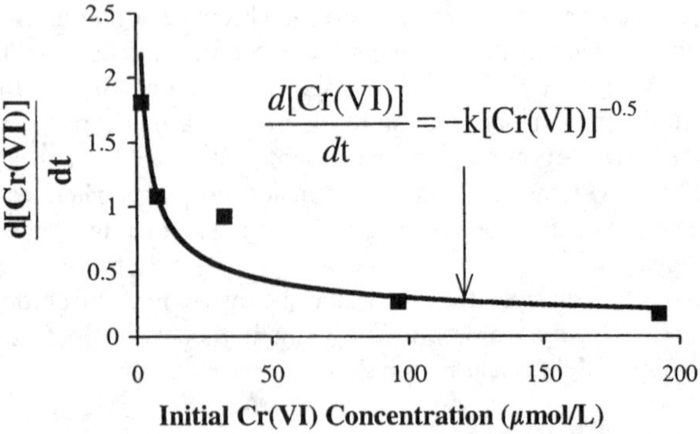

Fig. 5-2. Steady state removal rates of Cr(VI) by corroding iron wires in anaerobic 3 mM $CaSO_4$ background electrolyte solutions as a function of the initial Cr(VI) concentration. The fit of the kinetic model to the data points is shown by the solid curve. Data from Melitas et al. (2001).

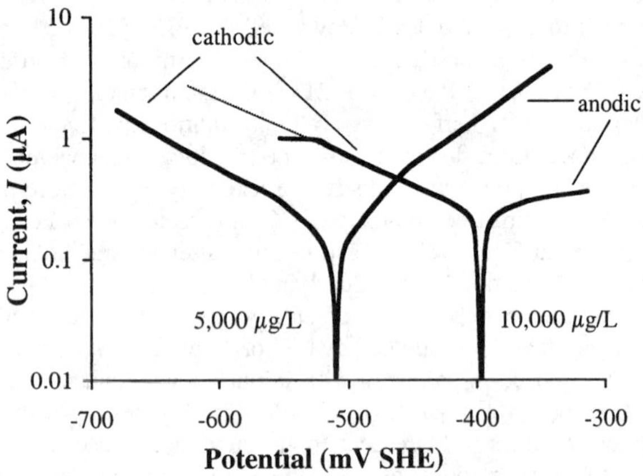

Fig. 5-3. Tafel polarization diagrams for iron wire electrodes in anaerobic 3 mM $CaSO_4$ background electrolyte solutions containing 5000 and 10,000 μg/L Cr(VI). To aid comparison of the cathodic currents, a dashed line has been used to extrapolate the linear Tafel region to current densities greater than the mass transfer limited value of ~1 $\mu A/cm^2$. Data from Melitas et al. (2001).

from 5000 to 10,000 μg/L resulted in a substantial decrease in the anodic current. The fact that Cr(VI) removal rates in Fig. 5-2 decreased with increasing Cr(VI) concentration indicates that rates of Cr(VI) reduction were controlled by the anodic reaction for iron corrosion. This conclusion is supported by the corrosion current data shown in Fig. 5-4. The trend of decreasing iron corrosion rates with increasing Cr(VI) concentrations indicates that higher Cr(VI) concentrations resulted in greater passivation of the iron surfaces. Thus, passivation effects can explain the -0.5 apparent reaction order for steady state Cr(VI) removal shown in Fig. 5-2.

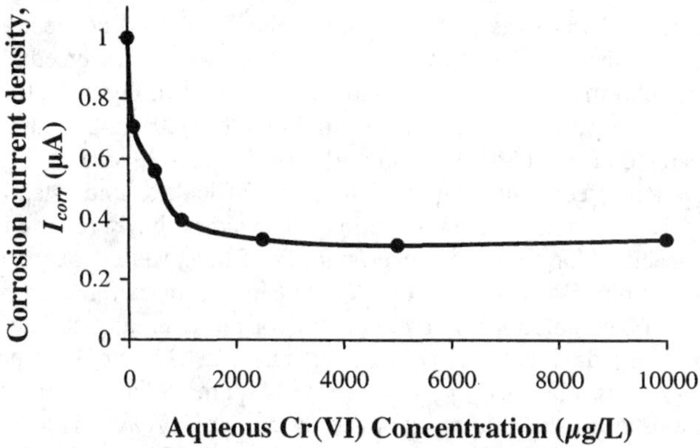

Fig. 5-4. Corrosion current densities for a corroding iron wire in anaerobic 3 mM $CaSO_4$ background electrolyte solutions as a function of the Cr(VI) concentration. Data from Melitas et al. (2002).

Surface passivation also contributes to the hysteretic performance of Fe^0 filings for Cr(VI) removal. Fig. 5-5 illustrates passivation induced hysteresis in a column reactor packed with *Master Builders Supply* Fe^0 filings. Over the first 12,500 pore volumes, the column was operated at influent Cr(VI) concentrations ≤5000 μg/L and complete Cr(VI) removal was observed (Melitas et al. 2001). Upon increasing the influent concentration from 5000 to 10,000 μg/L, breakthrough occurred after less

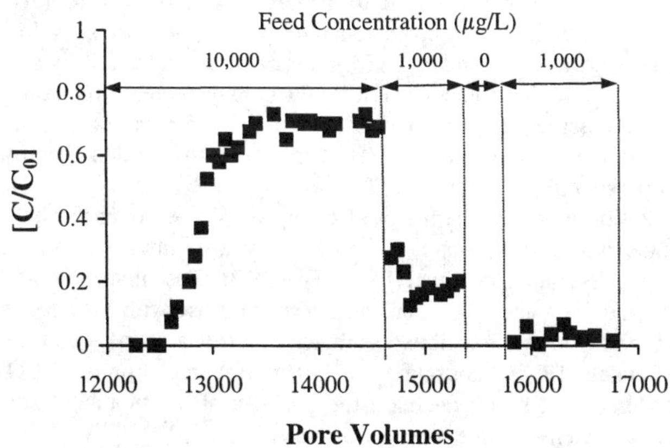

Fig. 5-5. Normalized effluent Cr(VI) concentrations for a column of *Master Builders Supply* Fe^0 filings operated with a hydraulic detention time of 25 min. At 12,500 pore volumes elapsed, the influent Cr(VI) concentration was increased from 5000 to 10,000 μg/L. Data from Melitas et al. (2001).

than 100 pore volumes, and a steady-state removal of ~3000 μg/L was observed for more than 1000 pore volumes. Upon decreasing the feed concentration to 1000 μg/L, a steady state removal of ~800 μg/L was attained. This behavior contrasts with the complete removal that was previously observed for a feed concentration of 1000 μg/L. The passivation induced hysteresis could be partially reversed by purging the column with chromate-free water, as shown by the data in Fig. 5-5 after 15,500 pore volumes. After resuming a feed concentration of 1000 μg/L, partial passivation reversal resulted in an additional ~150 μg/L of Cr(VI) removal.

Passivation effects may help explain some difficult to understand observations on Cr(VI) removal in column experiments. Studies have reported that Cr(VI) removal capacities for columns packed with Fe^0 filings were dependent on the flow rate. For example, Blowes et al. (1997) studied chromate removal from 342 μM (18,000 μg/L) feed solutions by two columns of Fe^0 operating at fluid velocities of 2.7 and 11 cm/day. In the column operated at 11 cm/day, near complete breakthrough was observed in less than 100 pore volumes. However, effluent concentrations from the column operated at 2.7 cm/day were below the detection limit for the duration of the experiment, which was 150 pore volumes. The early breakthrough in the column with the faster flow rate was attributed to mass transfer limitations. However, at a fluid velocity of only 11 cm/day, mass transfer limitations are not likely to account for the observed difference in column performance. Other studies have reported very fast Cr(VI) mass transfer and removal rates in columns packed with similar Fe^0 filings. Complete removal for over 500 bed volumes has been observed with a hydraulic detention time of 20 seconds for columns operated with influent Cr(VI) concentrations less than 5000 μg/L (Melitas et al. 2002). This suggests that improved performance at lower flow rates may be attributed to decreased surface passivation at lower Cr(VI) feed rates.

Decreased surface passivation in the presence of silicate minerals may also contribute to the ability of kaolinite to increase rates of Cr(VI) removal by corroding iron. Silica is a well-known corrosion inhibitor due to its ability to block reactive sites on Fe^0 surfaces by binding to iron corrosion products (Uhlig and Revie 1985). Adsorption of silica to iron corrosion products may decrease the ability of Cr(VI) to enforce the passivating ferric oxide. Thus, the main effect of silica may be to decrease the concentration of Cr(VI) at the iron surface and thereby decrease the extent of passivation.

In addition to water chemistry affecting Cr(VI) removal through its impact on Fe^0 surface passivation, water chemistry may also impact chromate removal by affecting the hydraulic conductivity of Fe^0 PRBs. Past investigators have reported that dissolved Ca^{2+} and Mg^{2+} concentrations decrease with distance into PRBs as a result of precipitation reactions promoted by the alkaline conditions within the barrier (Bennett, 1997; Yabusaki et al. 2001). Accumulation of $CaCO_3$, $MgCO_3$ and $Mg(OH)_2$ precipitates will decrease the porosity of the treatment zone and shorten the hydraulic detention time for Cr(VI) removal.

Several laboratory studies have investigated the effect of Ca^{2+} and Mg^{2+} concentrations on Cr(VI) removal in column reactors packed with Fe^0 filings. Karvonen (2004) reported that chromate removal decreased by 45% with increasing total hardness between 0 and 138 mg/L as $CaCO_3$. However, increasing the hardness between 138 and 826 mg/L decreased Cr(VI) removal by only an additional 15%. This result suggests that inhibition of Cr(VI) removal by the hardness ions cannot be attributable to a simple pore filling mechanism resulting from precipitation of calcium and magnesium containing minerals. Another study by Lo et al. (2003) found that Ca^{2+} and Mg^{2+} hardness had no effect on Cr(VI) removal for total hardness $\leq$ 200 mg/L, as $CaCO_3$. Since these studies were both conducted at Cr(VI) concentrations greater than those known to cause passivation, this disparity in the effects of hardness on Cr(VI) removal can likely be attributed to different effects of Ca^{2+} and Mg^{2+} on formation of the passivating film.

5.4 Other *In Situ* Cr(VI) Removal Methods

Although Fe^0 filings have been found to be effective for Cr(VI) removal in the laboratory, there have not been many field-scale applications of Fe^0 barriers for chromate remediation. In general, the use of Fe^0 PRBs for inorganic contaminant remediation has lagged behind that for chlorinated solvents (Morrison et al. 2002). The paucity of Fe^0 barriers for Cr(VI) remediation can likely be attributed to the high mobility of Cr(VI) that often results in deep contamination, and to the fact that there are other effective methods for *in situ* chromate remediation that have lower installation costs than PRBs filled with Fe^0 filings (Cummings and Booth 1997).

In situ redox manipulation (ISRM) using a variety of reducing agents is gaining in popularity for remediating Cr(VI) contamination. This idea was first proposed by Fruchter et al. (1992) and can be used for remediating a variety of redox active compounds. ISRM methods for chromate are based on creating reducing conditions in the contaminated zone via injection of chemical reducing agents, such as sulfite (SO_3^{2-}), dithionite ($S_2O_4^{2-}$), calcium polysulfide (CaS_x) or ferrous iron (Fe^{2+}). These reducing agents may precipitate Cr(III) compounds by direct reduction of Cr(VI), or by indirect reduction by first reducing structural Fe(III) in the aquifer materials to Fe(II) (USEPA 2000).

In addition to ISRM methods based on the injection of chemical reducing agents, there are also ISRM methods for Cr(VI) remediation based on microbiological activity. These methods involve injection of biodegradable organic substrates whose metabolism results in oxygen depletion and growth of anaerobic microorganisms in the contaminated zone. Compounds such as lactate (Tokunaga et al. 2003) and molasses (Suthersan 1997) have proven to be effective for this purpose.

Biological ISRM may also be combined with Fe^0 barriers. Henny et al. (2001) showed that addition of sulfate reducing bacteria (SRB) and lactate to a column reactor containing Fe^0 increased the amount of Cr(VI) removal in the column by 78%, as compared to Fe^0 alone. Additionally, seeding the Fe^0 column with SRB without lactate supplementation resulted in a 43% increase in Cr(VI) removal. The greater effectiveness for Cr(VI) removal in the SRB seeded column without added lactate can likely be attributed to microbial utilization of the hydrogen produced from water reduction. In the absence of microbes, the reducing equivalents of hydrogen gas are not available for Cr(VI) reduction (Gould 1982).

The structural Fe(II) in the aquifer minerals then serves to reduce Cr(VI) to Cr(III). In addition to dithionite oxidation to sulfite, some of the dithionate undergoes disproportionation to thiosulfate ($S_2O_3^{2-}$) and bisulfate (HSO_4^-). The solution injected for the pilot test consisted of 1874 liters of 0.05 M (8,700 mg/L) sodium dithionite containing a 0.025 M (3,000 mg/L) potassium bromide (KBr) tracer and a 0.1 M (10,000 mg/L) potassium bicarbonate ($KHCO_3$) buffer.

Injection of the test solution decreased the redox potential within the treatment zone from ~+400 to ~–400 mV (SHE), and decreased the dissolved oxygen levels in all monitoring wells to < 0.32 mg/L after 24 hours. Outside the 1 meter radius of the treatment zone, redox potentials remained close to their initial values. Sulfate concentrations within the treatment zone increased from ~150 mg/L, to as high as 5000 mg/L 24 hours after injection. Initial Cr(VI) concentrations ranging from 1200 to 7250 μg/L decreased to below the detection limit of 10 μg/L and remained there 48 weeks after injection. Concentrations of dissolved arsenic, nickel, lead and selenium inside and down gradient of the treatment zone were not altered by the ISRM. However, concentrations of dissolved barium, cadmium, and copper were slightly elevated inside the treatment zone during the first 8 weeks following the injection, but were not elevated outside the treatment zone. Based on the results of this pilot test, ISRM was approved for treating the entire source zone at the Elizabeth City site. Treatment of the source zone using ISRM is expected to eliminate the need for replacing the PRB at the site, and speed up the cleanup time from 30 to 10 years.

5.5.2 *Kolding, Denmark*

Another full scale 100% Fe^0 PRB for remediating TCE and chromate contamination has been installed at the former in Haardkrom electroplating facility site in Kolding, Denmark (Kjeldsen and Fuglsang 2000). Aqueous chromate concentrations at hot spots on the site were up to 110,000 μg/L. The Kolding PRB is 50 m long, 1 m wide and 1 to 3 m deep, and was installed in 1999 using an excavation box technique. The design costs for this PRB were $108,000 and the construction costs were $250,000. Performance results for the first year of operation indicated that the barrier did not effectively control the uneven distribution of Cr(VI) in the groundwater, and some areas of the wall had exhausted their capacity for chromate removal. This finding is consistent with the fact that Cr(VI) concentrations greater than 10,000 μg/L are known to rapidly passivate iron under groundwater treatment conditions (Melitas et al. 2001).

5.6 Conclusions

Both field and laboratory investigations indicate that Fe^0 filings are effective for removing soluble Cr(VI) from contaminated waters over extended periods of time for influent concentrations less than ~10,000 μg/L. However, concentrations above this level may sufficiently passivate the Fe^0 that steady state Cr(VI) removal cannot be sustained. In many cases, ISRM is able to precipitate Cr(III) as effectively as Fe^0 PRBs. The low capital costs for ISRM methods makes Fe^0 barriers an attractive remedial option only at sites containing both chlorinated solvent and Cr(VI) contamination. However, because of the widespread use of chlorinated solvents as degreasing agents prior to chrome electroplating, many of the chromate-impacted sites are also contaminated by chlorinated solvents. Thus, PRBs using Fe^0 remain an attractive option at many sites where Cr(VI) remediation is required.

Acknowledgements
Thanks to Ouatfa Chuffe-Moscoso and Nikos Melitas. Research support has been provided by the National Institute of Environmental Health Sciences of the National Institute of Health, with funds from the U.S. Environmental Protection Agency through grant 2P42ES04940-11.

5.7 References

Alowitz, M. J., and Scherer, M. M. (2002). "Kinetics of nitrate, nitrite and Cr(VI) reduction by iron metal." *Environ. Sci. & Technol.*, 36, 299-306.

Amonette, J. E., Szecsody, J. E. Schaef, H. T., Templeton, J. C., Gorby, Y. A., and Fruchter, J. S. (1994). "Abiotic reduction of aquifer materials by dithionate: A promising in situ remediation technology." In *Proceedings of the 33rd Hanford Symposium on Health and the Environment-In Situ Remediation: Scientific Basis for Current and Future Technologies*, Battelle Press, Columbus, OH, 851-882.

Baruthio, F., and Sferete, V. F. (1992). "Toxic effects of chromium and its compounds." *Biological Trace Element Research*, 32, 145-153.

Bennett, T. A. (1997). *An in situ reactive barrier for the treatment of hexavalent chromium and trichloroethylene in groundwater*, Master thesis, University of Waterloo, Ontario, Canada.

Blowes, D. W., Ptacek, C. J., and Jambor, J. J. (1997). "In-situ remediation of Cr(VI)-contaminated groundwater using permeable reactive walls: Laboratory studies." *Environ. Sci. & Technol.*, 31, 3348-3357.

Blowes, D. W., Gillham, R. W., Ptacek, C. J., Puls, R. W., Bennett, T. A., O'Hannesin, S. F., Bain, J. G., and Hantong-Fong, C. J. (1999a). *An in situ permeable reactive barrier for the treatment of hexavalent chromium and trichloroethylene in ground water, Vol. 1, Design and installation,* EPA/600/R-99/095a, United States Environmental Protection Agency, Washington, DC.

Blowes, D. W., Puls, R. W., Gillham, R. W., Ptacek, C. J., Bennett, T. A., Bain, J. G., and Hantong-Fong, C. J., and Paul, C. J. (1999b). *An in situ permeable reactive barrier for the treatment of hexavalent chromium and trichloroethylene in ground water, Vol. 2, Performance monitoring*, EPA/600/R-99/095b, United States Environmental Protection Agency, Washington, DC.

Cummings, M. A., and Booth, S. R. (1997). *Cost effectiveness of in situ redox manipulation for remediation of chromium contaminated groundwater,* LA-UR-97-165, Los Alamos National Laboratory, Los Alamos, NM.

Fendorf, S. E., and Li, G. (1996). "Kinetics of chromate reduction by ferrous iron." *Environ. Sci. & Technol.*, 30, 1614-1617.

Fendorf, S., Eick, M. J., Grossl, P., and Sparks, D. L. (1997). "Arsenate and chromate retention mechanisms on goethite. 1. Surface structure." *Environ. Sci. & Technol.*, 31, 315-320.

Fruchter, J. S., Zachara, J. M., Fredrickson, J. K., Cole, C. R., Amonette, J. E., Stevens, T. O., Holford, D. J., Eary, L. E., Black, G. D., and Vermeul, V. R. (1992). "Manipulation of natural subsurface processes: Field research and validation." *Pacific Northwest Laboratory Annual Report for 1991 to the DOE Office of Energy Research, Part 2: Environmental Sciences. PNL-8000 Pt. 2*, Pacific Northwest Laboratory, Richland, WA, 88-106.

Gould, J. P. (1982). "The kinetics of hexavalent chromium reduction by metallic iron." *Wat. Res.*, 16, 871-877.

Grossl, P. R., Eick, M. J., Sparks, D. L., Goldberg, S., and Ainsworth, C. C. (1997). "Arsenate and chromate retention mechanisms on goethite. 2. Kinetic evaluation using a pressure-jump relaxation technique." *Environ. Sci. & Technol.*, 31, 321-326.

Henny, C., Weathers, L. J., Katz, L. E., and MacRae, J. D. (2001). "Abiotic and biotic Cr(VI) reduction in a laboratory-scale permeable reactive barrier." In *Proceedings of the 6th International In Situ and On-Site Bioremediation Symposium*, Battelle Press, Columbus, OH, Vol. 8, 138-145.

Karvonen, A. (2004). "Cation effects on chromium removal in permeable reactive walls." *J. Environ. Eng. ASCE*, 130(8), 863-866.

Kjeldsen, P., and Fuglsang, I. A. (2000). "Demonstration program on reactive barrier technologies using zero-valent iron (Haardkrom, Denmark)." In *Proceedings of the 7th International FZK/TNO Conference on Contaminated Soil*, Forschungszentrum Karlsruhe (FZK) and the Netherlands Organization for Applied Scientific Research (TNO).

Lee, G., and Hering, J. G. (2003). "Removal of chromium from drinking water by redox assisted coagulation with iron(II)." *J. Water Supply Research and Tech.*, 52(5), 319-332.

Lingane, P. J., Anson, F. J., and Osteryoung, R. A. (1966). "Chronopotentiometric investigation of the reduction of chromate in alkaline solutions." *J. Electro. Anal. Chem.*, 12, 250-253.

Lo, I. M.-C., Lam, C. S.-C., and Chan, K. K. (2003). "Effects of groundwater characteristics on chromate removal by permeable reactive barriers." In *Proceedings of the 8th International FZK/TNO Conference on Contaminated Soil*, Forschungszentrum Karlsruhe (FZK) and the Netherlands Organization for Applied Scientific Research (TNO), 1657-1665.

Loyaux-Lawinczak, S., Lecomte, P., and Ehrhardt, J. J. (2001). "Behavior of hexavalent chromium in a polluted groundwater: Redox processes and immobilization in soils." *Environ. Sci. & Technol.*, 35, 1350-1357.

Miller, F. J., and Zittel, H. E. (1964). "Voltammetry of Ce(IV), Mn(VII), Cr(VI) and V(V) with the pyrolytic graphite electrode." *J. Electro. Anal. Chem.*, 7, 116-122.

Melitas, N., Chuffe, O., and Farrell, J. (2001). "Kinetics of soluble chromium removal from contaminated water by zerovalent iron media: Corrosion inhibition and passive oxide effects." *Environ. Sci. & Technol*, 35, 3948-3953.

Melitas, N., and Farrell, J. (2002). "Understanding chromate reactions with corroding iron media using Tafel analysis and electrochemical impedance spectroscopy." *Environ. Sci. & Technol.*, 36, 5476-5481.

Melitas, N., Conklin, M., and Farrell, J. (2002). "Understanding the mechanisms controlling the kinetics of arsenate and chromate removal from solution using zerovalent iron media, Chapter 12." *Innovative Strategies for the Remediation of Chlorinated Solvents and DNAPLs in the Subsurface*, American Chemical Society, Washington, DC.

Miehr, R., Tratnyek, P. G., Bandstra, J. Z., Scherer, M. M., Alowitz, M. J., and Bylaska, E. J. (2004). "Diversity of contaminant reduction reactions by zerovalent iron: Role of the reductate." *Environ. Sci. & Technol.*, 38, 139-147.

Morrison, S. J., Naftz, D. L., Davis, J. A., and Fuller, C. C. (2002). "Introduction to groundwater remediation of metals, radionuclides, and nutrients with permeable reactive barriers, Chapter 1." In *Handbook of groundwater remediation using permeable reactive barriers*, Academic Press, New York.

Palmer, C. D., and Wittbrodt, P. R. (1991). "Processes affecting the remediation of chromium-contaminated sites." *Environ. Health Persp.*, 92, 24-40.

Patterson, R. R., Fendorf, S., and Fendorf, M. (1997). "Reduction of hexavalent chromium by amorphous iron sulfide." *Environ. Sci. & Technol.*, 31, 2039-2044.

Paul, C. J., Khan, F. A., and Puls, R. W. (2002). "In situ reduction of chromium-contaminated soils and sediments by sodium dithionite, Chapter 16." In *Handbook of groundwater remediation using permeable reactive barriers*, Academic Press, New York.

Peterson, M. L., White, A. F., Brown, G. E., Jr., and Parks, G. A. (1997). "Surface passivation of magnetite by reduction with aqueous Cr(VI): XAFS and TEM results." *Environ. Sci. & Technol.*, 31, 1573-1576.

Pourbaix, M. (1966). *Atlas of electrochemical equilibria*, Pergamon Press, Oxford, UK.

Powell, R. M. (1994). Geochemical effects on chromate reduction and remediation utilizing the thermodynamic Instability of Zero-valence-state iron, Master Thesis, University of Oklahoma, Norman, OK.

Powell, R. M., Powell, P. D., and Puls, R. W. (2002). *Economic analysis of the implementation of permeable reactive barriers for remediation of contaminated ground water*, EPA/600/R-02/034, United States Environmental Protection Agency, Washington, DC.

Powell, R. M., Puls, R. W., Hightower, S. K., and Sabatini, D. A. (1995). "Coupled iron corrosion and chromate reduction: Mechanisms for subsurface remediation." *Environ. Sci. & Technol.*, 29, 1913-1922.

Pratt, A. R., Blowes, D. W., and Ptacek, C. J. (1997). "Products of chromate reduction on proposed subsurface remediation material." *Environ. Sci. & Technol.*, 31, 2492-2498.

Ritter, K., Odziemkowski, M. S., and Gillham, R. W. (2002). "An in situ study of the role of surface films on granular iron in the permeable iron wall technology." *J. Contam. Hydrol.*, 55, 87-111.

Ritter, K., Odziemkowski, M. S., Simpgraga, R., Gillham, R. W., and Irish, D. E. (2003). "An in situ study of the effect of nitrate on the reduction of trichloroethylene by granular iron." *J. Contam. Hydrol.*, *65*, 121-136.

Suthersan, S. (1997). *Remediation engineering design concepts*, Lewis Publishers, Boca Raton, FL.

Tokunaga, T. K., Wan, J., Firestone, M. K., Hazen, T. C., Herman, D. J., Sutton, S. R., and Lanzirotti, A. (2003). "In situ reduction of Cr(VI) in heavily contaminated soils through organic carbon amendment." *J. Env. Qual.*, 32(5), 1641-1649.

Uhlig, H. H., and Revie, R. W. (1985). Corrosion and corrosion control, 3rd Ed., John Wiley, New York.

USEPA (2000). *In situ treatment of soil and groundwater contaminated with chromium, Technical resource guide*, EPA/600/R-02/034, United States Environmental Protection Agency, Washington, DC.

Wilkin, R. T., Su, C., Ford, R. G., and Paul, C. J. (2005). "Chromium-removal processes during groundwater remediation by a zerovalent iron permeable reactive barrier." *Environ. Sci. & Technol.*, in press.

Williams, A. G. B., and Scherer, M. M. (2001). "Kinetics of Cr(VI) reduction by carbonate green rust." *Environ. Sci. & Technol.*, 35, 3488-3494.

Yabusaki, S., Cantrell, K., Sass, B., and Steefel, C. (2001). "Multicomponent reactive transport in an in-situ zerovalent iron cell." *Environ. Sci. & Technol.*, 35, 1493-1503.

Zachara, J. M., Girvin, D. C., Schmidt, R. L., and Resch, C. T. (1987). "Chromate adsorption on amorphous iron oxyhydroxide in the presence of major groundwater ions." *Environ. Sci. & Technol.*, 21, 589-594.

CHAPTER 6

Aqueous Nitrate Reduction by Zero-valent Iron Powder

Chih-Hsiang Liao[1], Shyh-Fang Kang[2], Jin Anotai[3], Chalermchai Ruangchainikom[4]

Abstract: This paper describes several aspects of nitrate removal by the two systems of zero-valent iron (Fe^0)/H_2SO_4 and Fe^0/CO_2, including (1) the water quality parameters (humic acid, n-propanol, Na_2CO_3, $CaCl_2$, NaCl, and H_2O_2) and operating parameters (pH, Fe^0 dosage, and batch/continuous mode); (2) the two side issues of concern, ferrous accumulation and ammonium formation; and (3) the proposing of a prototype of treatment system for complete nitrate removal. The results show that control of pH within 3-4 shortens the reaction time for 100% removal of nitrate in less than 30 min, given the Fe^0 dosage of 2 g/L. Background species such as humic acid, propanol, and calcium exhibit significantly inhibitory effects on nitrate removal. The presence of sodium shows insignificant effect on nitrate removal, whereas chloride ions enhance nitrate removal remarkably. In view of operating mode, it is recommended to use batch mode, rather than continuous flow mode, for a more efficient way to remove nitrate.

CE Database subject headings: Acid; Alkalinity; Ammonia; Carbon dioxide; Hardness; Iron; Nitrates; pH.

6.1 Introduction

Contamination of nitrate in natural waters has drawn widespread attention due to the public health and ecological concerns. Nitrate, if reduced into nitrite, was reported to cause methemoglobinemia in newborn infants or so-called "blue baby syndrome" (Walton, 1951). On top of this, the aforementioned nitrite species can also lead to the formation of N-nitrosamines, which is a possible carcinogen in stomach according to Mirvish (1985). Nitrate-rich nutrient streams, when discharged into surface water bodies especially lake and reservoir, can trigger an abnormal algae bloom, which will result in the difficulty and complexity in water purification processes. In general, nitrate contamination arises from such sources as agricultural fertilizers, high strength stock waste, and defected sanitary sewer system. For example, the use of

[1]Professor, Dept. of Environmental Resources Management, Chia Nan University of Pharmacy and Science, Tainan, Taiwan. (corresponding author). E-mail: chliao@mail.chna.edu.tw
[2]Professor, Dept. of Water Resource and Environmental Engineering, Tamkang University, Taipei, Taiwan.
[3]Associate Professor, Dept. of Environmental Engineering, King Mongkut's University of Technology Thonburi, Bangkok, Thailand.
[4]Ph.D. Candidate, National Research Center for Environmental and Hazardous Waste Management, Chulalongkorn University, Bangkok, Thailand.

poultry excrete as fertilizer for the tea trees plantation located at Nan-Tou, Taiwan has caused significant increase of nitrate content in groundwater over the years (Chang et al., 2002). Table 6-1 demonstrates that nitrate concentrations monitored in several wells in November 2001 have reached far beyond its criteria for drinking water quality. Because of serious degradation in groundwater quality within the contaminated area, biological denitrification was investigated to eradicate the posed problem by the responsible water treatment plant.

Table 6-1. Levels of Nitrate Monitored in the Wells Located at Min-Jen, Nan Tou, Taiwan

Well No.	Date of Well Construction	Initial NO_3^--N (mg/L)	NO_3^--N in Nov. 2001 (mg/L)	Well Depth (m)	Designed Pumping Rate (m^3/d)
Hsin-Je #1	Nov. 1979	-	**21.00**	120	4500
Hsin-Je #2	Mar. 1998	3.00	5.30	200	3000
Jen-Ho #1	Dec. 1998	0.64	5.60	150	3000
Jen-Ho #2	Jul. 2001	4.20	5.8 (Apr. 2002)	200	3000
Sha-Hsin #1	Apr. 1989	0.06	**22.0**	250	3000
Sha-Hsin #2	Aug. 1996	3.80	**10.00**	220	3500
Hsin-Min #1	Apr. 1997	0.96	1.10	300	4000
Hsin-Min #2	Apr. 1997	0.92	1.20	250	4000
Tan-Liaw	Feb. 1995	4.50	**18.00**	250	2000
Kam-Ka	Feb. 1997	4.60	7.70	200	3500
Hsin-Kwang	Mar. 1993	0.20	**23.0**(Sep. 1996)	-	2000
I-Tsan #1	Aug. 1990	0.04	**11.00**	200	2000
I-Tsan #2	Feb. 1998	1.80	-	-	2000
Er-Tsan	Jan. 1993	3.00	**28.00**	-	2000

Note: Taiwan drinking water quality standard for NO_3^--N is set at 10 mg/L.

In addition to biological method, the physicochemical processes for nitrate removal have been developed by different research groups. Of these methods, the Fe^0-based reduction process is now emerging as one of promising techniques to remove nitrate from contaminated waters. According to Kelly (1965), Fe^0, serving as electrons donor, provides its reducing power through the following reaction mechanisms:

$$Fe + H_2O \leftrightarrow Fe(H_2O)_{ads} \tag{6.1}$$
$$Fe(H_2O)_{ads} \leftrightarrow Fe(OH^-)_{ads} + H^+ \tag{6.2}$$
$$Fe(OH^-)_{ads} \leftrightarrow (FeOH)_{ads} + e^- \tag{6.3}$$
$$(FeOH)_{ads} \rightarrow (FeOH)^+ + e^- \tag{6.4}$$
$$(FeOH)^+ + H^+ \rightarrow Fe^{2+} + H_2O \tag{6.5}$$

Note that the subscript of "ads" in the above reaction equations represents the adsorbed form of species. Two electrons were released from iron metal in two stages, as shown in Eqs. (6.3) and (6.4), respectively. According to Eq. (6.5), the concentrations of $(FeOH)^+$ and Fe^{2+} are highly pH-dependent. Several researchers have attempted to reveal specifically the effect of pH on nitrate removal by Fe^0

(Cheng et al. 1997; Huang et al. 1998; Zawaideh and Zhang 1998; Liao et al. 2003a; Choe et al. 2004; Huang and Zhang 2004).

In view of application-oriented studies, some other researchers focused on the effect of background water quality on the process performance, including (1) natural organic matter, anthropogenic surfactants and quinines (Tratnyek et al. 2001); (2) n-propanol and hydrogen peroxide (Liao et al. 2003b); (3) formate, oxalate, citrate, chloride, sulfate, borate and phosphate (Su and Puls 2004). On the other hand, some others have emphasized on the reactor types for the treatment of nitrate such as Fe^0-packed columns (Westerhoff and James 2003) and bio-enhanced Fe^0-packed columns (Gandhi et al. 2002). In addition, kinetic study on nitrate reduction by Fe^0 was also reported under various initial nitrate and iron dosages (Choe et al. 2000), and different types of metal iron and pH buffers (Alowitz and Scherer 2002). In view of hybrid process for nitrate removal, Chew and Zhang (1998) investigated the use of electrokinetics coupled with a Fe^0 treatment wall for nitrate-contaminated soils, while Hsu et al. (2004) presented the aqueous nitrate removal by using Fe^0 combined with the bubbling of carbon dioxide.

In this paper, the authors attempt to present the earlier (Liao et al. 2003a; Liao et al. 2003b; Hsu et al. 2004) and ongoing studies on nitrate removal by using the two systems of Fe^0/H_2SO_4 and Fe^0/CO_2. Important process parameters investigated include the pH, iron dosage and size, and background species such as propanol, humic acids, inorganic ions (cations and anions). Based on the available information, experiments of different process operating modes were designed and conducted, and their interesting results are presented in this paper. In addition, a prototype of complete nitrate removal is proposed for the future field application, knowing that the NH_4^+ and Fe^{2+}, the byproduct of nitrate reduction by Fe^0, are undesired and need to be removed from the aqueous system as well.

6.2 Experimental Section

6.2.1 Material and Reagents

Zero-valent iron (Fe^0, 99.5%) purchased from *Merck* was used in all experiments without any pretreatment. Two different iron particle sizes (10 and 150 μm) were investigated for their effects on nitrate removals. Nitrate concentrations ranging from 10 to 100 mg/L were used in this study. The nitrate solution was prepared in the laboratory by dissolving predetermined amount of KNO_3 or $NaNO_3$ (*Merck KGaA*, Darmstadt, Germany) in the deionized water. To control the solution pH under a constant acidic condition, H_2SO_4 (*Kuan Heng*, Taiwan) was dosed automatically into solution as necessary. The stock H_2O_2 solution was obtained by diluting H_2O_2 reagent of 35% by weight (*Merck KGaA*, Darmstadt, Germany). With serial dilutions of this stock solution, different initial H_2O_2 concentrations were used in the reaction system. The reagent used for quantifying stock H_2O_2 solution was potassium permanganate (*Merck KGaA*, Darmstadt, Germany), while the chelating agent, potassium titanium (IV) oxalate (*Riedel-deHaen*, Seelze, Germany), was used for measurement of diluted H_2O_2 solution. In analysis of Fe^{2+}, the reagents used include ammonium acetate, 1,10-phenanthroline (*Merck KGaA*, Darmstadt, Germany), and 35% HCl (*Kuan Heng*, Taiwan). The CO_2 gas with purity greater than 99.5% was purchased from a local supplier. Humic acid (*Sigma-Aldrich*, St. Louis, USA) was prepared as organic species in solution. The $CaCl_2.2H_2O$, Na_2CO_3 and NaCl were purchased from *Merck.*

6.2.2 Reaction Systems and Operation

The experiments for the Fe^0/H_2SO_4 system were conducted in a cylindrical reactor of 2.3 L, the reactor of which is referred in detail elsewhere (Liao et al. 2003a; Liao et al. 2003b). Shaking of reactor or mixing of solution was reported to affect decomposition rate of target compounds significantly (Zawaideh and Zhang 1998; Choe et al. 2000). Therefore, the solutions to be treated in the Fe^0/H_2SO_4 system were mixed vigorously by a motor-driven stirrer with identical intensity for all experiments. Concerning the Fe^0/CO_2 system, the experiments were conducted in a cylindrical reactor of 1.3 L, as shown in Fig. 6-1. The liquid volume prepared was 1 L. The re-circulated flow was induced to achieve vigorous mixing of solution by using peristaltic pump. The CO_2 gas was introduced by passing through a disk diffuser of silicate material from the bottom of reactor.

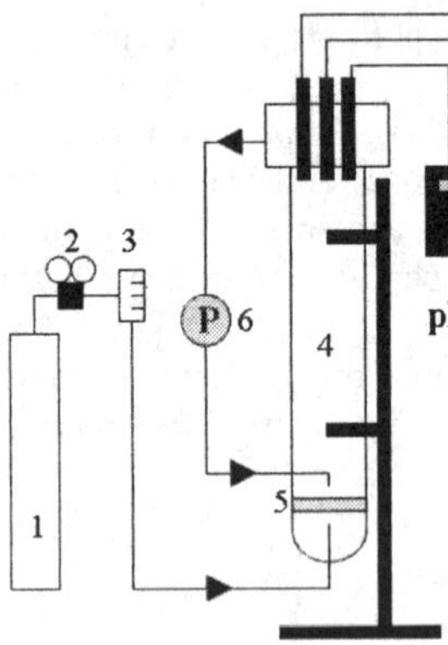

Fig. 6-1. Configuration of reactor for the Fe^0/CO_2 system. (1: CO_2 cylinder; 2: pressure gauge; 3: flow meter; 4: reactor; 5: disk diffuser; 6: recirculated water pump)

To investigate the effect of Fe^0 particle size on nitrate removal, the Fe^0 with 10 and 150 μm were tested, respectively. The Fe^0 dosages used were in the range of 1-4 g/L, and the pH values 3-5. In the presence of H_2O_2, the Fe^0 was not added into the working solution until the experiment was about to start; this was to avoid occurrence of Fenton reaction since considerable amount of Fe^{2+} might be released from Fe^0 before the experiment, especially under acidic pH condition. As the experiments were carried on, water samples were taken at different time intervals for subsequent analyses of nitrate, ammonium, H_2O_2 and ferrous concentration.

6.2.3 Instrumental Analyses

The residual nitrate for the Fe^0/H_2SO_4 system (Liao et al. 2003a; Liao et al. 2003b) was measured by the ultraviolet spectrophotometric method (standard method code 4500-NO_3^- B, APHA/AWWA/WEF 1998). 5 mL of water sample was taken from the reactor, filtered with a filter paper of 0.45 μm, diluted to a mark of 50 mL, and acidified by adding 1 mL of 1 M HCl. Through a nitrate calibration curve, the residual nitrate concentration was obtained by taking the two readings of ultraviolet absorption at 220 and 275 nm, the latter wavelength being used for interference correction purpose. In the Fe^0/CO_2 system, the residual nitrate, nitrite, and end product ammonium were analyzed by using Ion Chromatography (IC) (*DIONEX-120*, USA). In IC analyses, 4 drops of 15,000 mg/L H_2O_2 were spiked into the above filtrate to convert Fe^{2+} into ferric precipitate (Fenton's reaction product), and then the water sample was furthered filtered using a 0.45 μm membrane filter to remove the iron precipitate from solution. The stock H_2O_2 solution was quantified by using a potassium permanganate titration method (Vogel 1978). The residual H_2O_2 concentration in the reaction solution was determined by using the potassium titanium (IV) oxalate method (Sellers 1980). Under an acidic condition, the H_2O_2 reacts with Ti^{4+} to form a yellowish complex. Through the measurement of light absorption at 400 nm (*SHIMADZU*, UV-1201, Japan), the absorbance can be converted into equivalent H_2O_2 by reading the H_2O_2 value from a pre-determined linear calibration curve. Similarly, the Fe^{2+} is able to form a colored complex with 1,10-phenanthroline (standard method code 3500-Fe B, APHA/AWWA/WEF 1998), and its concentration can be determined through the measurement of light absorption at 510 nm (*SHIMADZU*, UV-1201, Japan), which is equivalent to a certain ferrous quantity. Humic acid was analyzed by TOC analyzer (*Elementar-liqui* TOC, Germany). In addition, the pH was monitored continuously by using a pH meter (*Suntex*, TS-1, Taiwan).

6.3 Results and Discussion

6.3.1 Fe^0/H_2SO_4 System

6.3.1.1 Effect of pH

As shown in Fig. 6-2, the reaction appears to stop almost completely at an elevated pH level of 5; the nitrate removal was not observed at all throughout the reaction period as the pH was controlled at 5. In contrast, the nitrate removals were significant as the pH values were controlled at 4, 3.5, and 3, with the most rapid removal of nitrate at pH 3, the next at pH 3.5, and the least at pH 4. In particular, the nitrate was removed by 90% approximately in 5 min at pH 3, and by 100% in 10 min at pH 3.5. In addition, the length of initial lag phase decreases with decreasing pH. For example, it lasts 15 min at pH 4, 10 min at pH 3.5, and far more less than 10 min at pH 3. As indicated in Eq. (6.5), the lower pH favors the acceleration of ferrous ejection from the iron metal surface, and in turn this results in faster release of one electron from $Fe(OH)_{ads}$. In addition, iron metal surface tends to form a protective film of iron oxide like Fe_3O_4 and γFe_2O_3 through anodic passivation process, which will eventually prevent further corrosion of iron metal. According to Cohen (1979), the breakdown of protective films is more rapid in acid solutions. It is also more rapid in solutions containing ions, which lead to acid formation in the anodic pores (e^--releasing zone). In summary, Fe^0 is capable of reducing nitrate at a relatively high reaction rate under acidic conditions.

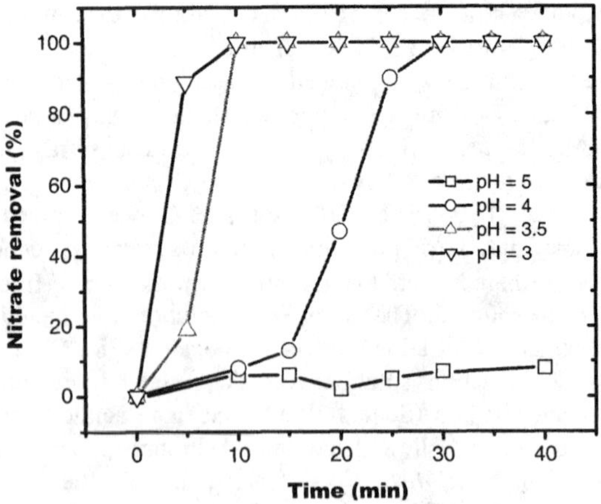

Fig. 6-2. Nitrate removal at various pH conditions (Fe^0 (10 μm) = 2 g/L; initial NO_3^- = 51 mg/L)

6.3.1.2 Effect of Fe^0 Dosage

Fig. 6-3 presents the profile of nitrate removals at various Fe^0 dosages. The Fe^0 dosages applied were 1, 2, and 3 g/L, and the iron particle size was 10 μm. The solution pH was controlled at 4 throughout the entire reaction period. It appears that there exists a lag phase over the initial period of 15 min, where the nitrate was removed only slightly. Following the lag phase, the nitrate removal was seen to increase rapidly. With the Fe^0 dosage of 1 g/L, the nitrate was removed by about 15% at 30 min, while the residue decreased to zero at 70 min. On the other hand, as

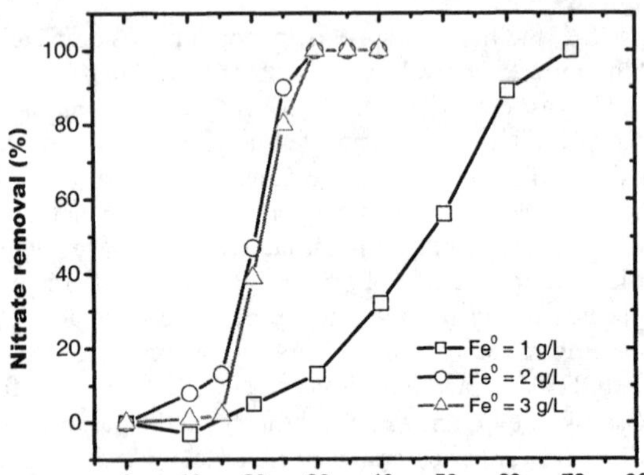

Fig. 6-3. Nitrate removal at various Fe^0 dosages (Fe^0 (10 μm) = 1, 2, 3 g/L; initial NO_3^- = 50 mg/L)

the Fe^0 dosages were 2 and 3 g/L, the nitrate was completely removed at time of 30 min. It is noticed here that the removal profiles of NO_3^- show no difference for the two dosages of 2 and 3 g/L applied, but the difference becomes quite significant as the dosage was reduced from 2 to 1 g/L. Such a result implies that the higher the Fe^0 dosage, the lower the NO_3^- residue in the reaction solution. However, the marginal benefit is decreasing as the Fe^0 is dosed in excessive amount.

6.3.1.3 Effect of Species with Hydroxyl Group

Eq. (6-1) shows that water molecule with hydroxyl group leads to the corrosion of Fe^0. Hence, organic (n-propanol) and inorganic (hydrogen peroxide) species with similar hydroxyl groups were investigated to reveal its individual effect on nitrate reduction by Fe^0, and the results of which were presented in Figs. 6-4 and 6-5, respectively.

In Fig. 6-4, the initial concentration of propanol used was 20.2 mg/L as DOC. The results show that propanol retards the removal of nitrate remarkably, and the solution DOC remains rather unchanged. With propanol, the nitrate removal was 10% at 25 min, compared to a 90% removal without propanol. It demonstrates that propanol exhibits significantly inhibitory effect on the nitrate removal, and the reason behind this might be due to the reactive sites of Fe^0 surface being occupied by propanol, leaving fewer sites available to nitrate for reduction. In Fig. 6-5, as iron metal with particle sizes of 10 and 150 μm were added, respectively, into H_2O_2-contained and pH-controlled solutions, the profiles of nitrate removal and H_2O_2 decomposition were quite different from each other. Surprisingly, with the use of particle size of 10 μm and with the extended reaction time as long as 145 min, there

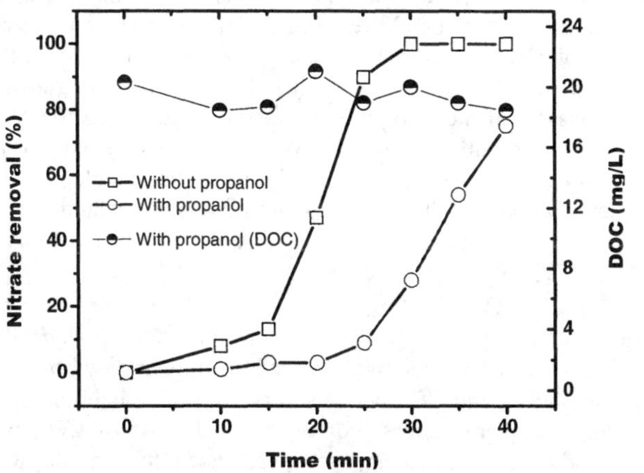

Fig. 6-4. Nitrate removal and DOC profile with and without propanol (pH = 4; initial NO_3^- = 53 mg/L; Fe^0 (10 μm) = 2 g/L; initial propanol concentration as DOC = 20.2 mg/L)

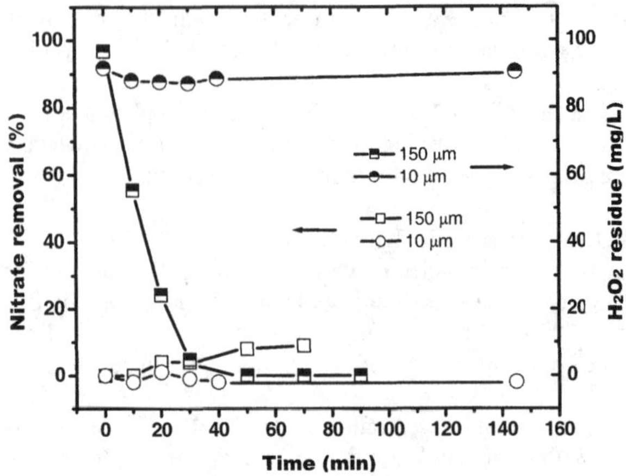

Fig. 6-5. Nitrate removal and residual profiles of H_2O_2 with two different particle sizes and in the presence of H_2O_2. Note that the H_2O_2 was present in the solution before the Fe^0 was spiked into the same solution.

was no nitrate removal at all, and the H_2O_2 remained almost unconsumed. In other words, the presence of H_2O_2 (92 mg/L) inactivates iron dissolving even in acid solution (pH = 4). However, by contrasting with the use of particle size of 150 μm, 10% nitrate was removed at time of 70 min, and H_2O_2 has been consumed almost completely at time of 50 min. As reported in detail elsewhere (Liao et al. 2003b), the consumption of H_2O_2 was through Fenton reaction, which infers that the Fe^{2+} has been dissolved from Fe^0 surface into bulk solution. In addition, the dissolved oxygen in solution was oversaturated due to the aforementioned Fenton reaction. The excessive oxygen in bulk solution can compete effectively with nitrate for electrons released from Fe^0 surface and consequently retard the removal of nitrate. This might be the reason why only slight removal of nitrate was observed in this case.

6.3.2 Fe^0/CO_2 System

6.3.2.1 Effect of CO_2 Bubbling

As mentioned earlier, hydrogen ion accelerates the rate of iron corrosion, leading to a higher efficiency of nitrate removal. Instead of supplying hydrogen ions with strong acids such as H_2SO_4 and HCl, acidic source from CO_2 bubbling is considered as a practice with many advantages. For example, the use of CO_2, which is a clear, colorless, odorless gas species, imposes no adverse effect on treated water quality. In addition, as the hydrogen ions are consumed in the reaction system, the resulting bicarbonate alkalinity can help remove background hardness and ferrous species through the formation of precipitates such as $CaCO_3$ and $FeCO_3$. In water purification processes, CO_2 is commonly used for stabilizing lime-softened water.

Depending on the pH level, the carbonated solution might be present with the species of H^+, HCO_3^- and CO_3^{2-}, according to Eqs. (6-6) to (6.9).

$$CO_2(g) \rightarrow CO_2(aq) \tag{6.6}$$
$$CO_2 + H_2O \rightarrow H_2CO_3 \tag{6.7}$$
$$H_2CO_3 \leftrightarrow HCO_3^- + H^+ \qquad pK_1 = 6.3 \tag{6.8}$$
$$HCO_3^- \leftrightarrow CO_3^{2-} + H^+ \qquad pK_2 = 10.3 \tag{6.9}$$

Fig. 6-6 presents the effect of CO_2 bubbling on the profiles of pH with and without Fe^0 in the solution. As can be seen, after only 10 min of CO_2 bubbling at a rate of 200 mL/min, the solution pHs dropped significantly down to a value around 4.5 (with 1-4 g/L of Fe^0) and 3.5 (without Fe^0), respectively. The pH values remained unchanged as the reaction time went beyond 10 min in the absence of Fe^0. In contrast, the pH profiles rebound in the presence of Fe^0. The rebounding pH can be reasonably explained by Eq. (6.10), in which the water molecule was reduced by Fe^0 to produce hydroxide ion, thus contributing the rising pH. In fact, the pH should be able to reach the steady-state condition after a certain time if the acidic supplying rate [Eqs. (6.6) to (6.9)] equals to the basic generating rate [Eq. (6.10)]. Therefore, if the CO_2 is bubbled into solution at a sufficiently high rate, the pH will remain unchanged, the Fe^0 will continue to corrode, and the NO_3^- will become reduced continuously.

$$Fe^0 + 2H_2O \rightarrow Fe^{2+} + H_2 + 2OH^- \tag{6.10}$$

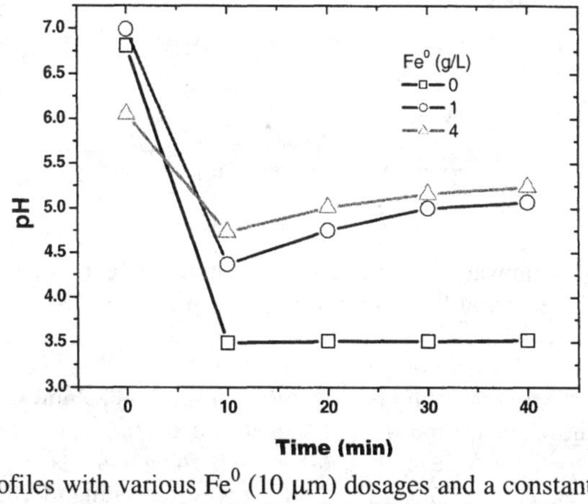

Fig. 6-6. pH profiles with various Fe^0 (10 μm) dosages and a constant CO_2 bubbling flow rate (solution volume = 1 L; solution recirculation flow rate = 1000 mL/min; CO_2 = 200 mL/min)

6.3.2.2 Effect of Initial Nitrate Concentration

Fig. 6-7 shows the remaining nitrate profiles with various initial nitrate concentrations (9.7-100.0 mg/L). Under the specified operating conditions, complete removal of nitrate can be achieved for the initials of 9.7 and 29.5 mg/L. The corresponding times of zero residues were around 20 and 30 min, respectively. For the initial of 46 mg/L, the final residue was kept at around 5 mg/L, while it was 20 mg/L for the initial of 100.0 mg/L. To achieve faster removal of nitrate, either Fe^0 dosage or the bubbling rate of CO_2 needs to be increased accordingly. In addition, it appears that the increase of reaction time is not an effective way to increase nitrate removal since the rebounding trend of pH at a longer reaction time can lead to retardation of Fe^0 corrosion or inhibition of nitrate reduction, as mentioned earlier. Furthermore, it was observed that the initial lag phase disappeared at the higher initial of 100.0 mg/L. As described earlier, decrease in the pH shortens the initial lag phase of nitrate removal profile (see Fig. 6-2). According to Eq. (6.11), the shortened initial lag phase seems to be justified as both nitrate and hydrogen ions increase in their molar concentrations, resulting in favorably forward reaction.

$$4Fe^0 + NO_3^- + 10H^+ \rightarrow 4Fe^{2+} + NH_4^+ + 3H_2O \tag{6.11}$$

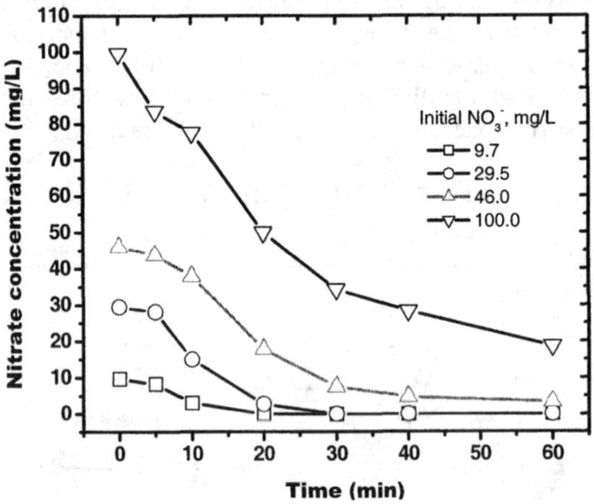

Fig. 6-7. Nitrate removal with various initial nitrate concentrations (Fe^0 (10 μm) = 2 g/L; solution recirculation flow rate = 1000 mL/min; CO_2 = 200 mL/min)

6.3.2.3 Effect of Humic Acid

Humic acid is a very common organic matter in soil and natural waters. It seems fit to investigate how this natural organic matter affects the reduction of nitrate by Fe^0. With and without humic acid, Fig. 6-8 presents the removal profiles with humic concentration range of 0-4.2 mg/L as TOC. It was interesting to find that humic acid content as small as only 0.6 mg/L imposed significant inhibitory effect on nitrate removal, as compared to the one without humic acid. With humic content more than 1.9 mg/L, no removal of nitrate was observed over the entire reaction period. Additionally, the initial lag phase of 30 min was extraordinarily long in the presence of either 0.6 or 1.9 mg/L as TOC. Initial strong adsorption or binding of humic acid onto Fe^0 surface seems to be a reasonable explanation for such a long lag phase. In addition, characteristics of humic acid showing a lower solubility at lower pH also helps understand such a phenomenon. Initially, the solution pH remains at around the neutral one, where humic acid is more soluble than at acidic pH. As the solution pH decreases because of CO_2 bubbling, the humic substances tend to precipitate and detach from the Fe^0 surface, thus allowing for the opportunity of nitrate reduction to occur. However, when humic acid exists in excessive amount like the ones of 1.9 and 4.2 mg/L as TOC, the equilibrium soluble concentration of humic substances might still dominate all over the reactive sites of Fe^0 surface, leaving no opportunity at all for the hydrogen ion to compete for the reactive sites of Fe^0 surface. Similar results were also reported on the inhibition of the reduction of carbon tetrachloride and trichloroethylene by natural organic matter (Tratnyek et al. 2001).

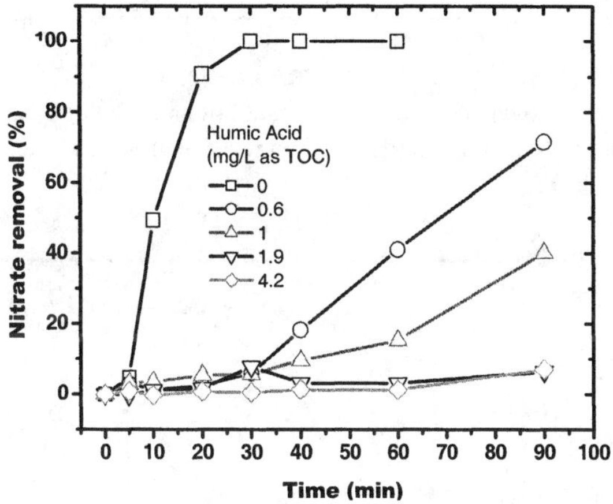

Fig. 6-8. Nitrate removal with various humic concentrations (Fe^0 (10 μm) = 2 g/L; NO_3^- = 31 mg/L; solution recirculation flow rate = 1000 mL/min; CO_2 = 200 mL/min)

6.3.2.4 Effect of Cations and Anions

In addition to humic acid, some other cations and anions commonly found in natural waters were also investigated for their impacts on nitrate reduction by Fe^0, including Na^+, CO_3^{2-}, Ca^{2+}, and Cl^-. The results were shown in Figs. 6-9 and 6-10, respectively. The use of sodium carbonate was intended to simulate the alkalinity of targeted waters, while the hardness was simulated by the presence of calcium ions, and possible seawater intrusion into groundwater was demonstrated by the presence of sodium chloride.

In Fig. 6-9, as sodium carbonate was increased from 50 to 150 mg/L, it appears that no significant differences were observed between the three nitrate removal profiles. By a closer look into these profiles, only very slight inhibitory effect on nitrate removal was imposed by the presence of Na_2CO_3; the higher the Na_2CO_3 concentration, the slower the nitrate removal. In the CO_2-water system, the CO_3^{2-} is present depending on the pH level, as depicted in Eq. (6.9). When Na_2CO_3 was present in the solution, the ionized CO_3^{2-} is obvious to assimilate itself into the

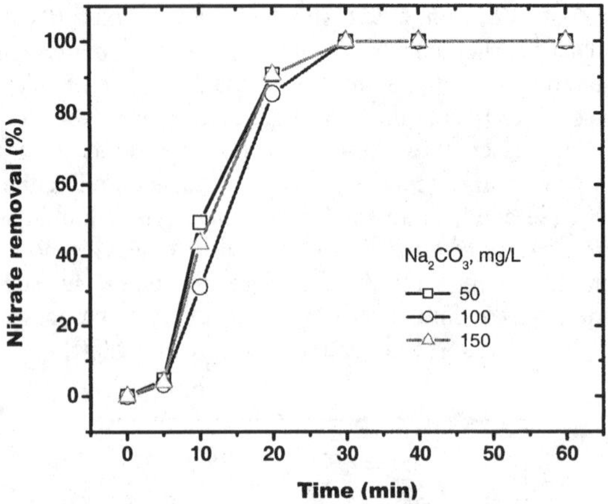

Fig. 6-9. Nitrate removal with various concentrations of sodium carbonate (Fe^0 (10 μm) = 2 g/L; initial NO_3^- = 29 mg/L; solution recirculation flow rate = 1000 mL/min; CO_2 = 200 mL/min)

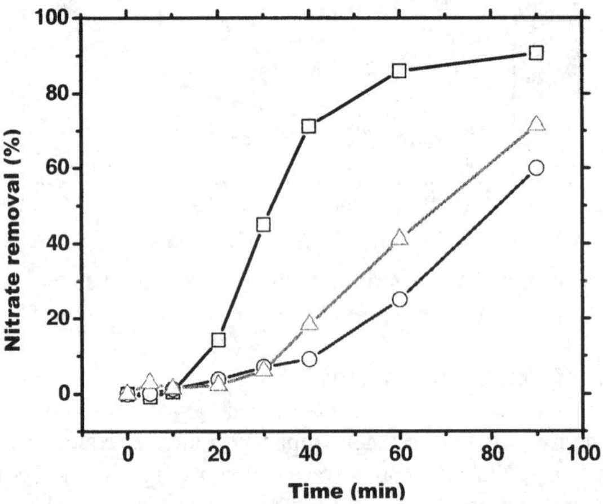

Fig. 6-10. Nitrate removal with various concentration combinations of Cl^-, Ca^{2+}, and Na^+. (-□-) With species of $NaNO_3$ (30 mg /L as NO_3^-), Na_2CO_3 (94 mg/L as $CaCO_3$), and NaCl (107 mg/L as Cl^-), which contain Cl^- = 107 mg/L, Ca^{2+}= 0 mg/L, and Na^+= 123 mg/L; (-o-) With species of $CaCl_2{\cdot}2H_2O$ (150 mg/L as $CaCO_3$), $NaNO_3$ (30 mg /L as NO_3^-), Na_2CO_3 (94 mg/L as $CaCO_3$), which contain Cl^- = 107 mg/L, Ca^{2+} = 60 mg/L, and Na^+ = 54 mg/L; (-Δ-)With species of $NaNO_3$ (30 mg /L as NO_3^-), which contains Cl^- = 0 mg/L, Ca^{2+}= 0 mg/L, and Na^+= 11 mg/L. Fe^0 (10 μm) = 2 g/L; initial NO_3^- = 30 mg/L; solution recirculation flow rate = 1000 mL/min; CO_2 = 200 mL/min; humic acid = 0.55 mg/L as DOC.

above CO_2-water system. Yet, it will take a longer time to bring down the pH to a certain acidic value by CO_2 bubbling for this alkaline species with higher concentration. Indeed, as illustrated in Fig. 6-9, there exist lag phases and slight differences between nitrate removal profiles. For the species Na^+ alone, it seems reasonable to infer that its effect on nitrate removal is insignificant, according to the previous statement.

In order to look into each individual effect of Cl^- and Ca^{2+}, reagents of $NaNO_3$, Na_2CO_3, NaCl, and $CaCl_2 \cdot 2H_2O$ were used to prepare different concentration levels of the aforementioned species. There are three sets of experiments conducted under these three conditions: (1) **(-□-)**: Cl^- = 107 mg/L, Ca^{2+}= 0 mg/L, and Na^+= 123 mg/L; (2) **(-o-)**: Cl^- = 107 mg/L, Ca^{2+} = 60 mg/L, and Na^+ = 54 mg/L; and (3) **(-Δ-)**: Cl^- = 0 mg/L, Ca^{2+}= 0 mg/L, and Na^+= 11 mg/L. As a result, Fig. 6-10 shows how these three conditions impose their effects on nitrate removals by Fe^0. Basically, the line with the legend of **(-Δ-)** is the result of the control experiment and it sits between the other two lines. In addition, it was inferred from the previous section that Na^+ imposes no effect on nitrate removal. With such perception, Ca^{2+} shows significantly inhibitory effect on nitrate removal by comparing the two lines with the legends of **(-□-)** and **(-o-)**. Also, by contrasting the two lines with the legends of **(-□-)** and **(-Δ-)**, it is concluded here that Cl^- accelerates the removal of nitrate in a significant way. The inhibitory property of Ca^{2+} might be owing to the mutual repelling factor between Ca^{2+} and Fe^{2+}. Note that the latter one is the resulting product of Fe^0 corrosion in bulk solution. As for the accelerating property of Cl^-, Choe et al. (2004) also reported that chloride ion in solution induces pitting corrosion of the Fe^0 surface, which could enhance surface reactivity or increase surface area of the Fe^0 for NO_3^- reduction.

6.3.2.5 Effect of Operating Modes

Two modes of operation were tested for the nitrate removal by Fe^0. Mode I was to treat nitrate-contained solution in a continuous way within a batch reactor, i.e., a predetermined amount of nitrate was spiked into the solution at different time points (60 and 120 min) without emptying the reactor or discarding the residual Fe^0 powder. Mode II was to treat the solution in a batch way; the treated solution was discharged and a new batch of nitrate-contained solution was treated by using the Fe^0 powder left from the previous run. In designing the experiments like Modes I and II, the durability or sustainability of Fe^0 can also be revealed. Consequently, the results show that Mode II outperforms Mode I, according to Fig. 6-11. In other words, it is more advantageous to operate in a batch way than in a continuous way when the Fe^0-contained batch reactor is employed in the treatment system. The reason is because the Fe^{2+} continues to build up in the reaction system in Mode I, and this will suppress the releasing of Fe^{2+} from Fe^0 surface, in view of decreasing concentration gradient between the Fe^0 surface and bulk solution. Additionally, the Fe^0 residue preserves its activity in a satisfactory way in the three batches in Mode II, with the third one becoming a slight degradation. In contrast, the final removal efficiency in each of the three batches is seen to decrease from 100%, 90% and to 70%, respectively, within the same reaction time, when Mode I was tested. In summary, it is recommended to operate the system in Mode II to prolong the sustainability of Fe^0 rather than Mode I.

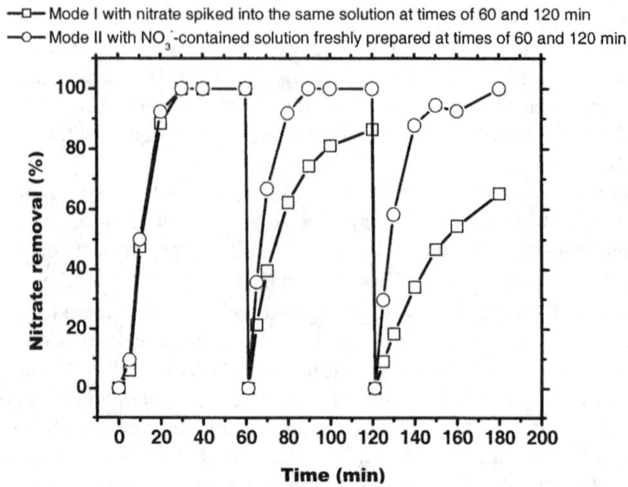

Fig. 6-11. Nitrate removal with two different operating modes (Fe^0 (10 μm) = 2 g/L; initial NO_3^- = 28 mg/L; solution recirculation flow rate = 1000 mL/min; CO_2 = 200 mL/min)

6.3.3 Issue of Undesired Byproducts and Its Resolution

In treating nitrate-contaminated water by Fe^0, two side issues need to be taken care of. One is the buildup of Fe^{2+} in the reaction system (Fig. 6-12), while another is the byproduct of ammonium from nitrate reduction (Fig. 6-13). These two parameters are usually regulated by the safe drinking water standards in most countries. In Fig. 6-12, the Fe^{2+} can accumulate up to as high as 275 mg/L under acidic conditions. In addition, as described in detail elsewhere (Liao et al. 2003b), the kinetic behavior of

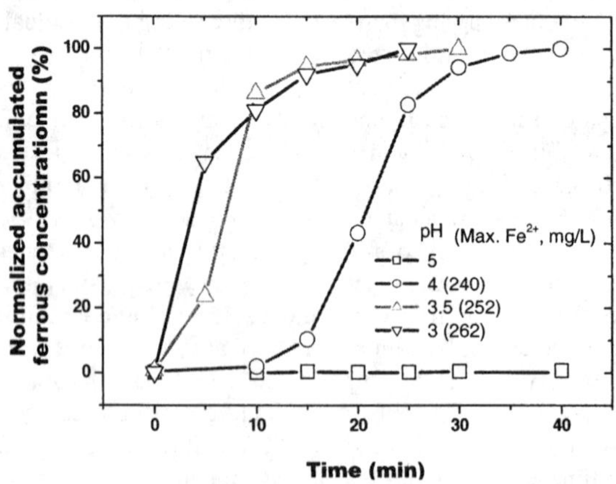

Fig. 6-12. Ferrous profiles with various pH conditions in the Fe^0/H_2SO_4 system (Fe^0 (10 μm) = 2 g/L; initial NO_3^- = 51 mg/L)

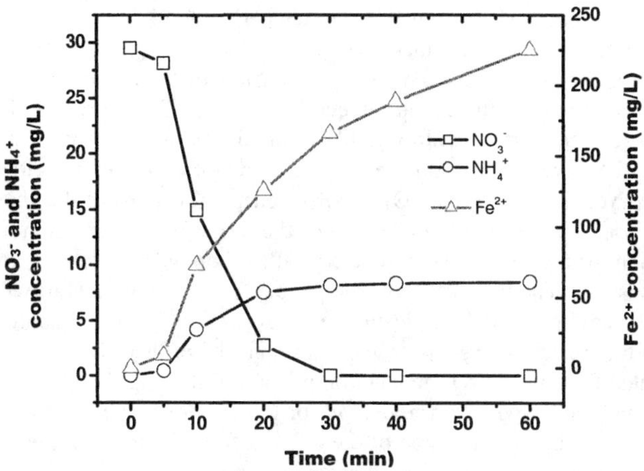

Fig. 6-13. Formation of ammonium and ferrous accumulation in the Fe^0/CO_2 system (Fe^0 = 2 g/L; initial NO_3^- = 30 mg/L; solution recirculation flow rate = 1000 mL/min; CO_2 = 200 mL/min)

ferrous accumulation can be expressed by a S-curve, including the initial lag phase, exponential phase, rate-decreasing phase and steady-state phase. At any rate, the cost-effective process needs to be developed to remove iron species from the system.

In Fig. 6-13, as the nitrate decomposes, the ammonium is seen to accumulate accordingly, in addition to the buildup of Fe^{2+}. Unfortunately, all nitrate treated in the Fe^0/CO_2 system are converted to the undesirable species of ammonium. This was also evidenced by Cheng et al. (1997). Though nitrogen gas was also reported as the end product of nitrate reduction by Fe^0, the particle size of Fe^0 used was in nanoscale (Choe et al. 2000). In view of nanoparticles preparation and field application, there still exist some practical issues to be resolved. According to Table 6-2 (Benjamin, 2002), reaction intermediates of nitrate in the Fe^0 system may include NO_2^-, NO, NH_4^+, and N_2. The reactions with respect to nitrite and ammonium have the lower values of E_H^0 (0.843, 0.89, 0.878 V), whereas the formation of nitrogen gas has the highest one (1.241 V). Thermodynamically, reaction for the formation of nitrogen gas should be the most favorable one to occur. However, the experimental results show that ammonium is the dominating species in the studied system. Further research needs to be conducted to reveal such phenomenon.

Table 6-2. Equilibrium Constants for Redox Half-cell Reactions of Nitrate

Reaction	Log K	pe^0	E_H^0, V
$NO_3^- + 2e^- + 2H^+ \Leftrightarrow \mathbf{NO_2^-} + H_2O$	28.57	14.29	**0.843**
$NO_3^- + 3e^- + 4H^+ \Leftrightarrow \mathbf{NO_{(g)}} + 2H_2O$	48.40	16.13	0.952
$2NO_3^- + 10e^- + 12H^+ \Leftrightarrow \mathbf{N_{2(g)}} + 6H_2O$	210.34	21.03	1.241
$NO_3^- + 8e^- + 10H^+ \Leftrightarrow \mathbf{NH_4^+} + 3H_2O$	119.08	14.89	**0.878**
$NO_2^- + 6e^- + 8H^+ \Leftrightarrow \mathbf{NH_4^+} + 2H_2O$	-	15.0	**0.890**

To achieve complete removal of nitrate, Fig. 6-14 shows a prototype of continuous treatment system proposed for further study, in terms of system performance and cost analysis. This system consists of three component units. The first unit is designed for the nitrate reduction by the Fe^0/CO_2 process. In this unit, nitrate will be converted mainly into ammonium, and the Fe^{2+} accumulates remarkably. To save the Fe^0 powder, the effluent of this unit is settled and Fe^0 residue is recycled back to the CO_2-bubbled reactor. To remove the Fe^{2+} from the aqueous phase, the second unit employs the process of air aeration for iron crystallization. Iron species will then be crystallized on the selected seeding material (e.g., quartz sand). The byproduct of iron pellets from the second unit will be tested and used for environmental application purpose, reutilizing the resources of solid waste from the treatment system. In the final unit, the effluent from the second unit will be treated first by raising the solution pH, and following the pH adjustment the byproduct ammonium will be stripped out of the system by air aeration. To ensure the final water quality, carbonation of the effluent from the third unit is to be carried out to bring down the pH to a required level.

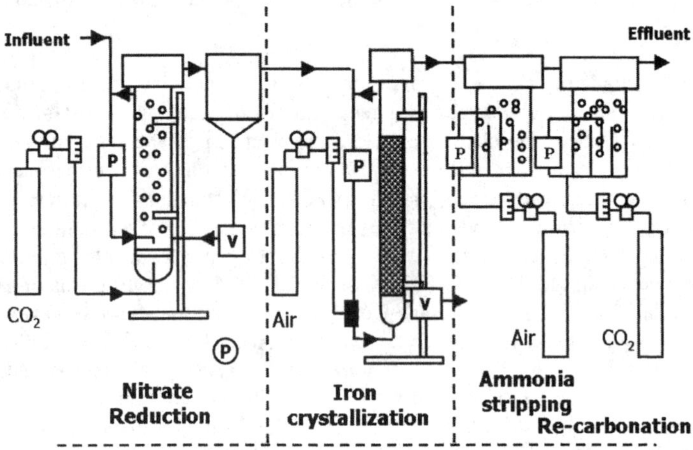

Fig. 6-14. Prototype of treatment process train for complete removal of nitrate

6.4 Conclusions and Recommendations

In this paper, several process parameters were identified for their impacts on nitrate removal by Fe^0. Control of pH within acidic range of 3-4 accelerates nitrate removal in a relatively short time period, e.g., less than 30 min. Appropriate iron dosages gain the most of marginal benefit in removing nitrate from waters. Background species of different kinds also play important roles in the process performance. The n-propanol with hydroxyl group exhibits retardation on nitrate removal significantly, while the presence of hydrogen peroxide (HO-OH) with two hydroxyl groups inactivates corrosion of Fe^0 of 10 μm size, but hydrogen peroxide itself decomposes when the Fe^0 of 150 μm size was applied. In the presence of higher sodium carbonate concentrations, the removal of nitrate was only slightly inhibited. As for the calcium species, it exhibits significant inhibition on the nitrate removal. But the presence of chloride improves the removal of nitrate remarkably.

In view of process operation, it is recommended that batch mode operation is employed as a more cost-effective way to treat nitrate-contaminated waters, rather than continuous flow mode. In view of the two side issues of concern, ferrous accumulation and end product of ammonium, a prototype of three-stage process for complete removal of nitrate is recommended for future study, including (1) nitrate reduction by Fe^0/CO_2 process; (2) ferrous crystallization by air aeration; and (3) ammonia stripping by air aeration.

Acknowledgement

The research material presented in this paper was the original work from the past and ongoing projects sponsored by the Taiwan National Science Council (NSC) under project No.: NSC89-2211-E-041-013 and NSC93-2211-E-041-001.

6.5 References

Alowitz, M. J., and Scherer, M. M. (2002). "Kinetics of nitrate, nitrite, and Cr(VI) reduction by iron metal." *Environ. Sci. & Technol.*, 36(3), 299-306.

APHA, AWWA and WEF (1998). *Standard methods for the examination of water and wastewater*, 20th Ed., American Public Health Association, American Water Works Association and Water Environment Federation, Washington DC, United States.

Benjamin M. M. (2002). *Water chemistry*, 1st Ed., McGraw-Hill Companies, Inc., New York.

Chang, H. L., Hwang, R. J., Gung, H. B., Horng, S. J., and Su, J. L. (2002). "Heterotrophic-denitrification pilot study in Hsin-Chia Water Treatment Plant on Nan-Tou operation station." In *Proceedings the 8th International Conference on Drinking Water Management and Treatment Technologies*, May 27-29, Chung Shan University, Kaohsiung, Taiwan, 3-14 – 3-30.

Cheng, I. F., Muftikian, R., Fernando, Q., and Korte, N. (1997). "Reduction of nitrate to ammonia by zero-valent iron." *Chemosphere*, 35(11), 2689-2695.

Chew, C. F., and Zhang, T. C. (1998). "In-situ remediation of nitrate-contaminated groundwater by electrokinetics/iron wall processes." *Wat. Sci. & Technol.*, 38(7), 135-142.

Choe, S., Chang, Y.-Y., Hwang, K.-Y., and Khim, J. (2000). "Kinetics of reductive denitrification by nanoscale zero-valent iron." *Chemosphere*, 41(8), 1307-1311.

Choe, S., Liljestrand, H. M, and Khim, J. (2004). "Nitrate reduction by zero-valent iron under different pH regimes." *App. Geochem.*, 19(3), 335-342.

Coehn, M. (1979). "Dissolution of iron." In *Corrosion chemistry*, American Chemical Society ACS, Washington, DC.

Gandhi S., Oh, B.-T., Schnoor, J. L., and Alvarez, P. J. J. (2002). "Degradation of TCE, Cr(VI), sulfate, and nitrate mixture by granular iron in flow-through columns under different microbial conditions." *Wat. Res.*, 36(8), 1973-1982.

Hsu, C.-Y., Liao, C.-H., and Lu, M.-C. (2004). "Treatment of aqueous nitrate by zero valent iron powder in the presence of CO_2 bubbling." *Groundwater Monitoring and Remediation*, 24(4), 82-87.

Huang Y. H., and Zhang T. C. (2004). "Effects of low pH on nitrate reduction by iron powder." *Wat. Res.*, 38(11), 2631–2642.

Huang, C.- P., Wang, H.-W., and Chiu, P.-H. (1998). "Nitrate reduction by metallic iron." *Wat. Res.*, 32(8), 2257-2264.

Kelly, E. J. (1965). "The active iron electrode: I. Iron dissolution and hydrogen evolution reactions in acidic sulfate solutions," *J. Electrochem. Soc.*, 112(2), 124-131.

Liao, C.-H., Kang, S.-F., and Hsu Y.-W. (2003a). "Characteristics of reductive removal of nitrate by suspended metallic iron powder." *J. Chinese Inst. of Environ. Eng.*, 13(4), 251-261.

Liao, C.-H., Kang, S.-F., and Hsu, Y.-W. (2003b). "Zero-valent iron reduction of nitrate in the presence of ultraviolet light, organic matter and hydrogen peroxide." *Wat. Res.*, 37(17), 4109-4118.

Mirvish, S. S. (1985). "Gastric cancer and salivary nitrate and nitrite." *Nature*, 315, 461-462.

Sellers, R. (1980). "Spectrophotometric determination of hydrogen peroxide using potassium titanium (VI) oxalate." *Analyst*, 105, 950-954.

Su, C., and Puls, R. W. (2004). "Nitrate reduction by zerovalent iron: effects of formate, oxalate, citrate, chloride, sulfate, borate, and phosphate." *Environ. Sci. & Technol.*, 38, 2715-2720.

Tratnyek, P. G., Scherer, M. M., Deng, B., and Hu, S. (2001). "Effects of natural organic matter, anthropogenic surfactants, and model quinones on the reduction of contaminants by zero-valent iron." *Wat. Res.*, 35(18), 4435-4443.

Vogel, A. (1978). *A textbook of quantitative inorganic analysis*, 4th Ed., Longman Scientific and Technical, Harlow, Essex, England.

Walton, G. (1951). "Survey of literature relating to infant methemoglobinemia due to nitrate-contaminated water." *Am. J. Public Health*, 41, 986-995.

Westerhoff, P., and James, J. (2003). "Nitrate removal in zero-valent iron packed columns." *Wat. Res.*, 37(8), 1818-1830.

Zawaideh, L. L., and Zhang, T. C. (1998). "The effects of pH and addition of an organic buffer (HEPES) on nitrate transformation in Fe^{o}-water systems." *Wat. Sci. & Technol.*, 38(7), 107-115.

CHAPTER 7

Removal of Nitrate from Water by a Combination of Metallic Iron Reduction and Clinoptilolite Ion Exchange Process

C. P. Huang[1], J. A., Yun[2], and S. W. Park[3]

Abstract: The effects of pH and nitrate surface loading on the reduction of nitrate by metallic iron and the removal of ammonium and ferrous ions by a natural ion exchanger, clinoptilolite, were investigated. The rate of nitrate reduction was observed to be a function of pH following the expression, $k_H = 6.6[H^+]^{0.58}$, where k is rate constant with a unit of min^{-1}. Nitrate surface loading plays an important role in the reduction of nitrate. The rate constant was related to surface loading (SL, in mM/m^2) by the expression: $k_{sl} = 2.00[SL]^{0.55}$. Based on results obtained in this study, there was a maximum surface loading above which nitrate reduction terminated. Clinoptilolite can be effective in removing ammonium ions in the presence of ferrous ions. The ion exchange reaction between ammonium ions and ferrous ions and clinoptilolite of the type, $Fe^{2+} + X - NH_4^+ \Leftrightarrow X - Fe^{2+} + NH_4^+ ; K_F^N$, was present. Results indicated that when the [Fe(II)] to ($[NH_3]$ + $[NH_4^+]$) molar ratio became greater than ca. 2, the above reaction will proceed backward rendering the clinoptilolite process infeasible. Overall, results indicated that the combination of metallic iron reduction and clinoptilolite ion exchange process can be an attractive alternative for the removal of nitrate from water.

CE Database subject headings: Ammonia; Iron; Ion exchange; Nitrates.

7.1 Introduction

Nitrate is a common contaminant in many United States water systems (Freeze and Cherry 1979). Many areas have nitrate levels above the standard of 10 mg/L NO_3^-N (Plumb 1998). Excessive use of nitrogen fertilizers, intensive use of irrigation, disposal of animal wastes, and septic systems are the major sources of nitrate in waters (Madison and Brunett 1985). Nitrate is an environmental pollutant, not only causing eutrophication of water but also threatening public health (Gunderloy et al. 1968). High nitrate concentration can cause a life-threatening methemoglobinemia in infants and may cause cancer in adults. The removal of nitrate is necessary to avoid these problems. Nitrate is a difficult contaminant to remove from water. The following processes are available for nitrate removal: biological denitrification (Richard 1989; Ainsworth 1990; Lyons 1990; Rogalla et al. 1990), ion exchange

[1]Professor, Dept. of Civil and Environmental Engineering, University of Delaware, Newark, DE 19716, United States (corresponding author). E-mail: huang@ce.udel.edu

[2]Research Assistant, Dept. of Civil and Environmental Engineering, University of Delaware, Newark, DE 19716, United States.

[3]Dagu Environmental Technology Development Center, Keimyung University, Daegu, Korea.

(Greeg 1973), chemical reduction (Szabo et al. 1951; Szabo et al. 1952; Szabo et al. 1957; Young et al. 1964; O'Brien 1968; Gunderloy et al. 1968; Gunderloy et al. 1970; Bruesh and Moraghan 1976; Sova 1986; Van Hecke et al. 1990; Huang et al. 1998; Hu et al. 2001; Hui et al. 2001; Hsu et al. 2004; Zhang and Huang 2005), a combined method of ion exchange and biological denitrification (Van der Hoek 1988a; Van der Hoek 1988b; Ainsworth 1990), and reverse osmosis (Eisenberg and Middlebrook 1986; Ainsworth 1990; Van der Hoek 1988a).

Biological processes, although capable of reducing nitrate, require a long process time, achieve only partial reduction, and sometimes generate nitrite, which is more toxic than nitrate. Ion exchange resins are not selective for nitrate, and the processes cause brine disposal problems. A combination of ion exchange and biological denitrification seems to ease these disadvantages, but this has not yet been investigated in detail. Reverse osmosis processes are too expensive and require disposal of voluminous concentrated waste.

Chemical reduction techniques using ferrous iron as a reductant have been developed, but some technical drawbacks limit their practicality. These processes produce large amounts of sludge for disposal, need a catalyst to drive the reaction forward, and should take place under anaerobic conditions. On the other hand, chemical reduction processes using iron particles do not need any catalyst and do not result in large amounts of sludge. These techniques are worth developing as a process for the practical removal of nitrate from water.

Huang et al. (1998) investigated the effects of pH and the iron-to-nitrate ratio on the rate of nitrate reduction using metallic iron and reported rapid nitrate reduction at pH less than or equal to 4. An iron-to-nitrate ratio of 120 m^2 Fe^0/mol NO_3 or higher was required to completely remove nitrate within an hour. Okazaki et al. (1998) studied the reduction of nitrate by iron powder coated with copper in neutral solutions. Nitrate was reduced to ammonium ion via nitrite as a first-order reaction. Hui et al. (2001) and Hu et al. (2001) studied the reduction of nitrate by metallic iron in aquatic solution and reported that the reduction of nitrate by metallic iron was a pseudo-first-order reaction. The production of nitrogen gas end product increased with increasing pH; no formation of nitrogen was found at pH = 2. Neutral condition enhanced the formation of nitrogen gas from nitrite reduction. Liao et al. (2003) investigated the reduction of nitrate using metallic iron in the presence of ultraviolet light (UV), organic matter, and hydrogen peroxide. The removal of nitrate increased with increasing iron dosage up to a concentration of 2 g/L. UV irradiation inhibited iron reduction of nitrate. The presence of propanol retarded nitrate reduction; the presence of H_2O_2 appeared to inactivate all reactions at as iron particles of the size of 10 μm were used. Westerhoff and James (2003) evaluated nitrate removal using zero-valent iron, Fe^0, packed bed columns and reported that in contrast to previous batch experiments with Fe^0 where nitrate was stoichiometrically converted to ammonium, only 70% of the applied nitrogen was recovered as nitrate, ammonium, or nitrite (<0.1 mg/L) during shorter-term column tests (2-20 bed volumes, BVs), and less than 25% of the applied nitrogen was recovered during longer-term field testing (500-1000 BVs) at elevated nitrate levels (similar to 25 mg N/L). Westerhoff (2003) studied the removal of three oxo-anions, i.e., nitrate, bromate, and chlorate, using Fe^0 and reported that ammonia was the primary reduction by-product from nitrate, chloride from chlorate, and bromide from bromate. Oxo-anion removal rates decreased as follows: bromate (BrO_3^-) > chlorate (ClO_3^-) > nitrate (NO_3^-). The atomic structure (atomic radii, electron orbital configuration, electron affinity) of nitrogen, chlorine, and bromine elements of the oxo-anions, and the bond dissociation energy between these elements and oxygen, were good indicators for the relative rates of reduction by Fe^0. Hsu et al. (2004) investigated the treatment of aqueous nitrate by Fe^0 powder in the presence of CO_2 bubbling and reported that CO_2 bubbling effectively created an acidic environment favorable to Fe^0 corrosion, which resulted in nitrate reduction. The end product in the reaction mixture was ammonium, which accounted for 90% to 104% of nitrate conversion with the presence of various Fe^0 dosages (0.5 to 2.0 g/L). Huang and Zhang (2004) studied the effects of low pH on nitrate reduction by iron powder and reported that nitrate could be rapidly reduced to ammonium at pH levels ranging from 2 - 4.5. The two major effects of pH on the kinetics of nitrate reduction are that (a) hydrogen ion (H^+) ions directly participate in the redox reaction of nitrate reduction following first-order kinetics; and (b) H^+ ions affect the nitrate adsorption onto reactive sites. Su and Puls (2004) investigated the effects of simple organic and inorganic acids such as formate, oxalate, citrate, chloride, sulfate, borate, and phosphate on nitrate reduction by Fe^0 and reported that nitrate reduction rates (pseudo-first-order) increased in the following order: phosphoric acid (H_3PO_4) < citric acid < boric acid (H_3BO_3) < oxalic acid < sulfuric acid (H_2SO_4) < formic acid < hydrochloric acid (HCl). The results are consistent with the sequence of strength of surface complexation of the inorganic ligands to iron oxides, which increases in the following order: chloride < sulfate < borate < phosphate. The blockage of reactive sites on the surface of Fe^0 and its corrosion products by specific adsorption of the inner-sphere complex forming ligands (oxalate, citrate, sulfate, borate, and phosphate) may be responsible for the decreased nitrate reduction by Fe^0 relative to the chloride system. Ginner et al. (2004) studied the influence of mass transport, temperature, and denitrifying microbes on nitrate and nitrite reduction by Fe^0 and reported that the activation energies for nitrate and nitrite reduction by granular Fe^0 were similar (21.7 ± 9.3 kJ mol^{-1} for nitrate and 23.8 ± 1.8 kJ mol^{-1} for nitrite). Addition of *P. denitrificans* to reactors resulted in faster nitrate removal compared to treatments with Fe^0 alone at all temperatures tested (5 to 50°C). Nitrite removal rates were typically 1.5 to 15 times faster than those of nitrate. Chi et al. (2004) investigated the reduction of nitrate by metallic iron under aerobic conditions in the pH range of 2.0 - 8.5 and reported that nitrate reduced predominantly to gaseous nitrogen at the beginning of the reduction reaction and ammonia/ammonium at the end of the reduction reaction and that pH value was an important parameter affecting the reduction of nitrate by metallic iron. At low pH values, ammonium was the main reduction product, while gaseous nitrogen was the dominant reduction product at high pH values. Choe et al. (2004) studied the reduction of NO_3^- by Fe^0 under anaerobic reductions in unbuffered solutions from pH 2 to greater than 10 and reported that the formation of green rusts divided the NO_3^- reduction process into two phases. Green rusts formed around a pH of 6.5 and contributed to the stabilization of pH. At low pH values, excessive accumulation of H_2 gas on the iron surface limited the availability of Fe^0 and retarded the NO_3^- reduction. Zhang and Huang (2005) studied the effect of pH on iron reduction of nitrate using three Good's pH buffers. Nitrate reduction by iron powder at near-neutral pH was negligible in an unbuffered system, but it was greatly enhanced with the presence of the buffer due in part to a direct reaction between the Good's pH buffer and the iron powder. Mishra and Farrell (2005) studied the nitrate reduction mechanism with Fe^0 media under conditions relevant to groundwater treatment using permeable reactive barriers (PRB) and reported that reaction with Fe^0 can be excluded as the mechanism for NO_3^- and NO_2^- reduction; reaction with Fe(II)-containing oxides coating the iron surface is the most likely reaction mechanism. This suggests that short-term batch tests requiring little turnover of reactive sites on the iron surface may overestimate long-term rates of NO_3^- removal because the effects of passivity are not apparent in batch tests conducted with high initial Fe(II) to NO_3^- ratios.

In summary, while it is clear that several factors such as pH, nitrate to iron ratio (surface loading) play an important role in nitrate reduction by metallic iron and that ammonium is the major product or metallic iron reduction under acidic conditions, little quantitative information is available regarding the nitrate reduction process. Moreover, while gas stripping is the most common practice in the removal of ammonia from water, the process must deal with the ultimate ammonia emission problem. There is need for alternative nitrate removal technology. The objectives of this research were (1) to establish the relationship between the rate of nitrate reduction and two important parameters, i.e., pH and surface loading, and (2) to assess the feasibility of ammonium removal by a natural rock mineral, clinoptilolite.

7.2 Materials and Methods

7.2.1 Chemicals

Ammonium chloride (NH_4Cl), sodium hydroxide (NaOH), sodium hypochlorite (NaClO, 5%), sulfuric acid (H_2SO_4), hydrochloric acid (HCl), sodium nitrate ($NaNO_3$), sodium nitrite ($NaNO_2$) and sodium sulfate ($NaSO_4$) were purchased from *Fisher Scientific* (Pittsburgh, PA). All chemicals used were analytical grade reagents, and high purity deionized water was used for preparation of all reagent solutions. Pyrogallol was from *Aldrich Chemical* Company. Hydrogen-reduced iron powder was purchased from *Alfa Aesar* (Ward Hill, MA). The iron powder had a purity of 99.5%, average diameter of 6 to 10 μm and specific surface area of 0.3125 m^2/g.

The source of clinoptilolite chosen in this experiment is *Idaho*, Preston PW31. The typical particle size is between 2 and 4 mm. According to the rock composition via X-Ray Diffraction, clinoptilolite content was 85-90%, and mineral content was Opal. The specific surface area (m^2/g) was 25.1304 ± 0.2300. Also, specific surface area by ethylene glycol monoethyl ether method was 93.5 m^2/g.

The clinoptilolite sample was ground and then sieved to different particle sizes. Five different particle size ranges were used: 0-0.177 mm, 0.177-0.841 mm, 0.841-1.0 mm, 1.0-1.7 mm, and 1.7-4.0 mm. The samples were washed by distilled water until the supernatant was cleaned, and then they were air-dried for 24 hrs before use.

7.2.2 Nitrate Reduction Experiments

To each of a series of test tubes, Pyrex glass, 0.3 g of iron powder and 15 mL of solution were added. The solution contained various amounts of sodium nitrate and sodium sulfate (0.05 M as ionic strength) and pre-adjusted pH values. The solutions were purged with nitrogen gas to expel free oxygen. The test tubes were placed on a shaker moving at 210 strokes per min. Samples were taken at various time intervals for the analysis of residual nitrate and by-products such as nitrite and ammonia. Samples were withdrawn with a syringe and filtered into a 30-mL beaker by 0.45-mm membrane filter (Gelman, Super-450, 25 mm). All runs were carried out in the laboratory at 25± 1 °C.

Residual nitrate concentrations in the samples were analyzed by ion chromatograph, model *Dionex* Bio LC, equipped with a conductivity detector. The sample loop had a volume of 50 mL. The *Dionex* Ionpace (HPIC-AS4A, 200x4.5 mm id) column preceded by *Dionex* Ionpac HPIC-AG4A guard column was thermostated at 25 °C. The eluent was a standard *Dionex* mixture of 1.8x10 M sodium carbonate and 1.7×10^{-3} M sodium bicarbonate. The eluent was pumped by an isocratic gradient pump at a flow-rate of 2.0 mL/min. The micro-membrane suppressor system (*Dionex* AMMS) was continuously regenerated by sulfuric acid (2.5×10^{-2} M) at a flow-rate of 3.0 mL/min.

7.2.3 Ion Exchange Experiments

Several clinoptilolite samples (0.5 g) were added to bottles containing NH_4Cl solutions (100 mL). The solutions were then mixed constantly over a shaker for a pre-selected period of time. For equilibrium studies, the samples were shaken for 24 hrs, and then filtered with 0.45-μm filters to collect the filtrate; after reagents were added, the residual ammonia concentration was analyzed. Ion exchange reaction was studied below pH 7 only because as long as pH is less than 9, i.e., pK_a, the distribution of NH_4^+ species is not affected by pH. Ammonia concentrations studied were 10, 20, 50, and 100 mg of NH_4^+ per liter using NH_4Cl salt. Temperature was kept constant throughout the experiments without controlling the ionic strength. The default conditions were: clinoptilolite concentration 5 g/L, 0.01 M Na_2SO_4 ionic strength, and temperature 20°C. The amount of ammonia ion was analyzed by the phenanthroline method (Abazic and Joseph 2001).

Experiments were conducted with clinoptilolite at a size of 1.35 mm and a concentration of 0.5 g per 100 mL sample. The equilibrium time was 24 hrs. The ferrous ion was made of iron (II) chloride tetrahydrate ($FeCl_2.4H_2O$, 99%), with the addition of 2 mL concentrated hydrochloric acid. The stock solution was preserved for one month. Ferrous ion concentration was determined by measuring the light absorbance at 510 nm after complexion with 1,10-phenanthroline (APHA/AWWA/WEF 1998). To avoid the interference from ferric ions, ammonium fluoride was chosen as a masking agent (Tamura et al. 1974). A one-milliliter sample was first diluted by 4 mL distilled water, and then the analyzing reagents were added in the following sequence: 1 mL of 3.6 M sulfuric acid, 1 mL of 2 M ammonium fluoride, 2 mL of 3.2 M ammonium acetate, and 1 mL of 1 % 1,10-phenanthroline. The solution was shaken vigorously after the addition of each reagent. Finally, it was analyzed by UV spectrophotometer at 510 nm light source. The standard solution was prepared as 1/20 mg of ferrous ion per liter. Dilution was made when necessary.

7.3 Results and Discussions

7.3.1 *Effect of pH*

Fig. 7-1a shows the effect of pH on the reduction of nitrate. The experimental conditions were as follows: 500 mg/L nitrate, 2 g/L iron and pH from 2.0 to 4.5. Results clearly showed that the reduction of nitrate occurred rapidly under low pH conditions. In the pH range of 2.0 to 3.5, the reduction of nitrate exhibited a rapid first stage followed by a slow second step reaction; the second-step reduction disappeared when the pH was in the range of > 4.0. Hu et al. (2001) reported that the reduction of nitrate by metallic iron was a pseudo-zero-order reaction. The reduction rate of nitrite increased when the pH of the reaction solution decreased, and the pseudo-zero-order reaction rate constants were 180, 130, 60, 15, 10, and 1 mM/h at pH = 2, 3, 4, 5, 6, and 7, respectively. Hui et al. (2001) reported that the reduction rate of nitrate by metallic iron increased with decreasing pH and the pseudo-first-

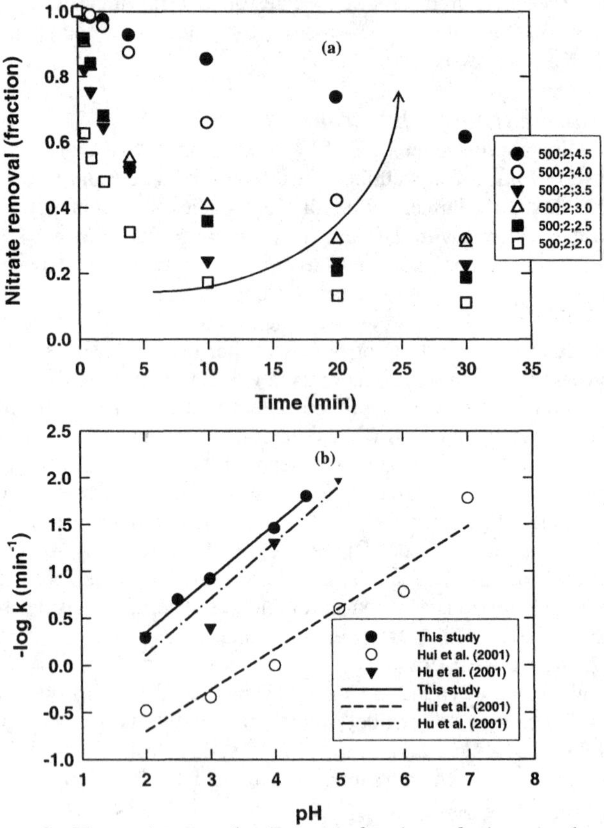

Fig. 7-1. Effect of pH on nitrate reduction: (a) fraction of nitrate reduced as a function of time; (b) relationship between rate constant and pH. Figures in the symbol-labeling box are concentration of nitrate in mg/L; iron concentration in g/L; pH of experiments.

order reaction rate constants were 0.49, 0.40, 0.05 and 0.01 h^{-1} at pH = 2, 3, 4 and 5, respectively.

Fig. 7-1b shows the plots of rate constant (k), -log k, versus pH. The following rate equations were obtained:

$$k_H = 6.6[H^+]^{0.58} \quad (k \text{ in 1/min; this study}) \tag{7.1}$$

$$k_H = 7.6 \bullet 10^3[H^+]^{0.54} \quad (k \text{ in mM/h; Hu et al. 2001}) \tag{7.2}$$

$$k_H = 12.3[H^+]^{0.60} \quad (k \text{ in 1/h; Hui et al. 2001}) \tag{7.3}$$

It is interesting to note that all three studies have the same order of proton effect, 0.58, 0.54 and 0.66 for this study, Hu et al. (2001) and Hui et al. (2001), respectively.

Huang et al. (1998) reported that acidity was the principal factor that controls the rate and the extent of nitrate removal by Fe^0. The rapid reduction of nitrate at low pH was most likely due to either direct reduction by Fe^0 or indirect reduction by surface hydrogen derived from proton. Ferrous species, Fe^{2+} and $Fe(OH)_2$, were probably not involved in this reaction. Apparently, as pH increases, the corrosion of iron decreases. Further, with increasing pH, the oxidized iron, i.e., Fe^{2+} species, becomes hydrolyzed with the formation of ferrous hydroxo species. Further increase in pH facilitates the further oxidation of ferrous ions with the formation of "green rust," which has been reported to inhibit the reduction of nitrate (Choe et al. 2004).

Huang and Zhang (2004) reported that as pH increased, a black coating, consisting of both Fe(II) and Fe(III), formed on the surface of the iron grains. The crystalline structure of the coating could not be matched with commonly known iron oxides/hydroxides/oxide hydroxides or green rust. The black coating does not inhibit the reactivity of Fe^0 at least at pH < 3. They also reported the two major effects of pH on the kinetics of nitrate reduction as (a) H^+ ions directly participating in the redox reaction of nitrate reduction following first-order kinetics; and (b) H^+ ions affecting the nitrate adsorption onto reactive sites.

7.3.2 *Effect of Nitrate Loading*

The effect of surface loading of nitrate on the rate of nitrate reduction is seen in Fig. 7-2a. Results indicate that as the surface loading increased, the extent and the rate of nitrate reduction decreased. At high surface loading, the fraction of nitrate removed reached a maximum value. As the surface loading decreased, that is, increase in iron dosage or decrease in nitrate concentration, the nitrate reduction reaction was rapid in the first 5 to 10 min and then became slow with further increase in reaction time.

Fig. 7-2b shows the relationship between the rate constant (k) and surface loading. The rate of nitrate reduction decreased with increasing surface loading. The rate constant dropped by one order of magnitude when the surface loading increased by approximately two orders of magnitude. The relationship between rate constant and surface loading (SL in mM nitrate per m^2 iron surface) can be described by the following equation:

$$k_{sl} = 2.00 \bullet [SL]^{0.55} \tag{7.4}$$

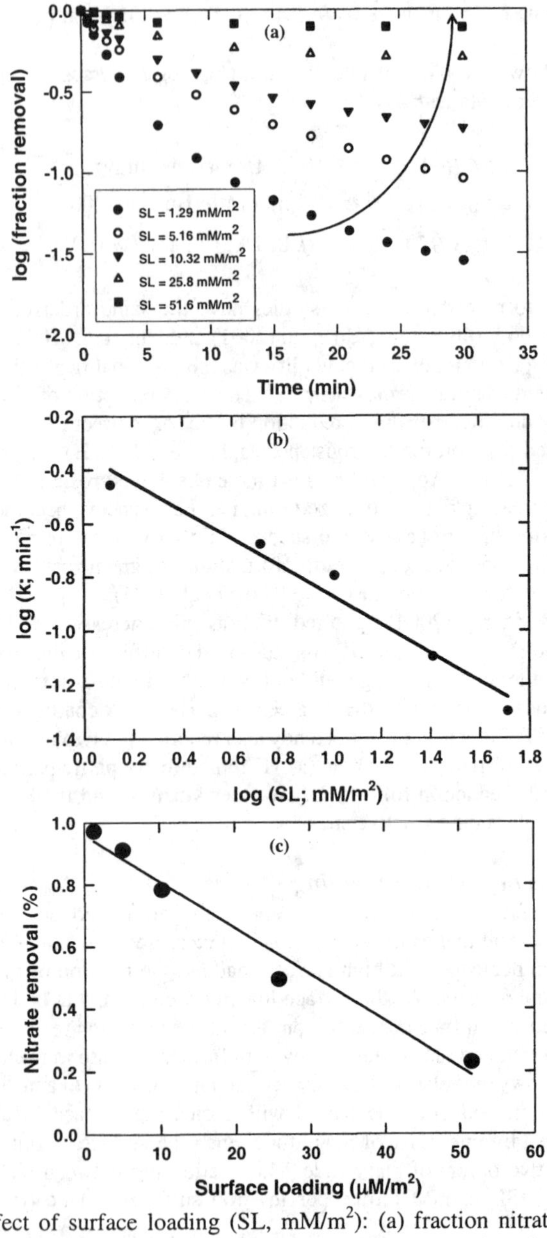

Fig. 7-2. Effect of surface loading (SL, mM/m^2): (a) fraction nitrate reduced as a function of time; (b) relationship between rate constant and surface loading; (c) nitrate reduction (30 min) as a function of surface loading (1 mM/m^2 = 14 mg as N/m^2)

This surface loading dependence of reaction rate and extent is indicative of the very nature of surface phenomenon. Apparently, at high surface loading, there are not enough active surface sites available for nitrate molecules to react, as there are a limited number of active surface sites. Once these sites are consumed, presumably on the first few molecular layers of the iron particles, the iron surface reaches corrosion passivity. As a result, no new site is generated and the reduction reaction is terminated.

Fig. 7-2c is a plot of the fraction of nitrate removal at 30 min as a function of surface loading. The following equation was obtained:

$$y = 0.959 - 0.015x \tag{7.5}$$

where y is the fraction of nitrate removal at 30 min and x is the surface loading in mM/m^2. The slope of this plot, 0.015 m^2/mM, gives the specific surface site of the iron particle, which is equivalent to 0.025 $Å^2$ per site (molecule). The equation also implies that there is a lowest limit possible for nitrate reduction by metallic iron. At zero nitrate removal, from Eq. (7.5), it is seen that the largest surface loading is 64 mM/m^2. Huang et al. (1998) reported that the maximum surface loading was 8 mM/m^2 at a reaction time of one hour.

7.3.3 Overall Nitrate Reduction Efficiency

Ideally, one would like to see a high percent utilization of the iron particles. This is more for consideration of the extra subsequent iron removal cost than the cost of the iron particle per se. If one can fully utilize the iron particles applied, one can minimize the generation of soluble iron species and the extra effort that is needed to get rid of the iron that is produced. The overall efficiency of the iron reduction process can be expressed in terms of end-product yield per mass of iron consumed (oxidized). Fig. 7-3 shows that as nitrate disappeared from the solution, ammonium appeared simultaneously. Furthermore, the results also demonstrate that the amount of iron consumed (oxidized) is equal to the amount of ammonia produced. The results shown in Fig. 7-3 were obtained under the following experimental conditions: pH = 4, Fe^0 = 0.94 m^2/L (or 3.0 g/L), total nitrate = 8.1 mM (113 mg N/mL) and ionic strength = 0.05 M Na_2SO_4. Chi et al. (2004) investigated the reduction of nitrate by metallic iron under aerobic conditions in the pH range of 2.0-8.5 and reported that pH value was an important parameter affecting the reduction of nitrate by metallic iron. At low pH values, ammonium was the main reduction product, while gaseous nitrogen was the dominant reduction product at high pH values. Gaseous nitrogen was 40.1, 56.0, 68.0, 77.8 and 81.0% after 150-min reduction at pH = 2.0, 3.5, 5.0, 7.0 and 8.5, respectively. In this study, we did not observe the release of nitrogen gas, however.

Hu et al. (2001) reported that the reduction products of nitrite were nitrogen gas and ammonium. The yields of nitrogen gas from nitrite reduction were 0.63, 0.74, 0.81, 0.87, 0.92, and 0.98 as molar ratios of nitrogen atom at pH 2, 3, 4, 5, 6, and 7, respectively. Neutral conditions enhanced the formation of nitrogen gas from nitrite reduction. Huang and Zhang (2004) reported that nitrate could be rapidly reduced to ammonium at a pH range of 2-4.5. Choe et al. (2004) reported that NH_3/NH_4^+ were

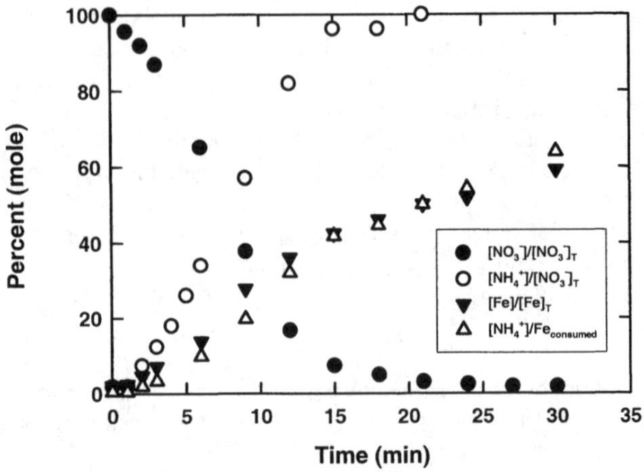

Fig. 7-3. Nitrate reduction efficiency in terms of ammonia yield and iron consumption (Note: $[Fe]/[Fe]_T$: ratio of soluble ferrous ion to total iron added; $Fe_{consumed}$: the amounts of powder iron oxidized; $[Fe]_T$: the total amounts of iron powder added.)

the only N products during the anaerobic reduction of nitrate using metallic iron. The formation of green rusts divided the NO_3^- reduction process into two phases. Green rusts formed around a pH of 6.5 and contributed to the stabilization of pH. With H_2SO_4, the available iron surface area was limited, initially by the excessive accumulation of H_2 gas at the interface, which inhibited NO_3^- reduction. The surface area normalized pseudo-first-order reaction rates for NO_3^- reduction at pH > 6.5 or after the formation of green rusts are consistent with those reported for buffered solutions.

7.3.4 Removal of Ammonia in the Presence of Fe(II) Ions

Fig. 7-4a shows the removal of ammonium and ferrous ions by clinoptilolite ion exchanger as a function of pH and various [Fe(II)] to ($[NH_4^+]$ + $[NH_3]$) molar ratios (or F/N). Results clearly indicated that the percent ammonium removal was dependent on the relative concentration of Fe(II) ions. Generally, the smaller the F/N ratio, the larger the ammonium removal percentage. As seen in Fig. 7-4a, at a F/N ratio of 8:1, the ammonium removal efficiency was between 60 and 70 % over a wide pH range, i.e., 1.0 to 10. As the F/N ratio increased, the ammonium removal percentage decreased, e.g., 69-75 % and 80-90 % at F/N ratio of 4:1, and 1:1, respectively. However, beyond the F/N ratio of 1:4, the removal of ammonium ion becomes lower than those of 1:1 and 1:4.

Fig. 7-4b shows the removal efficiency of ferrous ions. At pH greater than 4.0, the removal of ferrous ions was high, near 100%. The removal efficiency of ammonium ion fell sharply at pH > 8.0 due to acid-base equilibrium. The removal efficiency of ammonium ion in the presence of high concentration of Fe(II) ions was unfavorable. The removal efficiency of ferrous ion follows a similar trend as ammonium in all cases. It is seen that at above pH 4, most ferrous ion was removed. As the molar concentration of ferrous ion became larger than that of ammonium ion, i.e., high F/N ratio, the removal efficiency was better than other situations with a 60% removal.

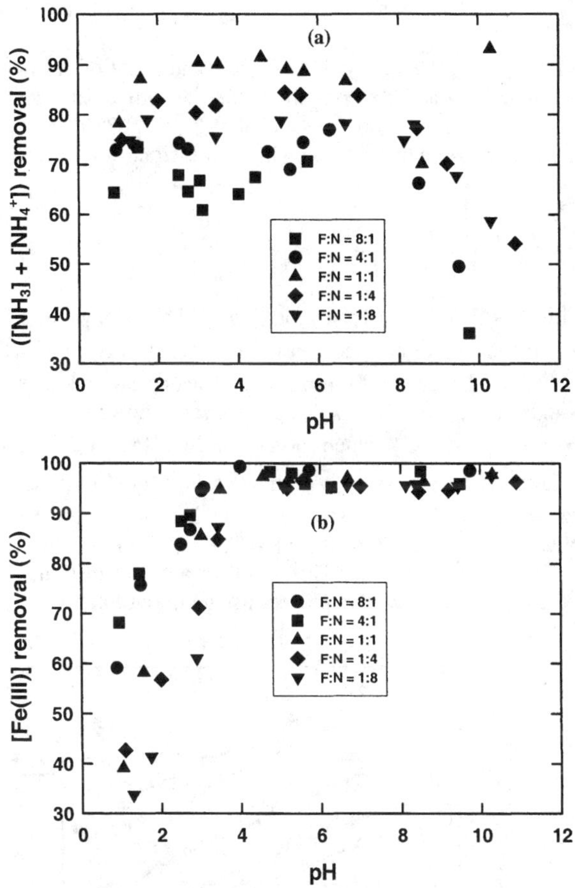

Fig. 7-4. Effect of Fe(II) ion on the removal of ammonium by clinoptitilotie at various total Fe(II) to total ammonia molar ratios (F/N): (a) removal of ammonium ion as a function of pH; (b) removal of Fe(II) ions as a function of pH (particle size of clinoptilolite = 1.35mm).

Apparently, there exists equilibrium between NH_4^+ and Fe^{2+} ions with respect to ion exchange reaction with clinoptilolite. That is:

$$Fe^{2+} + X - NH_4^+ \Leftrightarrow X - Fe^{2+} + NH_4^+ ; K_F^N \tag{7.6}$$

By applying the law of mass action, from Eq. (7.6), one has the equilibrium constant, K_F^N, expressed in the following equation:

$$K_F^N = \frac{[NH_4^+][Fe^{2+}]_s}{[Fe^{2+}][NH_4^+]_s} \tag{7.7}$$

where $[NH_4^+]$, $[Fe^{2+}]$, $[NH_4^+]_s$, and $[Fe^{2+}]_s$ are concentrations of free ammonium ion, free ferrous ion, adsorbed ammonium ion and adsorbed ferrous ions, respectively. Taking the logarithm on both sides of Eq. (7.7) and by rearranging there corresponding terms, one has the following expression:

$$\log\left(\frac{[Fe^{2+}]}{[Fe^{2+}]_s}\right) = \log\left(\frac{[NH_4^+]}{[NH_4^+]_s}\right) - \log K_F^N \tag{7.8}$$

A plot of log ($[Fe^{2+}]/[Fe^{2+}]_S$) versus log ($[NH_4^+]/[NH_4^+]_S$) yields slope and intercept, which gives the exchange constant between Fe^{2+} and NH_4^+ ions, K_F^N. A unity slope implies that the exchange equilibrium involves a single site. Results shown in Fig. 7-5a clearly indicated that the ion exchange reaction followed that described by Eq. 7.6. It must be noted that a K_F^N value larger than one favors the uptake of Fe^{2+} ions; exchange favors NH_4^+ when K_F^N is smaller than one as indicated in Fig. 7-5b. According to Fig. 7-5b, it is seen that the exchange removal of NH_4^+ ion by clinoptilolite occurs when the Fe(II) to $[NH_3]+[NH_4^+]$ molar ratio is less than ca. 2.0, the critical F/N, i.e., $(F/N)_{critical}$. This is important information with respect to the design of the ammonium removal process using clinoptilolite.

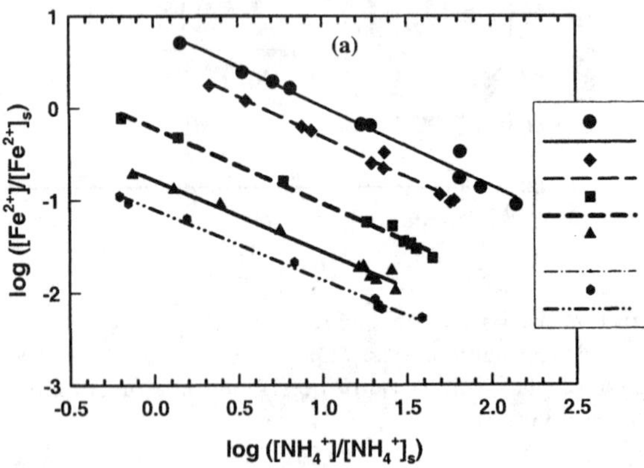

Fig. 7-5. (a) Plot of log([Fe]/[Fe]$_s$) versus log($[NH_4^+]/[NH_4^+]_s$) at various total Fe(II) to total ammonia molar ratios (F/N); (b) relationship between equilibrium exchange constant and total Fe(II) to total ammonia, ($[NH_3] + [NH_4^+]$), molar ratio (or F/N)

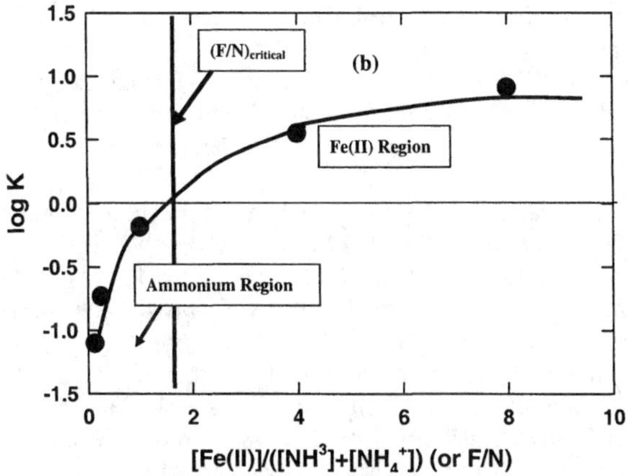

Fig. 7-5. (a) Plot of log([Fe]/[Fe]$_s$) versus log([NH_4^+]/[NH_4^+]$_s$) at various total Fe(II) to total ammonia molar ratios (F/N); (b) relationship between equilibrium exchange constant and total Fe(II) to total ammonia, ([NH_3] + [NH_4^+]), molar ratio (or F/N)

7.4 Summary

As shown above, a process design can be established by a combination of metallic iron reduction of nitrate and clinoptilolite ion exchange. The metallic iron reduction process will convert nitrate to ammonium ion under acidic pH conditions, i.e., pH < 5 with high yield. The effluent that contains ammonium and ferrous ions can be further treated with the clinoptilolite ion exchanger via which the ammonium and the ferrous ions will be retained by the ion exchanger. The effluent from the second treatment stage will be free of ammonium and ferrous ions. Based on the results shown above, it is clear that pH and the surface loading of nitrate are the two most important factors controlling the total treatment efficiency. A pH value of less than 5 will be adequate to achieve high percentages of nitrate reduction as long as the surface loading is kept at less than 64 mM/m^2 for a reaction time of 30 min. In order to partially remove ammonium ions using clinoptilolite, the iron to ammonia molar ratio, F/N, must be less than ca. 2.0; otherwise, the ion exchanger will selectively remove ferrous ions over ammonium ions.

Acknowledgement

This work was supported by a research grant from Dagu Environmental Technology Development Center, Keimyung University, Daegu, Korea. Mention of chemicals and products does not imply endorsement of the funding agency.

7.5 References

Abazic, S. D., and Joseph, N. R. (2001). "Particle release and permeability reduction in a natural zeolite (Clinoptilolite) and sand porous medium." *Environ. Sci. & Technol.*, 35(22), 4502-4508.

Ainsworth, R. G. (1990). "Water treatment for health hazards." *J. Inst. Water & Environ. Manag't*, 4(5), 489-493.

APHA, AWWA and WEF (1998). *Standard methods for the examination of water and wastewater*, 20th Ed., American Public Health Association, American Water Works Association and Water Environment Federation, Washington DC, United States.

Buresh, R. J., and Moraghan, J. T. (1976). "Chemical reduction of nitrate by ferrous iron." *J. Environ. Quality*, 5(3), 320-325.

Chi, I., Zhang, S. T., Lu, X., Dong, L. H., and Yao, S. L. (2004). "Chemical reduction of nitrate by metallic iron." *J. Water Supply & Techn.-Aqua*, 53(1), 37-41.

Choe, S. H., Ljestrand, H. M., and Khim, J. (2004). "Nitrate reduction by zero-valent iron under different pH regimes." *Applied Geochem.*, 19(3), 335-342.

Eisenberg, T. N., and Middlebrooke, E. J. (1986). *Reverse osmosis treatment of drinking water.* Butterworth Publishers, Stoneham, MA.

Freeze, R. A., and Cherry, J. A. (1979). *Groundwater.* Prentice-Hall, Englewood Cliffs, New Jersey.

Ginner, J. L., Alvarez, P. J. J., Smith, S. L., and Scherer, M. M. (2004). "Nitrate and nitrite reduction by Fe(0): Influence of mass transport, temperature, and denitrifying microbes." *Environ. Engineer. Sci.*, 21(2), 219-229.

Gregg, J. C. (1973). "Nitrate removed at water treatment plant." *Civil Engineering – American Society of Civil Engineers ASCE.*

Gunderloy, Jr. F. C., Cliff, Y., Dayan, F. V. H., and Gird, S. (1968). *Dilute solution reaction of the nitrate ion as applied to water reclamation*, Rept No. TWRC-1, FWPCA, Cincinnati, Ohio.

Gunderloy, Jr. F. C., Cliff, Y., and Dayan, F. V. H. (1970). *Development of a chemical denitrification process.* EPA Water Pollution Control Research Series. Rept. WPCR 17010 EEX10/70, United States Environmental Protection Agency, Cincinnati, Ohio.

Hsu, C. Y., Liao, C. H., and Lu, M. C. (2004). "Treatment of aqueous nitrate by zero valent iron powder in the presence of CO_2 bubbling." *Ground Water Monitoring & Remediation*, 24(4), 82-87.

Hu, H. Y., Goto, N., and Fujie, K. (2001). "Effect of pH on the reduction of nitrite in water by metallic iron." *Wat. Res.*, 35(11), 2789-2793.

Huang, C. P., Wang, H. W., and Chiu, P. C. (1998). "Nitrate reduction by metallic iron." *Wat. Res.*, 32(8), 2257-2264.

Huang, Y. H., and Zhang, T. C. (2004). "Effects of low pH on nitrate reduction by iron powder." *Wat. Res.*, 38 (11), 2631-2642.

Hui, H. Y., Goto, N., Fujie, K., Kasakura, T., and Tsubone, T. (2001). "Reductive treatment characteristics of nitrate by metallic iron in aquatic solution." *J. Chem. Engineer. Japan*, 34(9), 1097-1102.

Liao, C. H., Kang, S. F., and Hsu, Y. W. (2003). "Zero-valent iron reduction of nitrate in the presence of ultraviolet light, organic matter and hydrogen peroxide." *Wat. Res.*, 37(17), 4109-4118.

Lyons, T. A. (1990). *Management of nitrate-contaminated groundwater*, Master Thesis. University of Nebraska-Lincoln, Lincoln, NB.

Madison, R. J., and Brunett, J. O. (1985). "Overview of the occurrences of nitrates in groundwater of the United States." In *U.S. Geological Survey Water Supply Paper 2275, National Water Summary 1984*, 93-105.

Mishra, D., and Farrell, J. (2005). Understanding nitrate reactions with zero valent iron using Tafel analysis and electrochemical impedance spectroscopy." *Environ. Sci. & Technol.*, 39(2), 645-650.

O'Brien, W. J. (1968). *Chemical removal of nitrate from potable water supplies*, Contribution No. 23. Kansas Water Resources Research Institute, Manhattan, Kan. NTIS PB-193023.

Plumb, D. (1988). *Nitrates and groundwater: A public health concern*, Freshwater Foundation.

Okazaki, S., Sasaki, H., Asakura, S., Nakagawa, H., and Fukuda, K. (1998). "Reduction of nitrate ions by iron powder coated with copper in neutral solutions." *Denkikagaku*, 66(6), 620-624.

Richard, Y. R. (1989). "Operating experiences of full-scale biological and ion-exchange denitrification plants in France." *J. Inst. Water & Environ. Manag't.*, 3(2), 154-167.

Rogalla, F., Ravarini, P., DeLarminat, G., and Couttelle, J., (1990). "Large-scale biological nitrate and ammonia removal." *J. Inst. Water & Environ. Manag't.*, 4(4), 319-329.

Sova, R. J. (1986). *The chemical removal of nitrate from water supplies using ferrous sulfates and pickle liquor*, Master Thesis, University of Nebraska-Lincoln, Lincoln, NB.

Su, C. M., and Puls, R. W. (2004). "Nitrate reduction by zerovalent iron: Effects of formate, oxalate, citrate, chloride, sulfate, borate, and phosphate." *Environ. Sci. & Technol.*, 38(9), 2715-2720.

Szabo, Z. G., and Bartha, L. G. (1951). "A new titrimetric method for the determination of nitrate ion." *Analytica Chimica Acta*, 5(1), 33-45.

Szabo, Z. G., and Bartha, L. G. (1952). "The alkalimetric determination of nitrate by means of a copper-catalysed reduction." *Analytica Chimica Acta*, 6(5), 416-419.

Szabo, Z. G., and Bartha, L. G. (1957). "On the reducibility of nitrate ion." In *Recent aspects of the inorganic chemistry of nitrogen*, Special Publication No. 10, The Chemical Society, London, 131-136.

Tamura, H., Goto, K., Yotsuyanagi, T., and Nagayama, M. (1974). "Spectrophotometric deter-mination of iron(II) with 1,10-phenanthroline in the presence of large amounts of iron(III)." *Talanta.*, 21(4), 314-318.

Van Hecke, K., Van Cleemput, O., and Baert, L. (1990). "Chemo-denitrification of nitrate-polluted water." *Environ. Pollut'n.*, 63(3), 261-274.

Van der Hoek, J. P. (1988a). *Combined ion exchange/biological denitrification for nitrate removal from ground water*, Doctoral Thesis, Wageningen Agricultural University, Wageningen, The Netherlands.

Van der Hoek, J. P. (1988b). "Nitrate removal from ground water - Use of a nitrate selective resin and a low concentrated regenerant." *Water, Air, and Soil Pollut'n.*, 37(1/2), 41-53.

Westerhoff, P., and James, J. (2003). "Nitrate removal in zero-valent iron packed columns." *Wat. Res.*, 37(8), 1818-1830.

Westerhoff, P. (2003). Reduction of nitrate, bromate, and chlorate by zero valent iron." *J. Environ. Eng. ASCE*, 129(1), 10-16.

Young, G. K., Bungay, H. R., Brown, L. M., and Parsons, W. A. (1964). "Chemical reduction of nitrate in water." *J. Water Pollut' Control Fed.*, 36(3), 395-398.

Zhang, T. C., and Huang, Y. H. (2005). "Effects of selected Good's pH buffers on nitrate reduction by iron powder." *J. Environ. Eng. ASCE*, 131(3), 461-470.

CHAPTER 8

Utilization of Zero-valent Iron for Arsenic Removal from Groundwater and Wastewater

Chunming Su[1]

Abstract: Recently, increasing efforts have been made to explore the applicability and limitations of zero-valent iron (Fe^0) for the treatment of arsenic-bearing groundwater and wastewater. The experimental studies have demonstrated that arsenate, arsenite, and arsinilic acid are removed effectively from aqueous solution using Fe^0. Arsenic removal efficacy depends on the type of Fe^0, surface area and corrosion rate of the Fe^0, and water chemistry. Enhanced corrosion of Fe^0 in the presence of dissolved oxygen leads to an increase in surface area of the Fe^0 from pitting and neoformation of iron oxides and green rusts, which leads to greater arsenic removal. The iron-arsenic precipitates produced in the Fe^0 system seem to be stable in solution and the precipitation process seems to be irreversible. Toxicity Characteristic Leaching Procedure (TCLP) analysis of the spent media has shown that the arsenic concentration in the leachate is well below the 5 mg/L threshold for arsenic. Virtually all studies have demonstrated that Fe^0 is an effective sorbent for arsenic, implying that Fe^0 may be used as a medium in passive chemical permeable reactive barrier (PRB) technologies to immobilize arsenic from groundwater. Possible mechanisms for arsenic removal in the Fe^0 systems include surface adsorption, precipitation, co-precipitation, and redox transformation. Although experimental results are encouraging so far, there are still various gaps in the knowledge pertaining to detailed mechanisms of arsenic removal, to long-term stability of sequestered arsenic in the Fe^0 system, and to field-scale application of Fe^0. The influence of microbial activities in the presence of Fe^0 on arsenic fate and transport is not well understood. Further studies are needed to improve the technology for both *in situ* and *ex situ* arsenic removal from contaminated water.

CE Database subject headings: Adsorption; Arsenic; Barriers; Ground water; Iron; Transformations.

8.1 Introduction

There is an urgent need to develop efficient and cost-effective methods of arsenic (As) removal from water because of the enormity and severity of the problem of arsenic poisoning in groundwater resources in many areas of the world (Smedley and Kinniburgh 2002). Since arsenic naturally occurs in the soil, sediments, and rocks, it

[1]Soil Scientist, Ground Water and Ecosystems Restoration Division, National Risk Management Research Laboratory, Office of Research and Development, United States Environmental Protection Agency, 919 Kerr Research Drive, Ada, Oklahoma 74820, USA (corresponding author). E-mail: su.chunming@epa.gov.

can often be found in groundwater. With a growing number of people now drinking water from underground sources, numerous people are at serious risk worldwide. Tens of millions of people are exposed to dangerous levels of arsenic in their drinking water [up to 50 times the World Health Organization (WHO) threshold of 10 μg/L] in countries such as Bangladesh, India, Vietnam, and Argentina (e.g., McArthur et al. 2001; Nordstrom 2002). The United States Environmental Protection Agency (USEPA) has recently decreased the maximum contaminant level (MCL) of arsenic in drinking water from 50 to 10 μg/L. Adoption of this lower arsenic standard will require 3,000 community water systems, serving 11 million residents, to be corrected to lower the current levels of arsenic in their drinking water (USEPA 2000).

Arsenic occurs in groundwater predominantly as inorganic arsenite [As(III)] and arsenate [As(V)] (Korte and Fernando 1991). Arsenite (H_3AsO_3, $H_2AsO_3^-$) are more toxic and more mobile than arsenate (H_3AsO_4, $H_2AsO_4^-$, $HAsO_4^{2-}$, and AsO_4^{3-}). In the normal pH range of groundwater and drinking water, the dominant arsenite species is neutral (H_3AsO_3; $pK_{a,1} = 9.29$) and the dominant arsenate species are negatively charged ($H_2AsO_4^-$, $HAsO_4^{2-}$; $pK_{a,1} = 2.24$, $pK_{a,2} = 6.96$, and $pK_{a,3} = 11.50$) (Martel et al. 2003). Up until now, arsenic-contaminated groundwater, drinking water, and wastewater were treated away from its place of contamination, commonly referred to as *ex situ* treatment. Various physiochemical technologies such as anion exchange, chemical precipitation, adsorption, and reverse osmosis were used for the *ex situ* cleanup, but they show some disadvantages: they are expensive and generate high volumes of toxic sludge and brine, and the water recovers poorly. Consequently, new methods for *ex situ* and *in situ* (at the place, where the water occurs naturally) remediation are needed.

Recent research efforts have focused on using zero-valent iron (Fe^0) to remove arsenic from various types of water in laboratory batch, column, and field tests (Table 8-1). Previous studies have shown great potential of *in situ* remediation of certain organic and inorganic contaminants in groundwater using Fe^0 as a permeable reactive barrier (PRB) medium. A PRB is a passive treatment technology designed to clean up contaminated groundwater. A trench is dug down into the groundwater and filled with a permeable reactive material (granular Fe^0 or organic compost). When contaminated groundwater flows through the barrier, the contamination is removed by chemically reacting with the barrier material. Numerous Fe^0 based PRBs have been constructed at field scale to treat chlorinated hydrocarbon compounds in groundwater plumes via reductive dehalogenation (e.g., Scherer et al. 2000). Recent studies have also shown that Fe^0 effectively removes inorganic contaminants (CrO_4^{2-}, UO_2^{2+}, MoO_4^{2-}, TcO_4^-) from aqueous solution (Cantrell et al. 1995; Powell et al. 1995; Blowes et al. 1997; Gu et al. 1998; Puls et al. 1999; Blowes et al. 2000). The removal mechanism appears to be reductive precipitation for these metallic anions. In contrast, for arsenic, adsorption or precipitation appears to be the predominant removal mechanism for both As(V) and As(III) by Fe^0 (Lackovic et al. 2000; Su and Puls 2001a). In this paper, recent developments in the applications of Fe^0 to arsenic-bearing water treatment are reviewed, and possible future research needs are highlighted.

Table 8-1. Batch and Column Studies of Arsenic Removal Using Zero-valent Iron

Water type	Iron material	Reactors	Experimental conditions	References
Simulated wastewater containing Fe, Cu, Ni, Pb, Cd, Hg, and Cr(VI)	Unnamed	Batch and column	Initial pH 7.0-8.7, arsinilic acid at 0.66 – 5.5 mg As/L	Mayenkar and Heights (1986)
Berkely Pit acid mine drainage	ISP, Fluka	Batch	Initial pH 2.1 – 2.3, As 2 mg/L	Shokes and Möller (1999)
Berkely Pit acid mine drainage	Master Builder	Column	Initial pH 2.1 – 2.3, As 2 mg/L	Shokes and Möller (1999)
Groundwater from Bangladesh	Renwick Ironworks	3-pitcher	Initial pH 6.7-6.9, As 0.13 – 1.16 mg/L	Khan et al. (2000)
10 mM $NaNO_3$ ± 50 mg/L SO_4^{2-}	Baker, Master Builders	Lab column	Initial pH 6.3, As(V) or As(III) 1 mg/L	Lackovic et al. (2000)
Landfill leachate	Baker, Connelly-GPM	Field column	Initial pH 6.1-6.3, As(III) 0.3 mg/L	Lackovic et al. (2000)
10 mM NaCl	Fisher, Peerless, Aldrich, Master Builders	Batch	Initial pH 6.5, As(V) or As(III) or As(V + III) 2.0 mg/L	Su and Puls (2001a)
0.1 – 1 mM P, Si, B, Mo, Cr(VI)	Peerless	Batch	As(V) or As(III) 2.0 mg/L	Su and Puls (2001b)
7 mM NaCl + 0.86 mM $CaSO_4$	Peerless	Column	Initial pH 6.5, As(V + III) 2.0 mg/L	Su and Puls (2003)
Dilute K_2SO_4	Fisher	Batch	Initial pH 7, As(III) 0.2- 2 mg/L	Ramaswami et al. (2001)
3 mM $CaSO_4$	Aesar iron wire	Electrochemical cell	Initial pH 7, As(V) 0.1 or 5.0 mg/L	Farrell et al. (2001)
3 mM $CaSO_4$	Master Builders	Column	As(V) 0.1 – 20 mg/L	Farrell et al. (2001)
3 mM $CaSO_4$	Aesar iron wire	Electrochemical cell	Initial pH 7, As(V) 0.1 – 20 mg/L	Melitas et al. (2002a)
3 mM $CaSO_4$	Master Builders	Column	As(V) 0.1 – 50 mg/L	Melitas et al. (2002a)
3 mM $CaSO_4$	Aesar iron wire	Electrochemical cell	Initial pH 7, As(V) 0.1 - 1000 mg/L	Melitas et al. (2002b)
3 mM $CaSO_4$	Aesar iron wire	Electrochemical cell	As(V) 0.1 – 20 mg/L	Melitas et al. (2003)
3 mM $CaSO_4$	Master Builders	Column	As(V) 0.1 – 24 mg/L	Melitas et al. (2003)
Tap water	Iron wool	Column	As(V) 0.5 mg/L	Karschunke and Jekel (2002)
1 mM NaCl, tap water	Fisher	Batch	As(V) or As(III) 2, 15, or 150 mg/L	Manning et al. (2002)
Landfill leachate	Connelly-GPM	Large field column	Initial pH 6.0-6.3, As(III) 0.3 mg/L	Nikolaidis et al. (2003)
Bangladesh groundwater	Connelly-GPM	AsRT unit	Initial pH 6.8-7.1, As 0.2-0.6 mg/L	Nikolaidis et al. (2005)
Simulated acid mine drainage	Peerless, Fisher	Batch	Initial pH 2.3-4.5, As(V) 10 mg/L	Wilkin and McNeil (2005)
Simulated groundwater	Peerless	Batch and column	Initial pH 7.25, As(III) 50 mg/L	Lien and Wilkin (2005)
40 mM NaCl	U.S. Metals	Batch	pH controlled at 4 or 7 As(V) or As(III) 0.5 mg/L	Bang et al. (2005)
10 mM NaCl, groundwater	Nanoscale iron	Batch	Initial pH 3-12, As(III) 1 mg/L	Kanel et al. (2005)

8.2 Batch Tests with Non Mine-Impacted Waters

Before 1990, reports were sparse on utilization of Fe^0 to remove arsenic from contaminated water. Mayenkar and Heights (1986) obtained a patent for a process that removes dissolved heavy metals [Fe, Cu, Ni, Pb, Cd, Hg, Cr(VI)] and arsinilic acid using Fe^0 from a simulated metal cleaning wastewater that is typically generated in a coal-fired steam generating plant. The Fe^0 employed was not named; but it was said to be a "readily available iron in particulate form, e.g., decreased iron filings from the metalworking industry." One batch test they conducted showed significant removal of arsenilic acid from 0.66 to 0.021 mg As/L (initial pH 8.7) after a 211-h contact time with Fe^0 with concomitant decrease of dissolved Fe, Cu, Ni, Pb, Cd, and Cr(VI). In another batch test using a higher concentration of initial arsenilic acid (5.5 mg As/L) at an initial pH of 7.0, they were able to decrease arsenic concentration to 0.51 mg/L after 120 h. In both tests, the final pH increased to 9.2 at the end of each test. The removal mechanism was solely attributed to adsorption by these authors (later studies have shown that more complicated mechanisms could be involved). Nevertheless, the positive results reported in the patent document seem to be largely unnoticed in the scientific community for quite a few years as evidenced by a lack of citation of this patent in subsequently peer-reviewed papers.

More detailed studies gradually emerged. Su and Puls (2001a) conducted batch tests to evaluate the effectiveness of Fe^0 in removing arsenic from water. They tested four Fe^0 materials with a range of surface area values including *Fisher* electrolytic Fe^0 (99%, -100 mesh, *Fisher Scientific*, Fair Lawn, NJ, Cat. No. I60-3), *Peerless* Fe^0 (*Peerless* Metal Powders & Abrasive, Detroit, MI), *Master Builders* Fe^0 (*Master Builders Inc.*, Cleveland, OH), and *Aldrich* Fe^0 (-325 mesh, 97%, hydrogen reduced, Cat. No. 0930-9). Specifically, one gram of Fe^0 was added to a 50-mL polypropylene copolymer centrifuge tube (actual volume = 41.6 mL) that was filled without a headspace with a solution of 2 mg/L As(V), As(III), or As(V + III) (1:1) in 0.01 M NaCl. The pH of the suspension was allowed to drift freely. At preset time intervals, arsenic concentrations were determined and dissolved arsenic speciated using ion chromatography-hydride generation – graphite furnace atomic absorption spectrophotometry (IC-HG-GFAAS).

They showed that arsenic removal was affected largely by reaction time, the Fe^0 type, and to a lesser degree, by the initial arsenic species (Figs. 8-1 to 8-4). In the presence of Fe^0, arsenic concentration decreased exponentially with time. *Fisher, Peerless*, and *Master Builders* Fe^0 decreased concentrations of both As(III) and As(V) to less than 0.010 mg/L in 96 h. Pseudo-first-order reaction kinetics was found to describe the data well with R^2 values ranging from 0.85 to 0.98. The rate constant (k) for the removal of As(V), As(III), and As(V + III) by Fe^0 followed the order: Fisher > Peerless . Master Builders > Aldrich (Table 8-2). This sequence of reactivity is consistent with the results of reductive dechlorination of trichloroethene by these same Fe^0 samples from a previous study (Su and Puls 1999). The surface area normalized rate constants (k_{SA}), however, showed a different order: Fisher > Aldrich > Peerless . Master Builders. The removal of arsenic was not proportionally related to the surface area of the Fe^0 since Fisher Fe^0 had the least surface area, but it showed the greatest removal rates.

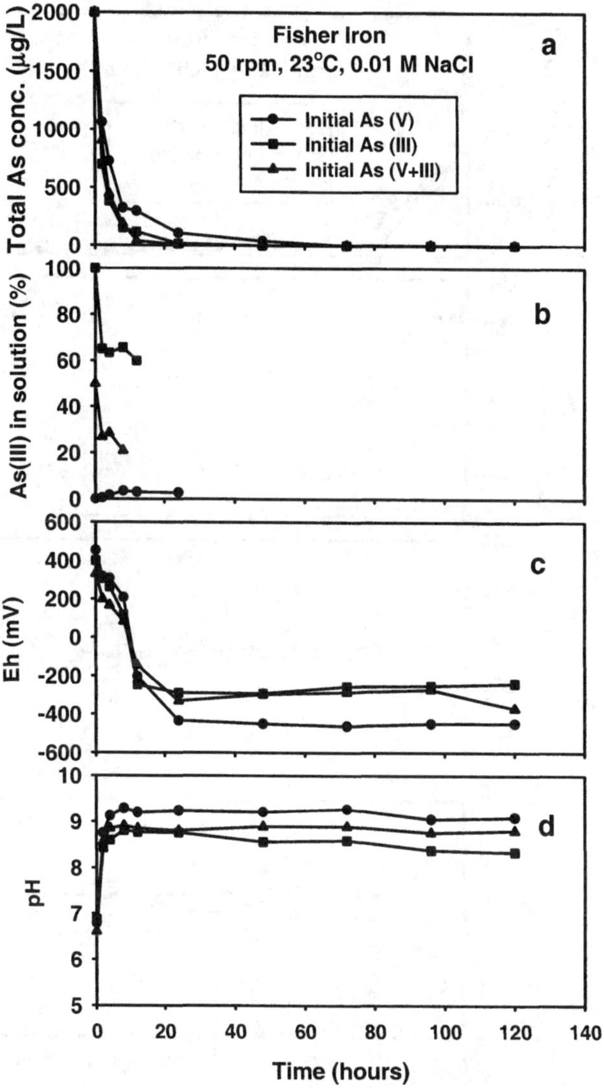

Fig. 8-1. Kinetics of As removal by *Fisher* Fe^0: (a) As concentration, (b) As(III) in solution, (c) Eh, and (d) pH. From Su and Puls (2001a), with permission.

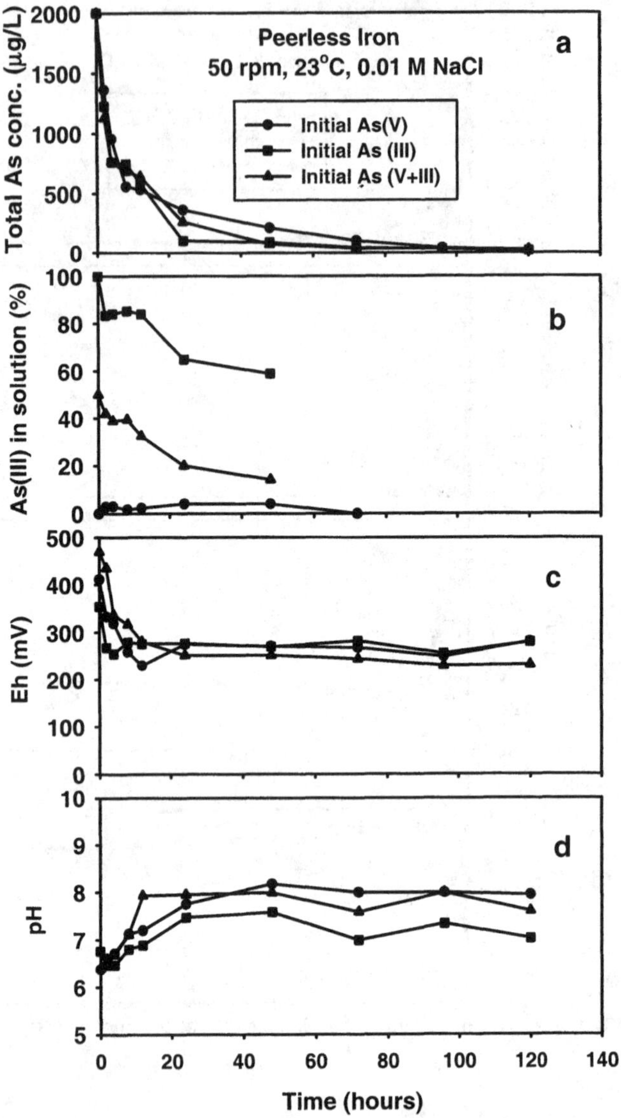

Fig. 8-2. Kinetics of As removal by *Peerless* Fe^0: (a) As concentration, (b) As(III) in solution, (c) Eh, and (d) pH. From Su and Puls (2001a), with permission.

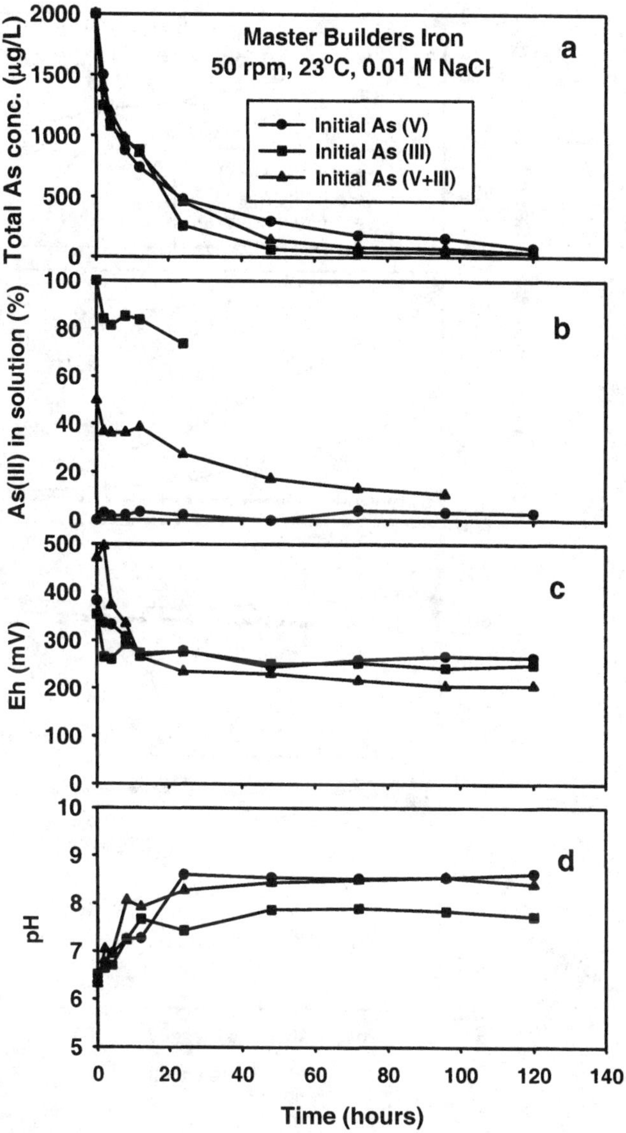

Fig. 8-3. Kinetics of As removal by *Master Builders* Fe^0: (a) As concentration, (b) As(III) in solution, (c) Eh, and (d) pH

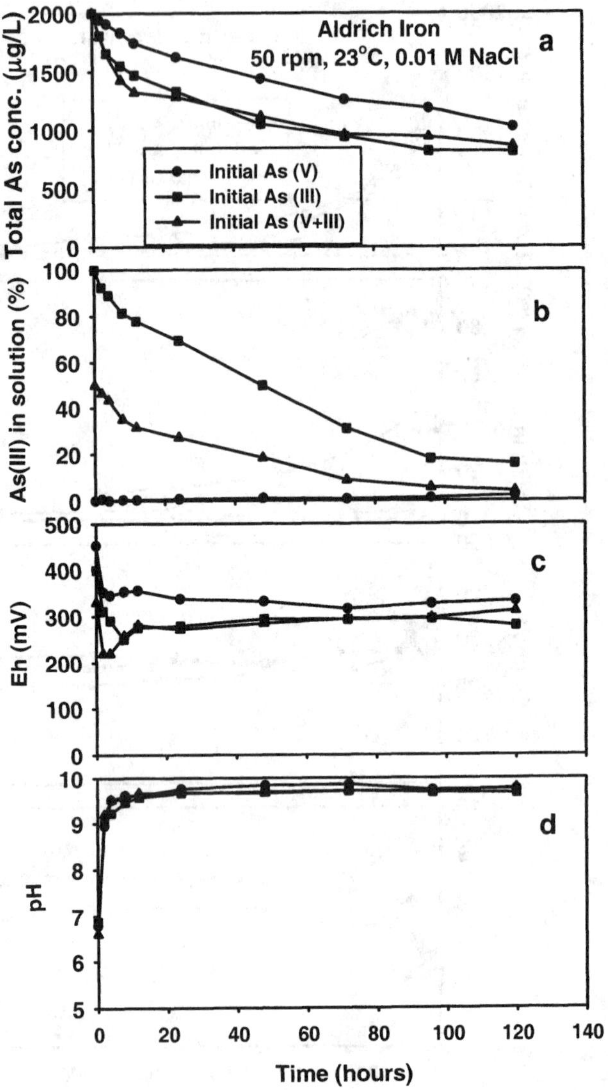

Fig. 8-4. Kinetics of As removal by *Aldrich* Fe^0: (a) As concentration, (b) As(III) in solution, (c) Eh, and (d) pH

Table 8-2. The Pseudo-first-order Rate Constants (k) and their Surface-area-normalized Rate Constants (k_{SA}) for Arsenic Removal by Zero-valent Iron

Iron material	Surface area m^2/g	$k \times 1000$ (h^{-1})			k_{SA} (h^{-1} m^{-2} mL)		
		As(V)	As(III)	As(V + III)	As(V)	As(III)	As(V + III)
Fisher	0.091 ± 0.005	77.8 ± 9.7	149 ± 4	149 ± 7	435.6 ± 4.4	68.2 ± 1.6	68.4 ± 3.2
Peerless	2.53 ± 0.44	34.9 ± 2.0	36.2 ± 0.5	35.7 ± 0.6	0.572 ± 0.033	0.594 ± 0.008	0.586 ± 0.033
Master Builders	2.33 ± 0.09	24.6 ± 1.9	36.1 ± 1.9	32.7 ± 1.8	0.439 ± 0.034	0.643 ± 0.034	0.582 ± 0.032
Aldrich	0.192 ± 0.001	5.31 ± 0.35	7.23 ± 0.69	6.08 ± 0.24	1.15 ± 0.08	1.56 ± 0.15	1.32 ± 0.05

Table 8-3. Arsenic Removal Capacities of Zero-valent Iron as Determined in Batch and Column Studies

Iron material	Removal capacity (mg/g)	Reactors	References	Comments
Peerless	1.77	Batch, Langmuir isotherm	Su and Puls (2001a)	As(III) removal
Peerless	0.73	Batch, Langmuir isotherm	Su and Puls (2001a)	As(V) removal
Peerless	7.5	Column	Lien and Wilkin (2005)	As(III), based on the breakthrough
Connelly-GPM	1.15	Column	Lackovic et al. (2000)	As(III) removal
Connelly-GPM	0.77-4.4	Column (Fe^0/sand mixture)	Nikolaidis et al. (2003)	As(III), based on mass of mixture
Baker	0.47	Column	Lackovic et al. (2000)	As(III) removal
Nanoscale iron	3.5	Batch	Kanel et al. (2005)	As(III) removal

Su and Puls (2001a) observed greater removal rates for As(III) than As(V), with As(V + III) having intermediate removal rates with Fe^0. They explained this phenomenon by a difference in the pH effect on adsorption of As(V) versus As(III) by the iron oxides. Adsorption of As(V) by amorphous iron oxide and goethite is more favorable than that of As(III) below pH 5-6; whereas, above pH 7-8, As(III) has a higher affinity for the solids (Dixit and Hering 2003). Thus, more As(III) adsorption is expected than As(V) adsorption at pH > 7. Since iron oxides are possible Fe^0 corrosion products, their reaction with As may be more predominant than the reaction with Fe^0. The oxidized layer on iron is likely the predominant sorption location for both arsenic species.

Su and Puls (2001a) showed that redox potential (Eh) in the *Fisher* Fe^0 system decreased with time from positive values of 300 to 400 mV to negative values of -200 to -400 mV; whereas, it was never below 200 mV in the *Peerless*, *Master Builders*, or *Aldrich* Fe^0 systems (Figs. 8-1 to 8-4). The pH increased with time, with the largest increases seen for the *Aldrich* Fe^0 system (up to pH 9.8). The As(III) gave the smallest pH increase compared to As(V) and As(V + III) (Figs. 8-1 to 8-4). The pH increase was explained by water decomposition by Fe^0 and by adsorption reactions of As(V) and As(III), which release OH^- groups from adsorbents via ligand exchange. Iron corrosion undergoes two pathways:

$$\text{Anaerobic corrosion} \quad Fe^0 + 2H_2O \rightarrow Fe^{2+} + H_2 + 2OH^- \tag{8.1}$$

$$\text{Aerobic corrosion} \quad 2Fe^0 + 2H_2O + O_2 \rightarrow 2Fe^{2+} + 4OH^- \tag{8.2}$$

A portion of the As retained by the *Peerless* Fe^0 in the amount of 2.88 to 12.2% of the initially sorbed As was desorbed by the first phosphate displacement, and a smaller portion (< 0.8%) was desorbed by the second phosphate treatment. The amount of the previously sorbed phosphate-extractable As decreased with residence time, suggesting that the As sorbed on the surface of the solids either formed stronger complexes (bidentate vs. monodentate), or that part of it diffused into the interior sites of the solids, making it less susceptible to phosphate displacement (Su and Puls 2001a).

A more recent study by Bang et al. (2005) generally confirmed the findings of Su and Puls (2001a) about arsenic removal kinetics. Bang et al (2005) demonstrated that As(III) removal rate was higher than that for As(V) when iron filings (80-120 mesh) were mixed with arsenic solutions purged with nitrogen gas in the pH range of 4-7. New results from Bang et al. (2005) showed XPS spectra of the reacted iron coupons that exhibited the reduction of As(III) to As^0. Soluble As(III) was formed when As(V) reacted with Fe^0 under anoxic conditions; however, no As^0 was detected on the iron coupons after 5 days of reaction in the As(V)-Fe^0 system. The removal of the arsenic species by Fe^0 was attributed to electrochemical reduction of As(III) to sparsely soluble As^0 and adsorption of As(III) and As(V) to iron hydroxides formed on the Fe^0 surface under anoxic conditions. When the solutions were open to atmospheric air, the removal rates of As(V) and As(III) were much higher than under the anoxic conditions, and As(V) removal was faster than As(III). The rapid removal of As(III) and As(V) was attributed to adsorption on ferric hydroxides formed readily through oxidation of Fe^0 by dissolved oxygen.

In other batch tests, inorganic arsenic has been treated using different Fe^0 materials in a variety of background electrolyte solutions and simulated waters (Table 8-1). They include dilute K_2SO_4 (Ramaswami et al. 2001), tape water (Manning et al. 2002; Karschunke and Jekel 2002), simulated groundwater (Lien and Wilkin 2005) and groundwater (Kanel et al. 2005).

Nanoscale Fe^0 has been suggested as an alternative means of exploiting the chemical reactivity of Fe^0 without the need to construct PRBs; it can be injected into the target zone of subsurface. No field application of nanoscale Fe^0 for arsenic remediation has been reported; however, encouraging results have been shown in laboratory bench tests. Nanoscale Fe^0 with a BET surface area of 24.4 m^2/g was tested for As(III) removal (Kanel et al. 2005) from Bangladesh groundwater. The reported pseudo-first-order rate constant values were 1000 times higher than those literature values for As(III) removal on micro size Fe^0. The maximum As(III) removal capacity in batch experiments calculated from Freundlich adsorption isotherm was 3.5 mg of As(III)/g of iron, which is also higher than those for the micro size Fe^0 (Table 8-3). So far, all the batch tests have shown effective removal of arsenic using Fe^0.

8.3 Batch Test with Acid Mine Drainage

Acid mine drainage (AMD) refers to drainage of water from areas that have been mined for coal or other mineral ores. The water is acidic and is loaded with toxic heavy metals because of its contact with sulfur-bearing materials that undergo chemical oxidation. AMD is harmful to aquatic organisms and leads to environmental degradation. Organic matter-based PRBs have been used to remediate AMD through sulfate reduction and metal-sulfide precipitation (e.g., Benner et al. 1997; Benner et al. 1999; Ludwig et al. 2002). A substrate composed of 40% municipal compost, 40% leaf compost, and 20 % wood chips, was mixed with pea gravel in a PRB to intercept and treat an AMD at the Nickel Rim mine site near Sudbury, Ontario, Canada (Benner et al. 1997). A field scale PRB consisted of 15% leaf compost, 84% peat gravel, and 1% limestone by volume was installed to treat a combination of heavy metals at an industrial site in British Columbia, Canada (Ludwig et al. 2002). Although arsenic was not a target contaminant in these studies, the approach may be applicable to arsenic-containing AMD in general.

The ability of Fe^0 to neutralize acidity and to reduce, adsorb, and co-precipitate dissolved heavy metal ions has generated interests in treatment of acid mine drainage using the metal. An electrochemical cell was constructed by Shelp et al. (1995) using a block of massive sulfide-graphite rock from a mine site as the cathode, scrap iron as the sacrificial anode, and acidic leachate collected from the same mine site as the electrolyte. The cell was effective at raising the pH of 41 L of leachate from 3.0 to 5.6. Concomitantly, the redox potential decreased from greater than 650 to less than 300 mV. Furthermore, iron sulfate precipitate formed, with a concomitant decrease of dissolved Al, Cd, Co, Cu, and Ni. Again, arsenic was not a target in this study; however, the test results demonstrated the proposed electrochemical approach to be a technically feasible and practical method of ameliorating AMD.

Arsenic-bearing AMD (2 mg As/L) from Berkeley Pit, MT was treated with Fe^0 (Shokes and Möller 1999). The dissolved metals in AMD are removed by several mechanisms proposed by the authors. Copper and cadmium cement onto the surface of the iron as zero-valent metals. Hydroxide forming metals such as aluminum, zinc, and nickel from complexes with iron and other metals precipitating from solution as the pH rises. Metalloids such as arsenic and antimony co-precipitate with iron. As metals precipitate from solution, various other mechanisms including co-precipitation, sorption, and ion exchange also enhance removal of metals from solution. Corroding iron also creates a reducing environment supportive for sulfate reducing bacteria (SRB) growth. Increases in SRB populations of 5000-fold were observed in iron-metal treated AMD solutions. Albeit a slow biological processes, sulfidogenesis is an additional pathway to further stabilize heavy metal precipitates.

A simulated AMD (initial pH 2.3 to 3.5) containing 10 mg As/L was treated with *Peerless* and *Fisher* Fe^0 in batch tests by Wilkin and McNeil (2003) who determined rates of acid-neutralization and of metal (Cu, Cd, Ni, Zn, Hg, Al, and Mn) and metalloid (As) uptake using simulated acid mine drainage (initial pH 2.3 to 4.5; total dissolved solids 14000 to 16000 mg/L). Metal removal from solution and acid-neutralization occurred simultaneously and were most rapid during the initial 24 h of reaction. Reaction half-lives ranged from 1.50 ± 0.09 h for Al to 8.15 ± 0.36 h for Zn. Based on geochemical model results, the authors concluded that metal removal is most effective in solutions that are highly undersaturated with respect to pure-metal hydroxides suggesting that adsorption is the initial and most rapid metal uptake mechanism; continued adsorption onto or co-precipitation with iron corrosion products are secondary metal uptake processes. Sulfate green rust was identified as the primary iron corrosion product, which is shown to be the result of elevated $[SO_4^{2-}]/[HCO_3^-]$ ratios in solution. Reversibility studies indicated that Fe^0 will retain metals after shifts in redox states are imposed, but that remobilization of metals may occur after the acid-neutralization capacity of the material is exhausted.

The role of microorganisms influencing the performance of Fe^0 in mine-impacted groundwater is significant. Enhanced metabolic activity of sulfate-reducing bacteria in Fe^0 has been well documented (Phillips et al. 2000; Roh et al. 2000; Wilkin et al. 2002). The presence of these organisms will offer additional pathways for metal removal, i.e., precipitation of insoluble metal sulfides. In addition, high total dissolved solids are expected to lead to higher rates of mineral precipitation and loss of pore space and permeability (Wilkin and McNeil 2003). The relatively low density calcium and magnesium carbonates, the most common authigenic precipitates in bicarbonate-dominated groundwater, are not expected in sulfate-dominated mine waste-impacted groundwater. Depending on input pH, significant accumulations of precipitates containing iron, aluminum, and manganese might be expected. Sulfate removal by Fe^0 in abiotic systems is incomplete. Limited formation of sulfate green rust and possible precipitation of high-density metal sulfides will act to counter the effect of high solute inputs to reactive media with respect to pore infilling (Wilkin and McNeil 2003).

8.4 Effects of Competing Inorganic Anions on Arsenic Removal by Zero-valent Iron

Some inorganic anions are known to be specifically adsorbed to iron and aluminum oxides, clay minerals, and soils. They include, but not limited to, phosphate, silicate, carbonate, borate, and to a less degree, sulfate. Competition of inner-sphere forming anions for adsorption sites, therefore, can have a significant effect on the fate and behaviour of the target anions of arsenate and arsenite. Studies have shown that phosphate decreases both As(V) and As(III) sorption by ferrihydrite, depending strongly on pH and phosphate concentration (Jackson and Miller 2000). The inhibiting effect of phosphate on As(V) sorption is greater at high pH than at low pH; whereas, the opposite trend is observed for As(III) (Jackson and Miller 2000; Jain and Loeppert 2000). Desorption of already-sorbed As(V) and As(III) on ferrihydrite and goethite by phosphate also shows the above trend towards pH and phosphate concentration. Application of phosphate fertilizers to lead arsenate-contaminated soils increases soil arsenic solubility and downward mobility (Peryea and Kammereck 1997). Competitive adsorption of As(V) with phosphate and molybdate has been reported to occur on goethite and gibbsite [$Al(OH)_3$] (Manning and Goldberg 1996). Dissolved silicate and phosphate negatively impact removal of arsenic via co-precipitation of iron (Roberts et al. 2004). Dissolved phosphate also decreases nitrate reduction by Fe^0 (Su and Puls 2004a).

Sulfate was reported not to influence As(V) sorption by ferrihydrite, but to result in a considerable decrease in As(III) sorption below pH 7, with the largest decrease at the lowest pH (Jackson and Miller 2000). Sorbed As(V) by ettringite [$Ca_6Al_2(SO_4)_3(OH)_{12}{\bullet}26H_2O$] was also not desorbable in the presence of concentrated sulfate and high ionic strength solutions (Myneni et al. 1997). On the contrary, sulfate was found to decrease both As(V) and As(III) sorption on hydrous ferric oxide in the pH range of 4 to 7 (Wilkie and Hering 1996). The release of As from a sandstone aquifer was found to be strongly and positively related to the bicarbonate concentration in the leaching solution (Kim et al. 2000). Concentrations of naturally occurring arsenic in alluvial aquifers of the Ganges Plain in Bangladesh and west Bengal was found to be correlated with concentrations of bicarbonate and were uncorrelated with concentrations of dissolved iron, which lead Nickson et al. (2000) to propose that the Ganges Plain is derived by reductive dissolution of iron oxyhydroxides in the sediment that is driven by microbial metabolism of sedimentary organic matter.

Su and Puls (2001b) conducted batch tests to evaluate the effects of inorganic anion competition on the kinetics of arsenate [As(V)] and arsenite [As(III)] removal by *Peerless* Fe^0 in aqueous solution. The oxyanions underwent either sorption-dominated reactions (phosphate, silicate, carbonate, borate, and sulfate) or reduction-dominated reactions (chromate, molybdate, and nitrate) with *Peerless* Fe^0 in the presence of As(V) or As(III), relative to chloride. Pseudo-first-order rate equations were found to describe satisfactorily both As(V) and As(III) removal kinetics in the presence of each competing anion. In the presence of phosphate, As removal rates ranged from 7.0×10^{-3} to 18.5×10^{-3} h^{-1}; whereas, As removal rates ranged from 34.9×10^{-3} to 36.2×10^{-3} h^{-1} in the presence of chloride. Silicate, chromate, and molybdate also caused strong inhibition of As removal, followed by carbonate and nitrate; whereas, borate and sulfate only caused slight inhibition to As(III) removal.

Relative to chloride at 10 mM, which is considered a non-specific anion in the solid-liquid interfacial reactions, the presence of phosphate at either 0.1 or 1.0 mM caused a significant decrease in both As(V) removal kinetics (Fig. 8-5) and As(III) removal kinetics (Fig. 8-6). Phosphate decreased both As(V) and As(III) removal by *Peerless* Fe^0 to a greater degree at 1.0 mM NaH_2PO_4 (P:As molar ratio 37:1) than at 0.1 mM NaH_2PO_4 (P:As molar ratio 3.7 :1) at each time interval (Figs. 8-5a and 8-6a).

The Eh generally decreased initially and approached steady state with positive values (Figs. 8-5c and 8-6c). The positive Eh values for *Peerless* Fe^0 were in contrast with those for *Fisher* electrolytic Fe^0 that gave negative (- 400 mV) Eh values within

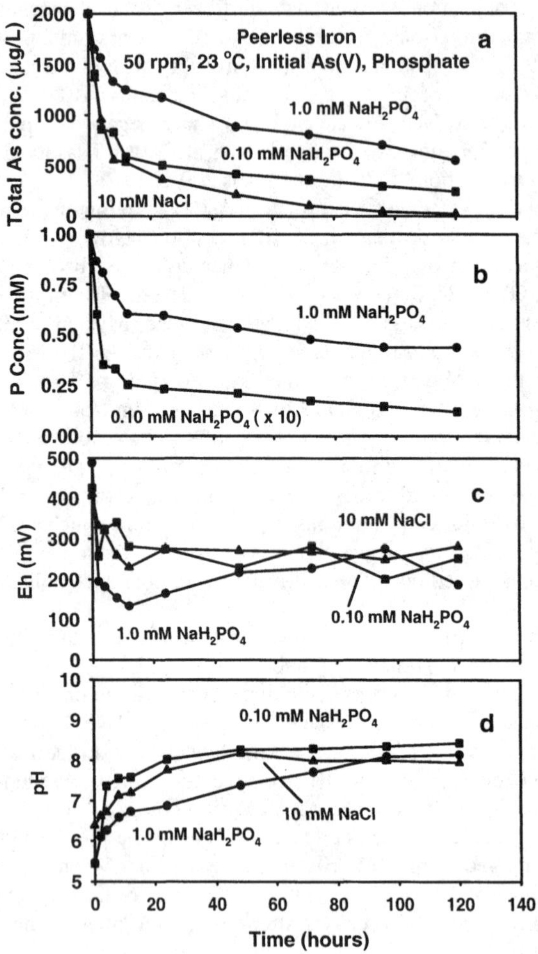

Fig. 8-5. Kinetics of As(V) removal by *Peerless* Fe^0 in 0.10 mM or 1.0 mM NaH_2PO_4 relative to 10 mM NaCl: (a) total As concentration, (b) phosphate concentration, (c) Eh, and (d) pH. From Su and Puls (2001b), with permission.

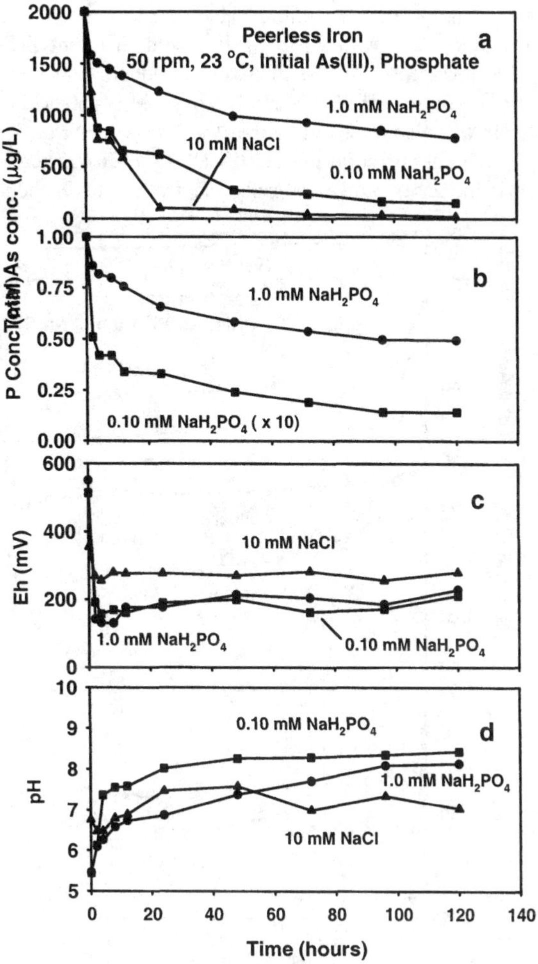

Fig. 8-6. Kinetics of As(III) removal by *Peerless* Fe^0 in 0.10 mM or 1.0 mM NaH_2PO_4 relative to 10 mM NaCl: (a) total As concentration, (b) phosphate concentration, (c) Eh, and (d) pH. From Su and Puls (2001b), with permission.

a few hours in 0.01 M NaCl (Su and Puls 2001a). The Eh values may be regarded as qualitative parameters that are not amenable to quantitative interpretation. In the *Peerless* Fe^0/anion systems, the redox potential is thought to be a mixed potential involving dissolved oxygen, the Fe^0/Fe^{2+} couple, the Fe^{2+}/Fe^{3+} couple, and oxyanion species. The Eh measurements in the batch tests may not be relevant to a PRB field situation where biological activities also influence redox reactions.

The effect of silicate at 1.0 mM on As(V) removal by *Peerless* Fe^0 was pH-dependent (Fig. 8-7a) in that an initial pH of 10.12 was far more effective in decreasing As(V) removal than an initial pH of 7.73. The Eh values for the two silicate solutions in contact with *Peerless* Fe^0 with different pH values showed convergence at 120 h (Fig. 8-7c). Less than 10% of added As(V) was removed from 1.0 mM silicate at an initial pH 10.12, and the pH was constant throughout the test (Fig. 8-7d). Silicate was also removed by the *Peerless* Fe^0 with greater removal at an initial pH of 7.73 than at an initial pH of 10.12 (Fig. 8-7b). Silicate adsorption is a function of pH; it increases with increasing pH from 4 to 9, then decreases with

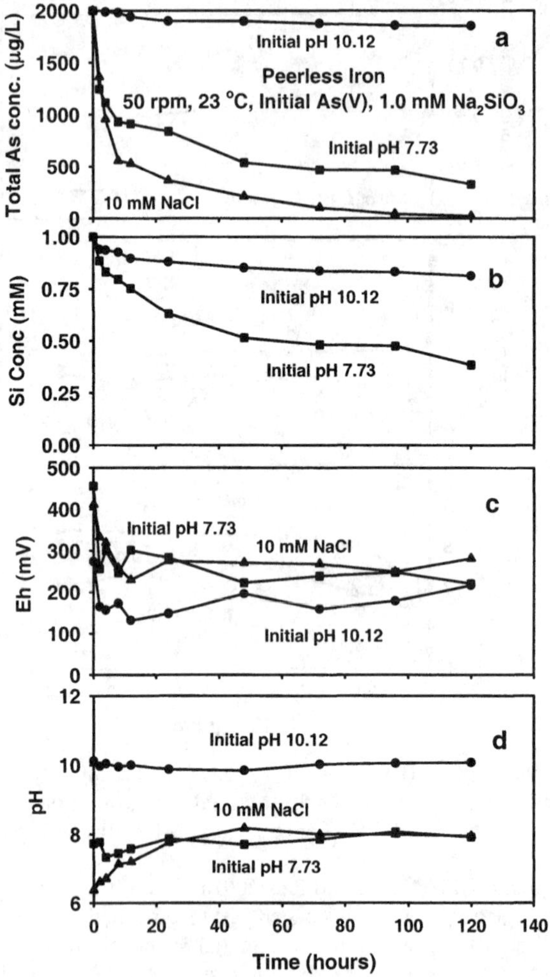

Fig. 8-7. Kinetics of As(V) removal by *Peerless* Fe^0 in 1.0 mM Na_2SiO_3 (initial pH 7.73, or 10.12): (a) total As concentration, (b) Si concentration, (c) Eh, and (d) pH. From Su and Puls (2001b), with permission.

increasing pH above 10 in soils, iron oxide minerals, and carbonate minerals (McKeague and Cline 1963; Hingston et al. 1972; Sigg and Stumm 1980). Evidence from x-ray photoelectron spectroscopy (XPS) experiments for silicate adsorption on ferrihydrite shows the presence of monomeric silicate or small units of polymerized silica at low solution silicate concentrations, and the presence of Si-rich precipitate at high silicate concentrations (Vempati et al. 1990).

The effects of competing anions (bicarbonate, silicate, and phosphate) on As(III) removal are also evident for nanoscale Fe^0 (Kanel et al. 2005). Although Fe^0 acts as a favorable sorbent for both phosphate and silicate, excess phosphate and silicate in groundwater may cause incomplete removal of As(V) and As(III). Other materials high in soluble silicate such as fly-ash should not be mixed with Fe^0 for arsenic removal. Alternatively, excess Fe^0 must be used to ensure the complete removal of arsenic in the presence of strong competing anions.

8.5 Column Tests and Field Applications

In an early laboratory column study of contaminant removal using Fe^0, Mayenkar and Heights (1986) showed that dissolved heavy metals [Fe, Ni, Cd, Cu, Pb, Zn, Hg, and Cr(VI)] and arsenilic acid in a simulated metal cleaning wastewater (initial pH 7.8 or 8.5) were effectively removed and longer residence time of the solution in contact with Fe^0 was found to be more effective for their removal than a shorter contact time.

A simple three-pitcher (locally known as '3-Kolshi') water filtration system made from locally available materials was tested for its efficiency in removing arsenic from the groundwater of Bangladesh (Khan et al. 2000). In a 3-Kolshi system, the first Kolshi has 3 kg of cast iron turnings and sand for arsenic removal, the second Kolshi has wood charcoal and sand for further arsenic removal and organic contaminants removal, and the third Kolshi is the collector for filtered water (Fig. 8-8). In their study, about 240 L of arsenic contaminated groundwater and

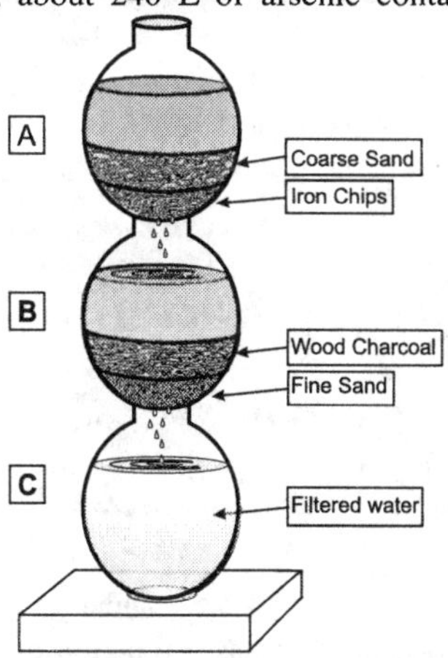

Fig. 8-8. Schematic diagram of a 3-Kalshi (pitcher) setup for removal of arsenic and organic contaminants from drinking water. Redrawn from Khan et al. (2000).

groundwater spiked with both As(III) and As(V) were filtered. More toxic As(III) was removed from 0.8 mg/L to below 0.002 mg/L; total As was lowered from 1.16 mg/L to below 0.016 mg/L. The final water quality meets and exceeds the guideline values set by USEPA, WHO, and Bangladesh.

Because in the presence of silicate the adverse effect of phosphate on As(V) adsorption by iron hydroxides had been shown to be magnified (Meng et al. 2002), recent column tests have been conducted on the combined effect of phosphate and silicate tests on As removal using Fe^0 (Su and Puls 2003). These column tests provided macroscale evidence for the continued corrosion of *Peerless* Fe^0 as the source of sorbents for As and confirmed previous batch tests (Su and Puls 2001b) showing the adverse effects of phosphate and silicate on As removal by Fe^0. Specifically, Su and Puls (2003) performed three column tests to study the behavior of *Peerless* Fe^0 to remove arsenic under dynamic flow conditions in the absence, as well as in the presence, of added phosphate and silicate. The three column tests were duplicated using a total of six columns: 1A and 1B (no added P and Si), 2A and 2B (low levels of added P and Si), and 3A and 3B (high levels of added P and Si). Each column consisted of a 10.3-cm depth of 50 : 50 (w : w, *Peerless* iron : sand) in the middle and a 10.3-cm depth of a sediment from Elizabeth City, NC in both upper and lower portions of the 31-cm long glass column (2.5 cm in diameter) with three side sampling ports. The flow velocity (upflow mode) was maintained at 4.3 m/d during the three- to four-month experiments. As expected, dissolved As concentrations in different positions of the column generally followed the reversed order of the longitudinal distance of these sampling ports from the influent end of the column: column influent > bottom port effluent > middle port effluent > top port effluent > column effluent (Figs. 8-9 to 8-11). The steady state As removal in the middle *Peerless* iron and sand mixture zone might be attributed to the continuous supply of corroded iron that served as the sorbents for both As(V) and As(III).

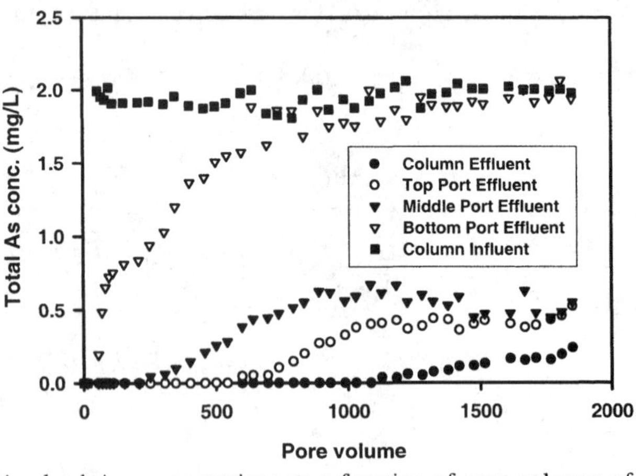

Fig. 8-9. Dissolved As concentrations as a function of pore volumes of the whole column. Column influent contained 1 mg As(V)/L + 1 mg As(III)/L in simulated Elizabeth City groundwater. From Su and Puls (2003), with permission.

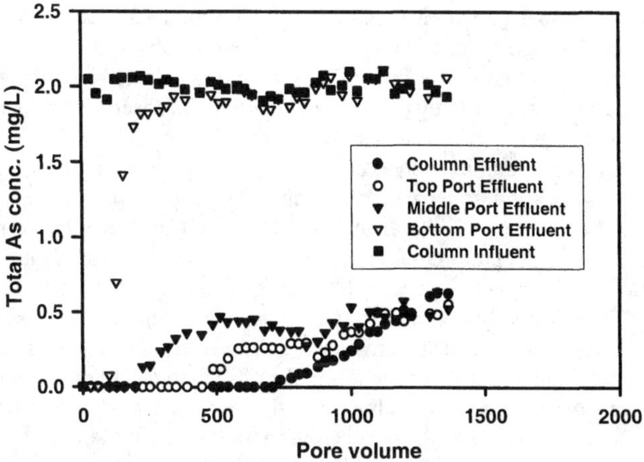

Fig. 8-10. Dissolved As concentrations as a function of pore volumes of the whole column. Column influent contained 1 mg As(V)/L + 1 mg As(III)/L in simulated Elizabeth City groundwater that also contained 0.5 mg P/L + 10 mg Si/L. From Su and Puls (2003), with permission.

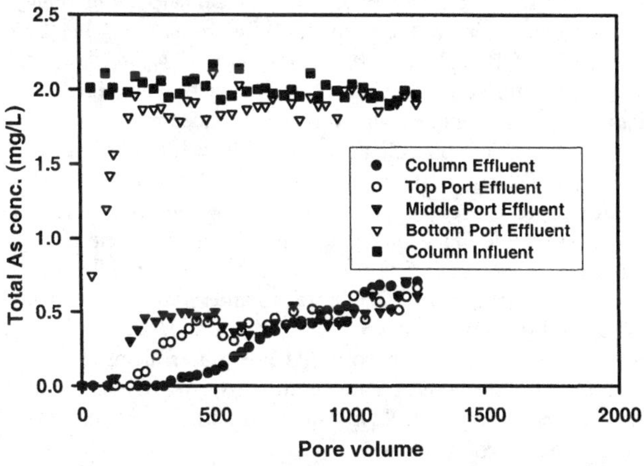

Fig. 8-11. Dissolved As concentrations as a function of pore volumes of the whole column. Column influent contained 1 mg As(V)/L + 1 mg As(III)/L in simulated Elizabeth City groundwater that also contained 1 mg P/L + 20 mg Si/L. From Su and Puls (2003), with permission.

Consistent with previous batch study findings, dissolved phosphate (0.5 or 1 mg P/L) and silicate (10 or 20 mg Si/L) showed strong inhibition for As(V) and As(III) (1 mg As(V)/L + 1 mg As(III)/L in 7 mM NaCl + 0.86 mM $CaSO_4$) removal by *Peerless* iron in the column tests (Figs. 8-10 and 8-11). The presence of added phosphate and silicate resulted in earlier breakthrough ($C = 0.5\ C_0$) and earlier complete breakthrough of dissolved arsenic relative to absence of added phosphate and silicate in the bottom port effluent. Phosphate and silicate were also removed by the PRB materials, especially by the *Peerless* iron. In the absence of added phosphate and silicate, the bottom port effluent showed breakthrough ($C = 0.5\ C_0$) at about 300 pore volumes, and complete breakthrough at 600 pore volumes (Fig. 8-9). As expected, total dissolved As concentrations in different positions of the upflow mode column followed an order: column influent > bottom port effluent > middle port effluent > top port effluent > column effluent. Dissolved As concentrations achieved steady state in the center of the middle zone (*Peerless* Fe^0 and Oil Creek sand mixture) at 900 pore volumes, and in the center of the top Elizabeth City sediment zone at 1100 pore volumes. The convergence of total dissolved As concentrations in the middle port and top port effluents in Fig. 8-9 is most likely attributed to the inactivation of the reactive materials within the zone of the top port effluent on As removal. The convergence of As concentration among the middle port, top port and column effluents observed in Figs. 8-10 and 8-11 further supports this hypothesis. No As was detected in the column effluent at pore volumes less than 1200, at which a total of 116 mg of As was removed by the whole column. This resulted in an average As concentration of 357 mg/kg of solids in the whole column.

The second column test (low P and Si) and third column test (high P and Si) results show that the presence of phosphate and silicate significantly changed the dissolved As concentrations in the effluents (Figs. 8-10 and 8-11). The bottom port effluent showed breakthrough ($C = 0.5\ C_0$) at only 170 pore volumes with 0.5 mg P/L + 10 mg Si/L, and at only 70 pore volumes with 1 mg P/L + 20 mg Si/L, as compared to 300 pore volumes without added phosphate and silicate. Complete breakthrough in the bottom port effluent was achieved at 400 pore volumes with 0.5 mg P/L + 10 mg Si/L (Fig. 8-10), and at 200 pore volumes (Fig. 8-11) with 1 mg P/L + 20 mg Si/L, as compared to 600 pore volumes without added phosphate and silicate (Fig. 8-9). At the higher phosphate and silicate concentrations, dissolved As concentrations achieved steady state in the center of the middle zone (*Peerless* Fe^0 and Oil Creek sand mixture) at 200 pore volumes, and in the center of the top Elizabeth City sediment zone at 400 pore volumes. Dissolved As concentrations converged in the middle port, top port, and column effluents at 700 pore volumes (Fig. 8-11). At 1000 pore volumes, concentrations of dissolved As in the middle of the *Peerless* Fe^0 zone were nearly the same (≈ 0.5 mg/L) for columns operated with and without Si and P (Figs. 8-9 to 8-11).

Nikolaidis et al. (2005) tested cartridges containing *Connelly-GPM* iron fillings mixed with sand in Bangladesh for their effectiveness in removing arsenic from groundwater. The set-up included two cartridges (A and B) connected in tandem for arsenic removal, and a downstream aeration vessel followed by a nylon filter to remove dissolved iron in the effluent. The tests were performed by pumping through units A and B, respectively, 1830 and 1140 L of groundwater over the course of a week from two wells containing 560 and 290 μg/L arsenic. During the first 3 to 4 days, arsenic concentrations in the outflow from unit A were at detectable (> 4 μg/L) levels, and in the outflow from unit B at ≈ 100 μg/L. Both units became effective during the following 3 days to maintain arsenic concentrations in the effluents at less than 10 μg/L. As pointed out by Nikolaidis et al. (2005), there are advantages of using Fe^0 fillings: a) readily available, b) without the addition of other elements of potential health concern, and c) no chemical reagents need to be routinely added.

The disposal of spent Fe^0 media in column tests may be a concern; however, recent studies show that the toxicity characteristic leaching test (TCLP) for the spent Fe^0 media show that the arsenic concentration in the leachate was two orders of magnitude lower than the 5 mg/L of TCLP for arsenic. The spent media are thus not hazardous and they can be disposed of in regular landfills (Nikolaidis et al. 2003).

8.6 Mechanisms of Arsenic Removal by Zero-valent Iron

So far virtually all the reported lab scale studies have shown that Fe^0 is effective in rapidly neutralizing acid and promoting removal and immobilization of dissolved heavy metals and metalloids (As). These studies indicate that rapid metal removal is due to adsorption onto the iron metal surface or onto iron corrosion products initially present on the unreacted metal surface. The mechanisms of arsenic removal by zero-valent iron fillings are complex, probably involving many processes such as adsorption, precipitation, co-precipitation, and redox transformation (Table 8-4). Adsorption appears to be a predominant process at the early stage of reaction and is expected for systems where solution composition is undersaturated with respect to known metal arsenate precipitates. For example, in a goethite suspension containing both Zn and arsenate, below As(V) surface saturation on goethite, As(V) was found to form bidentate binuclear bridging complexes on Fe and/or Zn octahedral, while Zn mainly formed edge-sharing complexes with Fe at the goethite surface. Above surface saturation, Zn was found to be increasingly complexed by arsenate, gradually forming an adamite-like [$Zn_2(AsO_4)OH$] surface precipitate on goethite (Gräfe et al. 2004).

To learn how arsenic is removed in the Fe^0 systems, it is important to characterize mineralogical changes associated with the corrosion of Fe^0. A few studies have reported detailed chemical and mineralogical characterization of iron corrosion products using a variety of analytical tools. They include scanning/transmission electron microscopy (e.g., Gu et al. 1999; Furukawa et al. 2002), X-ray diffraction (XRD) (e.g., Wilkin and McNeil 2003), infrared spectroscopy (e.g., Su and Puls 2004b), Raman spectroscopy (e.g., Odziemkowski et al. 1998), XPS (e.g., Su and Puls 2001a and 2004b; Bang et al. 2005), and X-ray absorption spectroscopy (e.g., Farrell et al. 2001; Melitas et al. 2002a; Manning et al. 1998 and 2002). A typical XRD pattern for pristine *Peerless* Fe^0 shows two strong peaks at d-spacings of 0.202 nm (2θ = 57.23) and 0.143 nm (2θ = 85.21) characteristic of the elemental iron and several smaller peaks for iron oxides (Fig. 8-12). The presence of iron oxides (magnetite, hematite, and lepidocrocite) and graphite is more clearly seen for the fines separated from the pristine *Peerless* Fe^0 (Fig. 8-12) (Su and Puls 2004b). Embedded graphite flakes are a minor but typical ingredient of cast iron. The presence of magnetite and hematite was reported via a Raman spectroscopic study for another pristine commercial zero-valent iron material, *Connelly GPM iron*, which is covered by an air-formed high temperature oxidation film, consisting of an inner layer of Fe_3O_4, and an outer, passive layer of Fe_2O_3 (Ritter et al. 2003).

Table 8-4. Possible Chemical Processes Related to Arsenic Removal and Fate in Zero-valent Iron Systems (Modified after Lien and Wilkin (2005))

Chemical Processes	Description	Examples	References
Adsorption	Binding of adsorbate onto the iron surface to form a complex at the solid-solution interface	Adsorption onto Fe^0 or Fe^0 corrosion products such as iron (oxydr)oxides to form inner-sphere, bidentate complexes	Mayenkar and Heights (1986) Lackovic et al. (2000) Su and Puls (2001a) Melitas et al. (2002b), and Manning et al. (2002)
Precipitation	Nucleation and growth of a solid phase with repeating molecular unit in three dimensions	Symplesite $Fe_3(AsO_4)_2 \cdot 8H_2O$, Realgar As_4S_4	Nikolaidis et al. (2003)
Co-precipitation	Incorporation of element as a minor constituent in a mineral structure as	Arsenic co-precipitation with: carbonate green rust $Fe_4^{2+}Fe_2^{3+}(OH)_{12}CO_3 \cdot nH_2O$, ferrous sulfide/disulfide FeS, Fe_3S_4, or FeS_2	Nikolaidis et al. (2003) Su and Puls (2004) Lien and Wilkin (2005)
Redox reactions	Reduction Oxidation	$As^{5+} + 2e^- \rightarrow As^{3+}$ (Insignificant) $As^{3+} + 3e^- \rightarrow As^0$ As^{3+} + oxidants $\rightarrow As^{5+}$ (Oxidants = e.g., manganese dioxide, carbonate green rust, magnetite, hematite, $OH^\cdot$ radical)	Farrell et al. (2001) Bang et al. (2005) Su and Puls (2001a, 2004b) Melitas et al. (2002b) Lien and Wilkin (2005) Kanel et al. (2005)

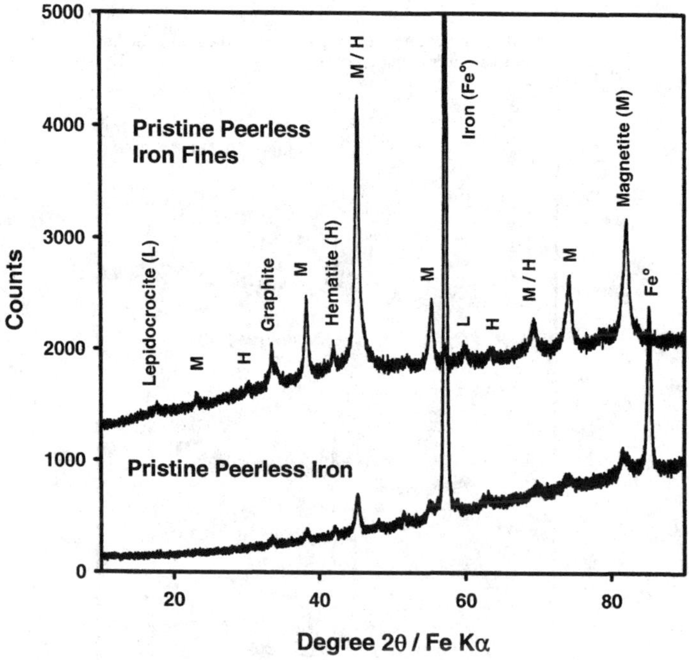

Fig. 8-12. X-ray diffraction patterns of the *Peerless* iron as received and its separated fines (< 0.05 mm). From Su and Puls (2004b), with permission.

Su and Puls (2004b) examined the corrosion products of *Peerless* Fe^0 used in three column tests for removing arsenic from water under dynamic flow conditions. XRD results of both 5 min sedimentation (Fig. 8-13) indicate that the dominant corrosion products were carbonate green rust (CGR) and magnetite. Green rust (GR) compounds are compositionally variable, mixed valence Fe(II)/Fe(III) layered hydroxides whose structural units consist of alternating positively charged tri-octahedral metal hydroxide sheets and negatively charged interlayers of anions (typically chloride, carbonate, or sulfate). GR compounds can be distinguished based on their characteristic d-spacings: GR1 shows a distance of 0.75 to 0.80 nm between hydroxide sheets as in carbonate green rust (CGR), and GR2 shows a distance of 1.1 nm as in sulfate GR (Taylor 1973; Hanson 1989; Genin et al. 1998). The four predominant d-spacings at 0.76, 0.38, 0.27, and 0.24 nm confirmed the formation of GR1 in the column tests. CGR and magnetite also have been found as the iron corrosion products in both laboratory tests and field PRBs (Gu et al. 1999; Roh et al. 2000; Furukawa et al. 2002).

Examination of iron corrosion in full-scale PRBs through solid phase characterization and geochemical modeling suggest that in non-mining impacted ground water, CGR precipitates preferentially over the sulfate green rust (SGR) in

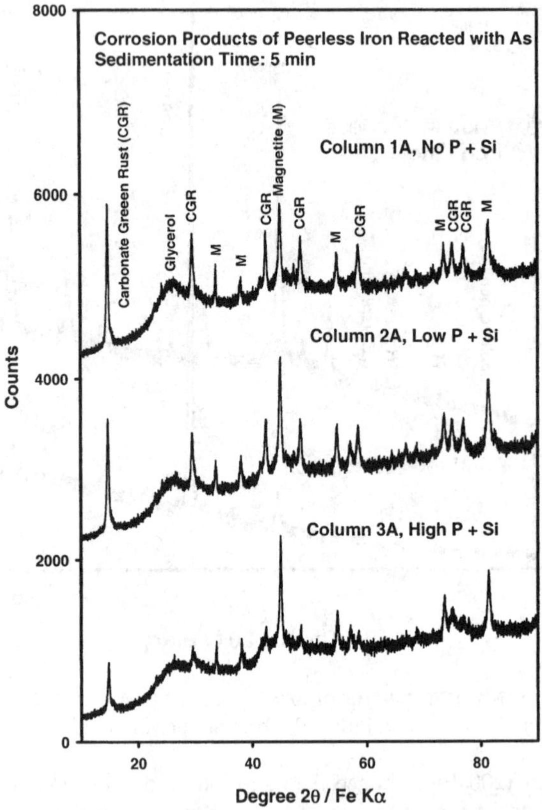

Fig. 8-13. X-ray diffraction patterns of the separated corrosion products of Peerless iron reacted with As in column tests that were collected after 5 min of sedimentation. Column influent contained 1 mg of As(V)/L + 1 mg of As(III)/L in simulated Elizabeth City ground water (ECGW) with a composition of 7 mM NaCl + 0.86 mM $CaSO_4$ (pH 6.5). Column 1A influent had no added P and Si, column 2A influent also contained 0.5 mg of P/L + 10 mg of Si/L in ECGW (pH 6.5), and column 3A influent also contained 1 mg of P/L + 20 mg of Si/L in ECGW (pH 6.5). From Su and Puls (2004b), with permission.

zero-valent iron systems (Wilkin et al. 2002). The equilibrium between SGR and CGR can be described by the following equation based on the anhydrous GR components:

$$Fe_4^{2+}Fe_2^{3+}(OH)_{12}SO_4 + CO_3^{2-} \; \Omega \; Fe_4^{2+}Fe_2^{3+}(OH)_{12}CO_3 + SO_4^{2-} \qquad (8.3)$$

An estimated equilibrium constant was obtained in a recent study (Wilkin and

McNeil 2003), assuming ideal mixing relations in the solids and taking thermodynamic data from Bourrie et al. (1999):

$$K_1 = [SO_4^{2-}/CO_3^{2-}]\ [Fe_4^{2+}Fe_2^{3+}(OH)_{12}CO_3/\ Fe_4^{2+}Fe_2^{3+}(OH)_{12}SO_4] = 10^{3.1} \quad (8.4)$$

The Fourier transform infrared - photo acoustic spectroscopy (FTIR-PAS) data of the sedimented fines separated from the corroded iron filings were composed of three main regions of interest: the O-H stretching region from 4000-2000 cm^{-1}, the combination band and overtone region from 2000 - 1200 cm^{-1}, and the lattice mode and molecular anion region from ca. 1200 - 400 cm^{-1} (Fig. 8-14). Each of the spectra contained the usual broad OH hump centered at 3404 cm^{-1} plus two additional narrow features on the high wavenumber side at 3612 and 3698 cm^{-1}. Such narrow bands usually are interpreted to mean distinct structural OH groups that are fairly well isolated from one another. The broad hump contains the signature of those OH groups that are sufficiently close to be interacting with one another. The relative size of this feature decreases in the order 1A > 2A > 3A.

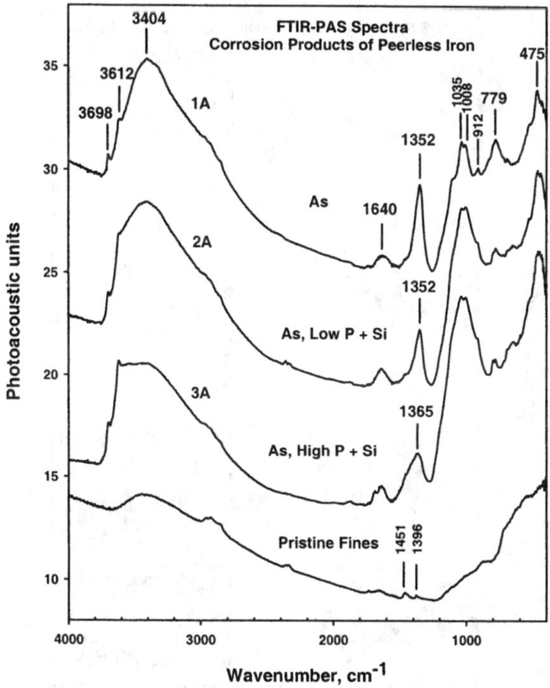

Fig. 8-14. FTIR-photo acoustic spectra of the separated corrosion products of *Peerless* iron reacted with As from column tests 1A, 2A, and 3A (5 min of sedimentation) and of the pristine *Peerless* iron fines. From Su and Puls (2004b), with permission.

Each of the spectra contained a fairly broad band near 1640 cm^{-1} assigned to the scissors vibration of molecular water. This band is characteristic of many hydrated minerals. Additionally, each spectrum contained a large band at 1352 or 1365 cm^{-1} that is assigned to CGR (Legrand et al. 2003). The normal modes for carbonate anions in crystalline carbonates such as calcite (1448 cm^{-1}) and siderite (1443 cm^{-1}) are not particularly close to either of these two bands (Johnson et al. 2002). Adsorbed carbonate onto amorphous iron oxide shows two bands at about 1480 cm^{-1} and 1350 cm^{-1}; whereas, free carbonate in solution show a ν_3 band at 1383 cm^{-1} (Su and Suarez 1997). The two weak bands at 1451 and 1396 cm^{-1} for the *pristine* Peerless Fe^0 fines may be attributed to adsorbed carbonate (Fig. 8-14).

The As 3d in the XPS spectra suggested an oxidation state closest to As(V) (Fig. 8-15). After a brief sputter, most of the carbon was removed and the As 3p 5/2 was shifted to lower binding energy of 143.7 eV. The As 3d region could be fit with two doublet sets, one at 44.2 eV for As(III) and another at 45.5 eV for As(V), which would mean that both As(III) and As(V) are present. After 1.0 min of sputtering similar results were obtained (Fig. 8-15).

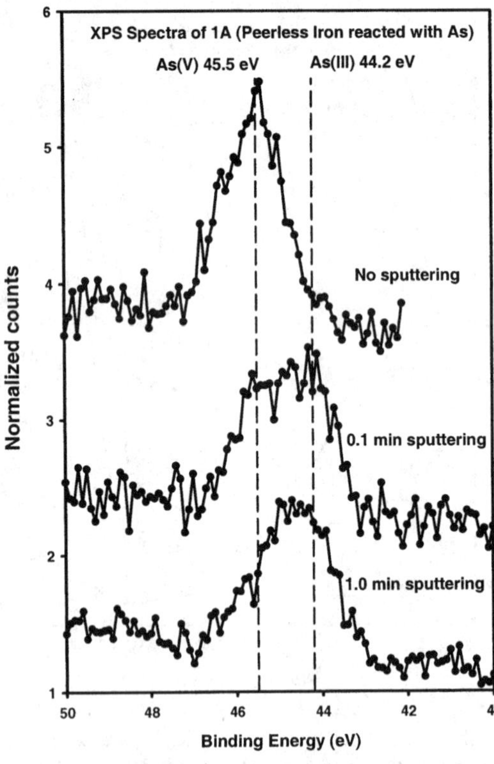

Fig. 8-15. XPS spectra of As 3d levels in corroded *Peerless* iron particles harvested from column test 1A. From Su and Puls (2004b), with permission.

It appears that the top layer of corroded iron surface contained As(V) despite the fact that the feeding solution in contact with *Peerless* iron contained more As(III) than As(V) as a result of preferential uptake of As(V) over As(III) by the Elizabeth City sediment (Su and Puls 2003). The XPS results for samples without sputtering agree with the chemical speciation results using IC-HG-AFS for 1.0 M HCl digested separates, which show that from 86 to 96% of the total extractable As (6.9 to 14.6 g/kg) was present in the form of As(V). Previous analyses of effluent solutions from laboratory column tests indicate that added As(III) was partially oxidized to form As(V) in the solution phase by the corrosion products of *Peerless* Fe^0 (Su and Puls 2003). This is consistent with the current XPS results.

Based on the mineralogical and compositional analyses of the Fe^0 system, Su and Puls (2004b) formulated chemical reactions to explain the dominant chemical reactions that could have taken place. Because the influent solution used in the column tests contained dissolved oxygen, aerobic corrosion of Fe^0 took place [Eq. (8.2)], resulting in elevated pH (pH = 8.5 to 9.5) within the *Peerless* Fe^0 zone. Dissolved oxygen subsequently oxidized portions of Fe^{2+} produced by corrosion of the Fe^0 [Eq. (8.5)]. The formation of CGR [Eq. (8.6)] is most likely to have occurred in the initial few centimeters of the middle zone of *Peerless* Fe^0 and sand mixture in the column; this is a transition zone where redox conditions changed from aerobic to anaerobic state. Measured Eh values in this zone ranged from 100 to 300 mV, which is the redox potential range at which both ferrous and ferric ions are produced to provide reactants (in addition to carbonate) for CGR formation. Sources of carbonate were NaOH used to adjust feeding solution pH that contained carbonate impurities and atmospheric CO_2 that reentered the feeding solution during column experiment plus the Elizabeth City sediment. Subsequently, CGR underwent partial degradation to form magnetite and to release carbonate ions to the solution [Eq. (8.7)]. Localized reducing environment may have also resulted in formation of iron sulfide [Eq. (8.8)].

$$4Fe^{2+} + O_2 + 2H_2O \ \Omega \ 4Fe^{3+} + 4OH^- \quad (8.5)$$

$$4Fe^{2+} + 2Fe^{3+} + CO_3^{2-} + 14H_2O \ \Omega \ Fe_4^{2+}Fe_2^{3+}(OH)_{12}CO_3 \cdot 2H_2O + 12H^+ \quad (8.6)$$

$$Fe_4^{2+}Fe_2^{3+}(OH)_{12}CO_3 \cdot 2H_2O + 0.5O_2 + 2OH^- \ \Omega \ 2Fe_3O_4 + CO_3^{2-} + 9H_2O \quad (8.7)$$

$$Fe^{2+} + HS^- \ \Omega \ FeS + H^+ \quad (8.8)$$

In carbonate-rich ground water, iron corrosion produces CGR (Gu et al. 1999; Bonin et al. 2000; Philips et al. 2003). Furthermore, this study shows that CGR also forms under the condition of much greater sulfate concentration in the feeding solution than the carbonate concentration. Because CGR seems to form before magnetite and because synthetic CGR has been shown to oxidize As(III) to As(V) (Su and Wilkin 2005), oxidation of As(III) by CGR may be more important than other iron corrosion products at the early stage of iron corrosion. Further studies are needed to separate As(III) oxidation by CGR versus oxidation by other ferric oxides. In addition to CGR, other mineral phases (ferrihydrite, magnetite, aragonite, calcite, mackinawite, greigite, and lepidocrcrocite) have been identified in the fine-grained fractions of PRB samples from the U. S. Coast Guard Support Center (Elizabeth City, NC) and the Denver Federal Center (Lakewood, CO) sites (Wilkin et al. 2002).

The most likely solid phases that host the sorbed As(V) and As(III) in a Fe^0 system are iron oxides and green rusts that are products of fast processes. Other slow processes may produce other strong precipitates and lead to the cementation of the column media. Infrared spectroscopic evidence supports the formation of inner-sphere complexes of As(V) and As(III) on ferrihydrite (Suarez et al. 1998). Analyses of As(III) and As(V) adsorption complexes in the Fe^0 corrosion products and synthetic iron oxides (goethite, hematite, lepidocrocite, maghemite, and magnetite) by extended X-ray absorption fine structure spectroscopy (EXAFS) indicate both As species form inner-sphere bidentate complexes (Manning et al. 2002). The experimental radial structure functions (RSF) produced from Fourier transformed EXAFS data are shown in Figs. 8-16 and 8-17 (solid lines). The fits of the theoretical EXAFS expression to the experimental data are shown with dotted lines. The predominant feature in Fig. 8-16 is a coherent shell of 4.0-4.2 oxygen atoms at an As-O interatomic distance of 1.69 Å. This distance is diagnostic for the As(V) species.

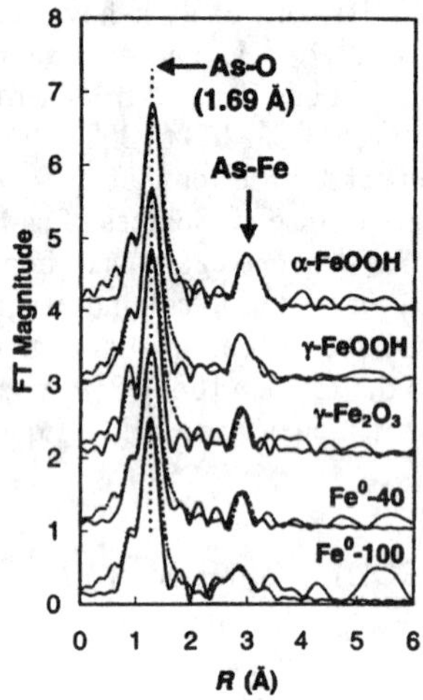

Fig. 8-16. Radial structure functions from EXAFS analysis of As(V)-treated iron oxides and Fe^0 materials. Solid lines are experimental data and dotted lines are the theoretical fits. From Manning et al. (2002), with permission.

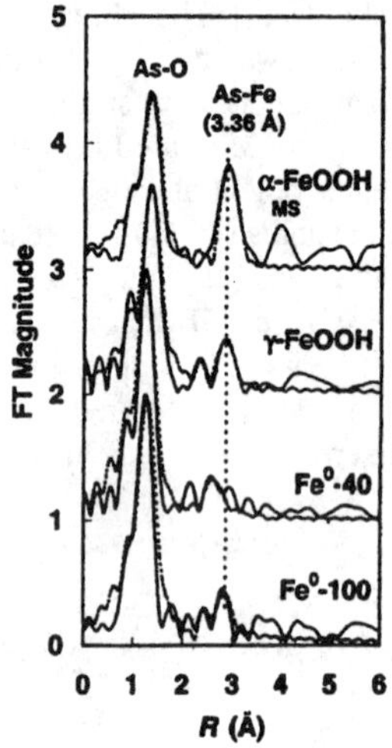

Fig. 8-17. Radial structure functions from EXAFS analysis of As(III)-treated iron oxides and Fe^0 materials. Solid lines are experimental data and dotted lines are the theoretical fits. From Manning et al. (2002), with permission.

A recent EXAFS study further demonstrated that As(V) forms predominantly bidentate corner-sharing complex on the iron oxides (ferrihydrite, goethite, hematite, and maghemite) (Sherman and Randall 2003). Other earlier spectroscopic studies also suggest that As(V) predominately forms inner-sphere bidentate surface complexes with ferrihydrite, although the nature of the bidentate complexes (corner-sharing vs. edge sharing) is controversial (Waychunas et al. 1993; Manceau 1995). As(V) was found to form inner-sphere surface complexes on amorphous Fe oxide while As(III) formed both inner- and outer-sphere surface complexes on an amorphous Fe oxide (Goldberg and Johnston 2001). As(V) also forms inner-sphere complex with carbonate green rust (Randall et al. 2001). Further studies are needed to explore if As(III) also forms similar complexes with carbonate green rust.

Oxidation of As(III) to As(V) in the Fe^0 systems has been observed by several investigators (Su and Puls 2001a and 2004b; Manning et al. 2002; Kanel et al. 2005). Examination of the As-O shells in the RSF data shown in Fig. 8-16 suggest that As(III) was oxidized to As(V) during corrosion of Fe^0-40 (mesh size) and Fe^0-100 but was not oxidized by goethite or lepidocrocite (Manning et al. 2002). This is consistent with the XANES data that show partial As(III) oxidation to form a mixed As(III)/As(V) speciation in the solids of As(III)- treated Fe^0 (Figs. 8-18a and 8-18b). The energy of the two first derivative peaks for As(III)-treated Fe^0-100 and Fe^0-40 correspond closely to the energy values for the As(III) and As(V) standards (Manning et al. 2002).

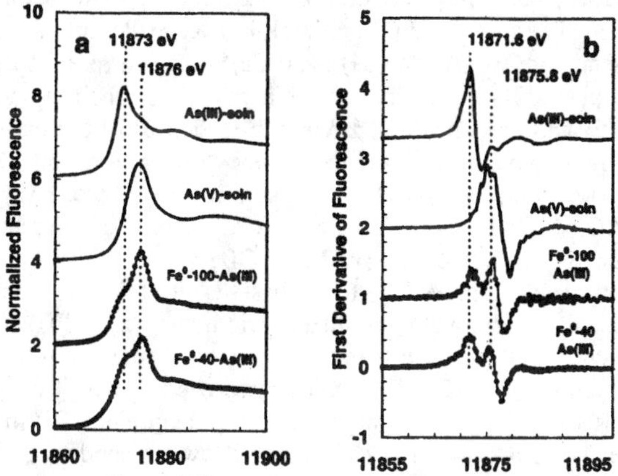

Fig. 8-18. XANES edges and first derivative plots of As(III)/As(III) standard solutions and As(III)-treated Fe^0 materials. From Manning et al. (2002), with permission.

Potential oxidants include manganese dioxide impurities in the Fe^0 sample (Su and Puls 2001a), green rusts and iron oxides (Su and Wilkin 2005), dissolved O_2 and Fe(II) (Hug and Leupin 2003), and $OH^•$ radical formed from corroding Fe^0 surfaces (Kanel et al 2005). It has been suggested that when Fe^0 corrodes, both ferrous iron and hydrogen peroxide forms, which in turn forms $OH^•$ radical (Voegelin and Hug 2003; Joo et al. 2004) that oxidizes As(III) to form As(V).

$$Fe^0 + O_2 + 2H^+ \ \Omega \ Fe^{2+} + H_2O_2 \quad (8.9)$$
$$Fe^{2+} + H_2O_2 \ \Omega \ Fe^{3+}OH^{2+} + OH^{\bullet} \quad (8.10)$$
$$2OH^{\bullet} + H_3AsO_3 \ \Omega \ H_2AsO_4^- + H_2O + H^+ \quad (8.11)$$

Oxidation of As(III) in the iron oxides alone systems also has been reported (De Vitre et al. 1991). From 77-85% of added As(III) at initial concentrations from 0.2 to 16 mg/L was converted to As(V) on intercalated Fe(III)-montmorillonite within 12 h as determined with XAFS combined with high-energy-resolution fluorescence spectrometry (Izumi et al. 2005). Regardless of the specific mechanisms, oxidation of As(III) is a desirable outcome because As(III) is considered to be more toxic and more mobile than As(V) in aqueous environments.

8.7 Alternative Materials of Iron and Aluminum Oxides for Arsenic Removal

Alternative materials of iron and aluminum oxides to Fe^0 have been tested for arsenic removal. Batch tests show that arsenic in mine tailing leachate can be effectively removed using steel manufacturing byproducts such as evaporation cooler dust, oxygen gas sludge, basic oxygen furnace slag, and electrostatic precipitator dust (Ahn et al. 2003). These by-products are high in iron oxides and are highly alkaline due to the presence of $CaCO_3$, $Ca(OH)_2$, and Ca_2SiO_4. In a recent batch and column study, Daus et al. (2004) tested five sorption materials for the removal of arsenic from water: activated carbon (AC), zirconium-loaded activated carbon (Zr-AC), a commercial sorption medium with the trade name "Absorptionsmittel 3" (AM3), Fe^0, and iron hydroxide granulates (GIH). Batch sorption kinetics of As(V) followed the sequence: Zr-AC >> GIH = AM3 > Fe^0 > AC; whereas, a different sequence was obtained for As(III): AC >> Zr-AC = AM3 = GIH = Fe^0. Column tests showed that As(III) was completely removed with the best results for GIH. Rapid small-scale column tests for As(V) removal in granular ferric hydroxide (GFH) could be used to aid in the selection and design of arsenic removal media for full-scale treatment facilities (Westerhoff et al. 2005). While fine-powdered iron hydroxide materials may be more effective due to their larger surface area, granular Fe^0 may be preferred in a PRB application because it does not slow down groundwater flow due to its larger particle size.

Attempts have been made to construct portable household water purification systems that can be used in the rural area of the developing countries. Meng et al. (2001) conducted laboratory and field tests to evaluate the effectiveness of a household filtration process and to investigate the effects of phosphate and silicate on the removal of arsenic from Bangladesh groundwater by ferric hydroxides. The household filtration process included co-precipitation of arsenic by adding a packet (approximately 2 g) of ferric and hypochlorite salts to 20 L of well water and subsequent filtration of the water through a bucket sand filter. A field demonstration study was performed to test the treatment system in seven households in Bangladesh in March and April 2000. Experimental results showed that the household treatment process removed arsenic from approximately 300 μg/L in the well water to less than 50 μg/L. The field demonstration results showed that the household filtration process could effectively remove arsenic from Bangladesh well water. The household co-precipitation and filtration process was proved reliable and easy to use. Based on a daily consumption of 50 L of filtered water, it is estimated that the chemical costs are less than US $4 annually for a family. The co-precipitation results indicate that elevated phosphate and silicate concentrations in Bangladesh well water dramatically decreased adsorption of arsenic by ferric hydroxides. A Fe/As mass ratio of greater than 40 was required to reduce arsenic concentration to less than 50 μg/L in the well water.

A similar arsenic removal system at a household level was used in Bangladesh by six households for four months to treat well water containing 190-750 μg/L As as well as 0.4-20 mg/L Fe and 0.2-1.9 mg/L P (Cheng et al. 2004). The system removes As from a 16-L batch of water in a bucket by filtration through a sand bed following the addition of about 1.5 g of ferric sulfate and 0.5 g of calcium hypochlorite. Arsenic concentrations in all but 1 of 72 samples of treated water were below the Bangladesh drinking water standard of 50 μg/L for arsenic. About half of the samples also met the WHO guideline of 10 μg/L. The bucket system requires daily handling and attention to support the water consumption of a typical household and it does not appreciably remove Mn that ranged from 0.730 to 5.080 mg/L in the well water. In contrast, the 3-Kalshi system decreased Mn concentration in Bangladesh groundwater from 0.723 mg/L to less than 0.001 mg/L (Khan et al. 2000).

Activated alumina also has been used as a sorbent for arsenic. Since 1997, over 135 well-head (above ground) arsenic removal units containing activated alumina have been installed in remote villages in the Indian state of West Bengal bordering Bangladesh (Sarkar et al. 2005). Every component of the arsenic removal treatment system including activated alumina sorbent is procured indigenously. Each unit serves 200–300 households and contains about 100 L of activated alumina. No chemical addition, pH adjustment or electricity is required for operating these units. The villagers are responsible for the day to day operation and the upkeep of the units. The arsenic concentration in the influent varies from around 100 μg/L to greater than 500 μg/L. In the treated water, arsenic concentration is consistently below 50 μg/L. The units are capable of removing both As(III) and As(V) from the contaminated groundwater for several months, often exceeding 10,000 bed volumes. In the top portion of the column, the dissolved iron present in ground water is oxidized by atmospheric oxygen into hydrated Fe(III) oxides or HFO particles which in turn selectively bind both As(III) and As(V). Upon exhaustion, these units are regenerated by caustic soda solution followed by acid wash. The arsenic-laden spent regenerant is converted into a small volume sludge (less than 500 g) and contained over a coarse sand filter in the same premise requiring no disposal. Many units have been operating for several years without any significant operational difficulty. The treated water is used for drinking and cooking.

8.8 Knowledge Gaps and Research Needs

Great progress has been made in recent years in research pertaining to Fe^0 for arsenic removal from groundwater and wastewater; yet, much remains to be known. The stability of initially sorbed arsenic in the Fe^0 system over extended periods of time (years to decades) is not known. No information is available about possible later release of sequestered As due to subsequent mineralogical transformation of initially formed corrosion products. Zero-valent iron corrodes at first to form green rusts which are generally not thermodynamically stable under most field conditions; green rusts transform to magnetite and other iron oxides over time (Odziemkowski et al. 1998; Wilkin et al. 2002). Since green rusts have higher sorption capacity for As than does magnetite, what happens to arsenic that is initially sorbed by green rusts when green rusts decompose to form magnetite?

Dissolved solutes may have positive or negative effects on arsenic removal and their role merits further investigation. Dissolved fulvic acid displaces arsenate from and inhibits its adsorption to ferrihydrite and gibbsite (Simeoni et al. 2003). Natural organic matter delays the attainment of sorption equilibrium and diminishes the extent of sorption of both As(V) and As(III) onto hematite (Redman et al. 2002). Dissolved organic matter thus may negatively impact the performance of Fe^0 for arsenic removal. Dissolved carbonate results in an enhanced and/or suppressed extent of As(V) adsorption on hematite depending on reaction conditions [e.g., available surface sites, initial As(V) concentrations, and reaction times] (Arai et al. 2004). On the other hand, surface complexation model calculation shows that sorption of carbonate at common soil and groundwater concentrations reduce the sorption of capacity of arsenic on ferrihydrite significantly (Appelo et al. 2002).

Another issue is related to the porosity loss of Fe^0-based PRB. The working life of Fe^0-based PRBs is limited by corrosion of Fe^0 and precipitation of secondary minerals (Liang et al. 2000). Quantification of mineral precipitation and the long-term impact of corrosion products on Fe^0-PRB longevity need to be further studied. Extensive porosity loss due to high levels of corrosive species in groundwater may have a significant negative impact on long-term performance of PRBs. A 2-yr field column study showed high porosity reduction (40.6 % to 45.1% for the inlet sand/Fe^0 interface and 7.4% to 25.6% for effluent samples) of two test columns subject to high nitrate groundwater (Kamolpornwijit et al. 2004). This study reported the formation of goethite, magnetite, aragonite, and calcite as major precipitated minerals, with $Fe_2(OH)_2CO_3$ and green rust as minor phases.

Microbial aspect of arsenic in the aqueous environment is an important component of arsenic biogeochemical cycling (Cullen and Reimer 1989). It is not clear how microbial processes within and around iron PRBs impact arsenic removal and transformation. It has been shown that hydrogen gas generation from Fe^0 corrosion stimulates the growth of microbial populations and leads to sulfate reduction (Gu et al. 1999; Wilkin et al. 2003). The biogeochemical coupling of iron, sulfur, and arsenic in Fe^0 based PRBs needs further investigation.

8.9 Conclusions

We face the international challenge to design or create a workable, sustainable, efficient point-of-use water treatment system for arsenic-contaminated groundwater in Bangladesh, India, Nepal, and other developing countries. We also face the domestic task to improve PRB technologies for groundwater remediation and drinking water/wastewater treatment at field scale in the United States to meet the new USEPA MCL for arsenic. Zero-valent iron has shown its effectiveness in laboratory bench-scale and a few pilot-scale field settings for arsenic removal on a short term basis. Full-scale implementation of the technology is expected. Long-term PRB performance is likely to be affected by mineral precipitation, gas production, and microbial activities. More work remains to be done before zero-valent iron-based systems become technically robust, reliable, maintainable, socially acceptable, and affordable.

Acknowledgments

Although the research described in this report has been funded wholly by the United States Environmental Protection Agency, it has not been subjected to Agency review and, therefore, does not necessarily reflect the views of the Agency, and no official endorsement should be inferred.

8.10 References

Ahn, J. S., Chon, C. M., Moon, H. S., and Kim, K. W. (2003). "Arsenic removal using steel manufacturing byproducts as permeable reactive materials in mine tailing containment systems." *Wat. Res.*, 37, 2478-2488.

Appelo, C. A. J., van der Weiden, M. J. J., Tournassat, C., and Charlet, L. (2002). "Surface complexation of ferrous iron and carbonate on ferrihydrite and the mobilization of arsenic." *Environ. Sci. & Technol.*, 36, 3096-3103.

Arai, Y., Sparks, D. L., and Davis, J. A. (2004). "Effects of dissolved carbonate on arsenate adsorption and surface speciation at the hematite-water interface." *Environ. Sci. & Technol.*, 38, 817-824.

Bang, S., Johnson, M. D., Korfiatis, G. P., and Meng, X. (2005). "Chemical reactions between arsenic and zero-valent iron in water." *Wat. Res.*, 39, 763-770.

Benner, S. G., Blowes, D. W., and Ptacek, C. J. (1997). "A full-scale porous reactive wall for prevention of acid mine drainage." *Ground Water Monit. Rem.*, 17, 99–107.

Benner, S. G., Blowes, D. W., Gould, W. D., Herbert Jr., R. B., and Ptacek, C. J. (1999). "Geochemistry of a permeable reactive barrier for metals and acid mine drainage." *Environ. Sci. & Technol.*, 33, 2793–2799.

Blowes, D. W., Ptacek, C. J., and Jambor, J. L. (1997). "In-situ remediation of chromate contaminated groundwater using permeable reactive walls." *Environ. Sci. & Technol.*, 31, 3348-3357.

Blowes, D. W., Ptacek, C. J., Benner, S. G., McRae, C. W. T., Bennet, T. A., and Puls, R. W. (2000). "Treatment of inorganic contaminants using permeable reactive barriers." *J. Contam. Hydrol.*, 45, 123-137.

Bonin, P. M. L., Odziemkowski, M. S., Reardon, E. J., and Gillham, R. W. (2000). "In situ identification of carbonate-containing green rust on iron electrodes in solutions simulating groundwater." *J. Solut. Chem.*, 29, 1061-1074.

Bourrié, G., Trolard, F., Génin, J., Jaffrezic, A., Maître, V., and Abdelmoula, M. (1999). "Iron control by equilibria between hydroxy-Green Rusts and solutions in hydromorphic soils." *Geochim. Cosmochim. Acta,* 63, 3417-3427.

Cantrell, K. J., Kaplan, D. I., and Wietsma, T. W. (1995). "Zero-valent iron for the in situ remediation of selected metals in groundwater." *J. Hazard. Mater.*, 42, 201–212.

Cheng, Z., van Geen, A., Jing, C., Meng, X., Seddique, A., and Ahned, K. M. (2004). "Performance of a household-level arsenic removal system during 4-month deployments in Bangladesh." *Environ. Sci. & Technol.*, 38, 3442-3448.

Cullen, W. R., and Reimer, K. J. (1989). "Arsenic speciation in the environment." *Chem. Rev.*, 89, 713-764.

Daus, B., Wennrich, R., and Weiss, H. (2004). "Sorption materials for arsenic removal from water: A comparative study." *Wat. Res.*, 38, 2948-2954.

De Vitre, R., Belzile, N., and Tessier, A. (1991). "Speciation and adsorption of arsenic on diagenetic iron oxyhydroxides." *Limnol. Oceanogr.*, 36, 1480-1485.

Dixit, S., and Hering, J. G. (2003). "Comparison of arsenic (V) and arsenic (III) sorption onto iron oxide minerals: Implications for arsenic mobility." *Environ. Sci. & Technol.*, 37, 4182-4189.

Farrell, J., Wang, J., O'Day, P., and Conklin, M. (2001). "Electrochemical and spectroscopic study of arsenate removal from water using zero-valent iron media." *Environ. Sci. & Technol.*, 35, 2026-2032.

Furukawa, Y., Kim, J.-W., Watkins, J., and Wilkin, R. T. (2002). "Formation of ferrihydrite and associated iron corrosion products in permeable reactive barriers of zero-valent iron." *Environ. Sci. & Technol.*, 36, 5469-5475.

Genin, J-M. R., Bourrie, G., Trolard, F., Abdelmoula, M., Jaffrezic, A., Refait, P., Maitre, V., Humbert, B., and Herbillon, A. (1998). "Thermodynamic equilibria in aqueous suspensions of synthetic and natural Fe(II)-Fe(III) green rusts: Occurrences of the mineral in hydromorphic soils." *Environ. Sci. & Technol.*, 32, 1058-1068.

Goldberg, S., and Johnston, C. T. (2001). "Mechanisms of arsenic adsorption on amorphous oxides evaluated using macroscopic measurements, vibrational spectroscopy, and surface complexation modeling." *J. Colloid Interface Sci.*, 234, 204-216.

Gräfe, M., Nachtegaal, M., and Sparks, D. L. (2004). "Formation of metal-arsenate precipitates at the goethite-water interface." *Environ. Sci. & Technol.*, 38, 6561-6570.

Gu, B., Liang L., Dickey, M. J., Yin, X., and Dai, S. (1998). "Reductive precipitation of uranium(VI) by zero-valent iron." *Environ. Sci. & Technol.*, 32, 3366-3373.

Gu, B., Phelps, T. J., Liang, L., Dickey, M. J., Roh, Y., Kinsall, B. L., Palumbo, A. V., and Jacobs, G. K. (1999). "Biogeochemical dynamics in zero-valent columns: Implications for permeable reactive barriers." *Environ. Sci. & Technol.*, 33, 2170-2177.

Gu, B., Watson, D. B., Wu, L., Philips, D. H., White, D. C., and Zhou, J. Z. (2002). "Microbiological characteristics of in a zero-valent iron reactive barrier." *Environ. Monit. Assess.*, 77, 293-309.

Hanson, H. C. B. (1989). "Composition, stabilization, and light absorption of Fe(II)Fe(III) hydroxy carbonate (green rust)." *Clay Mineral.*, 24, 663-669.

Hingston, F. J., Posner, A. M., and Quirk, J. P. (1972). "Anion adsorption by goethite and gibbsite. I. The role of proton in determining adsorption envelopes." *J. Soil Sci.*, 23, 177-192.

Hug, S. J., Canonica, L., Wegelin, M., Gechter, D., and von Gunten, U. (2001). "Solar oxidation and removal of arsenic at circumneutral pH in iron containing waters." *Environ. Sci. & Technol.*, 35, 2114-2121.

Hug, S. J., and Leupin, O. (2003). "Iron-catalyzed oxidation of arsenic(III) by oxygen and by hydrogen peroxide: pH-dependent formation of oxidants in the Fenton reaction." *Environ. Sci. & Technol.*, 37, 2734-2742.

Izumi, Y., Masih, D., Aika, K., and Seida, Y. (2005). "Characterization of intercalated iron(III) nanoparticles and oxidative adsorption of arsenite on them monitored by x-ray absorption fine structure combined with fluorescence spectrometry." *J. Phys. Chem. B*, 109, 3227-3232.

Jackson, B. P., and Miller, W. P. (2000). "Effectiveness of phosphate and hydroxide for desorption of arsenic and selenium species from iron oxides." *Soil Sci. Soc. Am. J.*, 64, 1616-1622.

Jain, A., and Loeppert, R. H. (2000). "Effect of competing anions on the adsorption of arsenate and arsenite by ferrihydrite." *J. Environ. Qual.*, 29, 1422-1430.

Johnson, T. E., Martens, W., Frost, R. L., Ding, Z., and Kloprogge, J. T. (2002). "Structured water in hydrotalcites of formula $Mg_xZn_{6-x}A_{12}(OH)_{16}CO_3 \bullet 4H_2O$: A Raman microscopic study." *J. Raman Spectrosc.*, 33, 604-609.

Joo, S. H., Feitz, A. J., and Waite, T. D. (2004). "Oxidative degradation of the carbothioate herbicide, molinate, using nanoscale zero-valent iron." *Environ. Sci. & Technol.*, 38, 2242-2247.

Kamolpornwijit, W., Liang, L., Moline, G. R., Hart, T., and West, O. R. (2004). "Identification and quantification of mineral precipitation in Fe^0 filings from a column study." *Environ. Sci. & Technol.*, 38, 5757-5765.

Kanel, S. R., Manning, B., Charlet, L., and Choi, H. (2005). "Removal of arsenic(III) from groundwater by nanoscale zero-valent iron." *Environ. Sci. & Technol.*, 39, 1291-1298.

Karschunke, K., and Jekel, M. (2002). "Arsenic removal by iron hydroxides, produced by enhanced corrosion of iron." *Water Sci. Technol: Water Supply*, 2, 237-245.

Khan, A. H., Rasul, S. B., Munir, A. K. M., Habibuddowla, M., Alauddin, M., Newaz, S. S., and Hussam, A. (2000). "Appraisal of a simple arsenic removal method for groundwater of Bangladesh." *J. Environ. Sci. Health*, A35, 1021-1041.

Kim, M., Nriagu, J., and Haack, S. (2000). "Carbonate ions and arsenic dissolution by groundwater." *Environ. Sci. & Technol.*, 34, 3094-3100.

Korte N. E., and Fernando Q. (1991). "A review of arsenic (III) in groundwater." *Crit. Rev. Environ. Contr.*, 21, 1-39.

Lackovic, J. A., Nikolaidis, N. P., and Dobbs, G. M. (2000). "Inorganic arsenic removal by zero-valent iron." *Environ. Engin. Sci.*, 17, 29-39.

Legrand, L., Maksoub, R., Sagon, G., Lecomte, S., Dallas, J. P., and Chaussé, A. (2003). "Electroanalytical and kinetic investigations on the carbonate green rust-Fe(III) redox system. *J. Electrochem. Soc.*, 150, B45-B51.

Liang, L., Korte, N., Gu, B., Puls, R., and Reeter, C. (2000). "Geochemical and microbial reactions affecting the long-term performance of in situ iron barriers." *Adv. Environ. Res.*, 4, 273–286.

Liang, L., Sullivan, A. B., West, O. R., Moline, G. R., and Kamolpornwijit, W. (2003). "Predicting the precipitation of mineral phases in permeable reactive barriers." *Environ. Eng. Sci.*, 20, 635-653.

Lien, H. L., and Wilkin, R. T. (2005). "High-level arsenite removal from groundwater by zero-valent iron." *Chemosphere*, 59, 377-386.

Ludwig, R. D., McGregor, R. G., Blowes, D. W., Benner, S. G., and Mountjoy, K. (2002). "A permeable reactive barrier for treatment of heavy metals." *Ground Water*, 40, 59-66.

Manceau, A. (1995). "The mechanism of anion adsorption on iron oxides: Evidence for the bonding of arsenate tetrahedral on free $Fe(O, OH)_6$ edges." *Geochim. Cosmochim. Acta*, 59, 3647-3653.

Manning, B. A., and Goldberg, S. (1996). "Modeling competitive adsorption of arsenate with phosphate and molybdate on oxide minerals." *Soil Sci. Soc. Am. J.*, 60, 121-131.

Manning, B. A., Fendorf, S. E., and Goldberg, S. (1998). "Surface structures and stability of As(III) on goethite: Spectroscopic evidence for inner-sphere complexes." *Environ. Sci. & Technol.*, 32, 2383-2388.

Manning, B. A., Hunt, M. L., Amrhein, C., and Yarmoff, J. A. (2002). "Arsenic(III) and arsenic(V) reactions with zerovalent iron corrosion products." *Environ. Sci. & Technol.*, 36, 5455-5461.

Martel, A. E., Smith, R. M., and Motekaitis, R. J. (2003). *NIST Critically Selected Stability Constants of Metal Complexes Database;* NIST Standard Reference Database 46, Version 7. NIST: Gaithersburg, MD.

Mayenkar, K. V., and Heights III, G. (1986). "Removal of dissolved heavy metals from aqueous waste effluents." United States Patent 4,565,633.

McArthur, J. M., Ravenscroft, P., Safiulla S., and Thirlwall M. F. (2001). "Arsenic in groundwater: Testing pollution mechanisms for sedimentary aquifers in Bangladesh." *Water Resou. Res.*, 37, 109-117.

McKeague, J. A., and Cline, M. G. (1963). "Silica in soil solutions. II. The adsorption of monosilicic acid by soil and by other substances." *Can. J. Soil Sci.*, 43, 83-96.

Melitas, N., Wang, J., Conklin, M., O'Day, P., and Farrell, J. (2002a). "Understanding soluble arsenate removal kinetics by zerovalent iron media." *Environ. Sci. & Technol.*, 36, 2074-2081.

Melitas, N., Conklin, M., and Farrell, J. (2002b). "Electrochemical study of arsenate and water reduction on iron media used for arsenic removal from portable water." *Environ. Sci. & Technol.*, 36, 3188-3193.

Melitas, N., Conklin, M., and Farrell, J. (2003). "Understanding the mechanisms controlling the kinetics of Arsenate and chromate removal from solution using zerovalent iron." *ACS Symp. Ser.*, 837, 165-180.

Meng, X., Korfiatis, G. P., Bang, S., and Bang, K. W. (2002). "Combined effects of anions on arsenic removal by iron hydroxides." *Toxicol. Lett.*, 133, 103-111.

Meng, X., Korfiatis, G. P., Christodoulatos, C., and Bang, S. (2001). "Treatment of arsenic in Bangladesh well water using a household co-precipitation and filtration system." *Wat. Res.*, 35, 2805-2810.

Morrison, S. J., Metzler, D. R., and Dwyer B. P. (2002). "Removal of As, Mn, Mo, Se, U, V, and Zn from groundwater by zero-valent iron in a passive treatment cell: Reaction progress modeling." *J. Contaminant Hydrol.*, 56, 99-116.

Myneni, S. C. B., Traina, S. J., Logan, T. J., and Waychunas, G. A. (1997). "Oxyanion behavior in alkaline environments: Sorption and desorption of arsenate in ettringite." *Environ. Sci. & Technol.*, 31, 1761-1768.

Myneni, S. C. B., Traina, S. J., Waychunas, G. A., and Logan, T. J. (1998). "Experimental and theoretical vibrational spectroscopic evaluation of arsenate coordination in aqueous solutions, solids, and at mineral-water interfaces." *Geochim. Cosmochim. Acta,* 62, 3285-3300.

Nickson, R. T., McArthur, J. M., Ravenscroft, P., Burgess, W. G., and Ahmed, K. M. (2000). "Mechanism of arsenic poisoning of groundwater in Bangladesh and West Bengal." *Appl. Geochem.*, 15, 403-413.

Nikolaidis, N. P., Dobbs, G. M., and Lackovic, J. A. (2003). "Arsenic removal by iron fillings: Field, laboratory and modeling studies." *Wat. Res.*, 37, 1417-1425.

Nikolaidis, N. P., Cheng, Z., and van Geen, A. (2005). "Removal of arsenic from Bangladesh Groundwater with Zero-valent Iron." *Amer. Chem. Soc. Symp. Ser.*, 915, in press.

Nordstrom, D. K. (2002). "Worldwide occurrences of arsenic in ground water." *Science*, 296, 2143-2145.

Odziemkowski, M. S., Schuhmacher, T. T., Gillham, R. W., and Reardon, E. J. (1998). "Mechanism of oxide film formation on iron in simulating groundwater solutions: Raman spectroscopic studies." *Corros. Sci.*, 40, 371-389.

Peryea, F. J., and Kammereck, R. (1997). "Phosphate enhanced movement of arsenic out of lead arsenate-contaminated topsoil and through uncontaminated subsoil." *Water, Air, and Soil Pollut'n.*, 93, 243-254.

Phillips, D. H., Gu, B., Watson, D. B., Roh, Y., Liang, L., and Lee, S. Y. (2000). "Performance evaluation of a zerovalent iron reactive barrier: Mineralogic characterization." *Environ. Sci. & Technol.*, 34, 4169-4176.

Phillips, D. H., Watson, D. B., Roh, Y., and Gu, B. (2003). "Mineralogical characteristics and transformation during long-term operation of a zerovalent iron reactive barrier." *J. Environ. Qual.*, 32, 2033-2045.

Powell, R. M., Puls, R. W., Hightower, S. K., and Sabatini, D. A. (1995). "Coupled iron corrosion and chromate reduction: Mechanisms for subsurface remediation." *Environ. Sci. & Technol.*, 29, 1913-1922.

Puls, R. W., Blowes, D. W, and Gillham, R. W. (1999). "Long term performance monitoring for a permeable reactive barrier at the US Coast Guard Support Center, Elizabeth City, North Carolina." *J. Hazard. Mater.*, 68, 109-124.

Ramaswami, A., Tawachsupa, S., and Isleyen, M. (2001). "Batch-mixed iron treatment of high arsenic waters." *Wat. Res.*, 35, 4474-4479.

Randall, S. R., Sherman, D. M., and Ragnarsdottir, K. V. (2001). "Sorption of As(V) on green rust ($Fe_4(II)Fe_2(III)(OH)_{12}SO_4 \cdot 3H_2O$) and lepidocrocite ($\gamma$-FeOOH): Surface complexes from EXAFS spectroscopy." *Geochim. Cosmochim. Acta,* 65, 1015-1023.

Redman, A. D., Macalady, D. L., and Ahmann, D. (2002). "Natural organic matter affects arsenic speciation and sorption onto hematite." *Environ. Sci. & Technol.*, 36, 2889-2896.

Roh, Y., Lee, S. Y., and Elless, M. P. (2000). "Characterization of corrosion products in the permeable reactive barriers." *Environ. Geol.*, 40, 184-194.

Ritter, K., Odziemkowski, M. S., Simpgraga, R., Gillham, R. W., and Irish, D. E. (2003). "An in situ study of the effect of nitrate on the reduction of trichloroethylene by granular iron." *J. Contamin. Hydrol.*, 65, 121-136.

Roberts, L. C., Hug, S. J., Ruettimann, T., Billah, M. M., Khan, A. W., and Rahman, M. T. (2004). "Arsenic removal with iron(II) and iron(III) in waters with high silicate and phosphate concentrations." *Environ. Sci. & Technol.*, 38, 307-315.

Sarkar, S., Gupta, A., Biswas, R. K., Deb, A. K., Greenleaf, J. E., and SenGupta, A. K. (2005). "Well-head arsenic removal units in remote villages of Indian subcontinent: Field results and performance evaluation." *Wat. Res.*, 39, 2196-2206.

Scherer, M. M., Richter, S., Valentine, R. L., and Alvarez, P. J. J. (2000). "Chemistry and microbiology of permeable reactive barriers for in situ groundwater clean up." *Crit. Rev. Environ. Sci. & Technol.*, 30, 363-411.

Shelp, G. S., Chesworth, W., and Spiers, G. (1995). "The amelioration of acid mine drainage by an *in situ* electrochemical method-I. Employing scrap iron as the sacrificial anode." *Applied Geochem.*, 10, 705-713.

Sherman, D. M., and Randall, S. R. (2003). "Surface complexation of arsenic(V) to iron(III) (hydr)oxides: Structural mechanism from ab initio molecular geometries and EXAFS spectroscopy." *Geochim. Cosmochim. Acta,* 67, 4223-4230.

Shokes, T. E., and Möller, G. (1999). "Removal of dissolved heavy metals from acid rock drainage using iron metal." *Environ. Sci. & Technol.*, 33, 282-287.

Simeoni, M. A., Batts, B. D., and McRae, C. (2003). "Effect of groundwater fulvic acid on the adsorption of arsenate by ferrihydrite and gibbsite." *Applied Geochem.*, 18, 1507-1555.

Sigg, L., and Stumm, W. (1980). "The interaction of anions and weak acids with the hydrous goethite (α-FeOOH) surface." *Collid. Surf.*, 2, 101-117.

Smedley, P. L, and Kinniburgh, D. G. (2002). "A review of the source, behavior and distribution of arsenic in natural waters." *Appl. Geochem.*, 17, 517-568.

Su, C., and Puls, R. W. (1999). "Kinetics of trichloroethene reduction by zerovalent iron and tin: Pretreatment effect, apparent activation energy, and intermediate products." *Environ. Sci. & Technol.*, 33, 163-168.

Su, C., and Puls, R. W. (2001a). "Arsenate and arsenite removal by zerovalent iron: Kinetics, redox transformation, and implications for in situ groundwater remediation." *Environ. Sci. & Technol.*, 35, 1487-1492.

Su, C., and Puls, R. W. (2001b). "Arsenate and arsenite removal by zerovalent iron: Effects of phosphate, silicate, carbonate, borate, sulfate, chromate, molybdate, and nitrate, relative to chloride." *Environ. Sci. & Technol.*, 35, 4562-4568.

Su, C., and Puls, R. W. (2003). "In situ remediation of arsenic in simulated groundwater using zerovalent iron: Laboratory column tests on combined effects of phosphate and silicate." *Environ. Sci. & Technol.*, 37, 2582-2587.

Su, C., and Puls, R. W. (2004a). "Nitrate reduction by zerovalent iron: Effects of formate, oxalate, citrate, chloride, sulfate, borate, and phosphate." *Environ. Sci. & Technol.*, 38, 2715-2720.

Su, C., and Puls, R. W. (2004b). "Significance of iron(II, III) hydroxycarbonate green rust in arsenic remediation using zerovalent iron in laboratory column tests." *Environ. Sci. & Technol.*, 38, 5224-5231.

Su, C., and Suarez, D. L. (1997). "In situ infrared speciation of adsorbed carbonate on aluminum and iron oxides." *Clays Clay Miner.*, 45, 814-825.

Su, C., and Wilkin, R. T. (2005). "Arsenate and arsenite sorption on and arsenite oxidation by iron(II, III) hydroxycarbonate green rust." *Amer. Chem. Soc. Symp. Ser.*, 915, in press.

Suarez, D. L., Goldberg, S., and Su, C. (1998). "Evaluation of oxyanion adsorption mechanisms on oxides using FTIR spectroscopy and electrophoretic mobility." *Amer. Chem. Soc. Symp. Ser.*, 715, 136-178.

Taylor, R. M. (1973). "Crystal structures of some double hydroxide minerals." *Mineral. Mag.*, 39, 377-389.

USEPA (2000). *Arsenic in drinking water rule: Economic analysis*, EPA/815/R-00/026, United States Environmental Protection Agency, Washington, DC, 2000.

Voegelin, A., and Hug, S. (2003). "Catalyzed oxidation of arsenic (III) by hydrogen peroxide on the surface of ferrihydrite: An situ ATR-FTIR study." *Environ. Sci. & Technol.*, 37, 972-978.

Vlassopoulos, D., Pochatila, J., Lundquist, A., Andrews, C. B., Rafferty, M. T., Chiang, K., Sprel, D., and Nikolaidis, N. (2002). "An elemental iron reactor for arsenic removal from groundwater." In *Proceedings of the 3rd International Conference on Remediation of Chlorinated and Recalcitrant Compounds*, Paper 2H-05.

Vempati, R. K., Loeppert, R. H., Dufner, D. C., and Cocke, D. L. (1990). "X-ray photoelectron spectroscopy as a tool to differentiate silicon-bonding state in amorphous iron oxides." *Soil Sci. Soc. Am. J.*, 54, 695-698.

Waychunas, G. A., Rea, B. A., Fuller, C. C., and Davis, J. J. (1993). "Surface chemistry of ferrihydrite. I. EXAFS studies of the geometry of coprecipitated and adsorbed arsenate." *Geochim. Cosmochim. Acta*, 57, 2251-2269.

Westerhoff, P., Highfield, D., Badruzzaman, M., and Yoon, Y. (2005). "Rapid small-scale column tests for arsenate removal in iron oxide packed bed columns" *J. Environ. Eng. ASCE*, 131, 262-271.

Wilkie, J. A., and Hering, J. G. (1996). "Adsorption of arsenic onto hydrous ferric oxide: Effects of adsorbate/adsorbent ratios and co-occurring solutes." *Colloid. Surf. A.*, 107, 97-110.

Wilkin, R. T., Puls, R. W., and Sewell, G. W. (2002). *Environmental research brief - long-term performance of permeable reactive barriers using zero-valent iron: An evaluation at two sites*, EPA/600/S-02/001, Office of Research and Development, United States Environmental Protection Agency, Washington, D. C. 20460, United States.

Wilkin, R. T., and McNeil, M. S. (2003). "Laboratory evaluation of zero-valent iron to treat water impacted by acid mine drainage." *Chemosphere*, 53, 715-725.

Wilkin, R. T., Puls, R. W., and Sewell, G. W. (2003). "Long-term performance of permeable reactive barriers using zero-valent iron: Geochemical and microbiological effects." *Ground Water*, 41, 493-503.

CHAPTER 9

Removal of Arsenic from Groundwater – Mechanisms, Kinetics, Field/Pilot and Modeling Studies

Nikolaos P. Nikolaidis, P.E.[1], and Konstantina Tyrovola[2]

Abstract: Arsenic presence in natural waters is associated to natural and anthropogenic mobilization processes. Due to its carcinogenic effects, the World Health Organization (WHO), the United States Environmental Protection Agency (USEPA) and the European Commission, have proposed a new guideline for arsenic in drinking water (10 μg/L). Many methods have been developed to remove arsenic from drinking water. A few of the most commonly used methods include anion exchange, activated alumina, and reverse osmosis. This chapter highlights the potential of using zero-valent iron (Fe^0) as filter media for the removal of arsenic from groundwater. Specifically, this chapter presents the mechanism of arsenic removal by Fe^0 and the effect of competing ions on the kinetics of the removal process. In addition, a comparison of the experimental conditions and arsenic removal efficiencies of several field/pilot studies as well as the approaches used to model the arsenic removal processes from aqueous solutions using Fe^0 are also presented.

CE Database subject headings: Adsorption; Arsenic; Kinetics; Iron; Models; Precipitation.

9.1 Introduction

Arsenic contamination in groundwater has gained great attention in recent years because of its toxicity to humans. Arsenic is a toxic metalloid that causes cancer (Azcue and Nriagu 1994; Pontius et al. 1994; Yamauchi and Fowler 1994). Its toxicity to humans depends on its concentration and the time of exposure. The early symptoms are various skin disorders and chronic arsenicosis can cause skin and liver cancer (Rahman 1999; Nriagu 1994a).

Arsenic occurrence in groundwater is attributed to natural and anthropogenic sources. There are four general categories of arsenic contaminated groundwater. These include:

- *Sedimentary Deposits* – River deltas are typical areas of high arsenic contamination. The most calamitous occurrence of arsenic contamination is that of Bangladesh (Dhar et al. 1997) and West Bengal (Chatterjee et al. 1995; Das et al. 1995; Mandal et al. 1996) where arsenic is derived from natural deposits in

---

[1]Associate Professor, Dept. of Environmental Engineering, Technical University of Crete, 73100 Chania, Greece (corresponding author). E-mail: nnikolai@mred.tuc.gr.

[2]Ph.D. Candidate, Dept. of Environmental Engineering, Technical University of Crete, 73100 Chania, Greece.

- the Ganges River delta (Dhar et al. 1997). Other places such as the Axios River delta in Greece have also been reported having arsenic contamination (Meladiotis et al. 2002). Arsenic mobilization in these reducing environments results in high concentrations in groundwater (Smedley and Kinniburgh 2002; Duker et al. 2005).
- *Volcanic Deposits, Geothermal Fields and Neotectonic active fault areas* – Arsenic contamination in geothermal systems have been reported in the western United States (Morton 1976; Feinglass 1976; Matisoff et al. 1982; Welch et al. 1988), Mexico (Del Razo et al. 1990), central America, Japan, New Zealand, Papua New Guinea, Chile, Philippines, Indonesia, Kamchatka, Alaska, Iceland, France, Hungury (Smedley and Kinniburgh 2002; USEPA 2000a; Welch et al. 2000; Webster and Nordstrom 2003) and Greece (Tyrovola et al. 2005). The most known case of arsenic contamination due to geothermal fields is the Yellowstone National Park in western USA (Webster and Nordstrom 2003). In geothermal systems, arsenic is discharged mainly as As(III) (Ballantyne and Moore 1988; Wilkie and Hering 1998; Hering 1997; Ball et al. 1998) and is derived from the dissolution of the host rock (Webster and Nordstrom 2003). As(III) is oxidized to As(V) during the ascension of geothermal hot waters and their mixing with shallow oxidized groundwater.
- *Regions that are near lacustrine or marine deposits* – These are regions where lake or marine deposits have created anoxic groundwater conditions that mobilize arsenic. Such places exist in New Hamsphire, Maine, and Oxio States in the USA (Zuena and Keane 1985; Schlottmann and Breit 1992; Marvinney et al. 1986; Boudette et al. 1985).
- *Industrial Contamination and Mine Runoff* – Arsenic has been used in medicine, the cosmetics industry, and agriculture (Azcue and Nriagu 1994; Mackie and Zdepski 1998) throughout the ages. It has also been used as insecticide, rodenticide, and herbicide. Industrial activities such as mining and smelting have caused arsenic contamination (Crecelius et al. 1974; Ragaini et al. 1977). Naturally occurring arsenic is found as a main component of several minerals such as arsenopyrite (Boyle and Jonasson 1973; Bhumbla and Keefer 1994), as well as in coal and coal combustion by-products (Kusnierova 1994; Germani and Zoller 1988).

Other areas affected by arsenic contamination are Argentina (Borgono et al. 1977), Taiwan (Chen et al. 1994), Northern China, Vietnam, Hungary, Romania, Mexico, and Chile (Smedley and Kinniburgh 2002).

The predominant species of arsenic in soils and natural waters are the inorganic arsenic of As(III) – arsenite (H_3AsO_3, $H_2AsO_3^-$, $HAsO_3^{2-}$), and As(V) – arsenate (H_3AsO_4, $H_2AsO_4^-$, $HAsO_4^{2-}$). The dominant arsenite species is neutral ($H_3AsO_3^0$) and the dominant arsenate species are the negatively charged anions ($H_2AsO_4^-$ and $HAsO_4^{2-}$) in the pH range of natural waters. Both inorganic species are subject to microbially mediated redox and methylation reactions and produce the organic arsenic species of monomethylarsonic acid (MMA) and dimethylarsinic acid (DMA) (Andreae and Klumpp 1979; Compeau and Bartha 1984; Freeman et al. 1986). The scientific literature contains a significant number of review papers on the chemistry, fate and transport of arsenic in the environment (Cullen and Reimer 1989; WHO

1981; Nriagu 1994b; Frankenberger 2002; Smedley and Kinniburgh 2002; Bissen and Frimmel 2003a; Welch and Stollenwerk 2003).

The World Health Organization (WHO), the United States Environmental Protection Agency (USEPA) and the European Commission, in order to minimize the health effects of arsenic, have proposed a new guideline for arsenic (10 μg/L). In USA, all public water supply systems have to comply with the drinking water standard for arsenic (10 μg/L) by 2006. The USEPA has estimated that 4,000 community water systems will have to take measures to lower the arsenic concentrations in the drinking water. The USEPA has estimated that communities will have to spend between 195 and 675 million dollars in order to comply to the new standard. The cost to each household will be between 58 and 327 dollars per year.

Many methods have been developed to remove arsenic from drinking water. A few of the most commonly used methods include anion exchange, activated alumina, and reverse osmosis (Frey et al. 1998). Ferric iron hydroxides have also been used to remove arsenic from drinking water systems (Davis and Kent 1990; Wilkie and Hering 1996; Hering et al. 1996; Elkhatib et al. 1984; Waychunas et al. 1993; Fuller et al 1993; Edwards 1994; Raven et al. 1998; Joshi and Chaudhuri 1996; Maeda et al. 1990; Huang and Vane 1989). These methods are typically effective for arsenate removal and require a pre-oxidation step for arsenite removal which makes them more expensive. A detailed comparison of the main available technologies for arsenic removal has been conducted by USEPA illustrating the availability and performance of the technologies (USEPA 2000b; USEPA 2001; USEPA 2002a; USEPA 2002b; USEPA 2003). During 2003 and 2004, the USEPA conducted demonstrations of low-cost treatment technologies. The first round of 12 demonstrations was carried out in 2003 and the second round of 29 demonstrations was in 2004. Case studies have been conducted in Scottsdale (Arizona), Tucson (Arizona), Paramount (California), and Faibanks (Alaska). Additional research papers have been published presenting the efficiency of various arsenic removal technologies (Daus et al. 2004; Bissen and Frimmel 2003b; Jiang 2001). One technology for arsenic removal from aqueous solution used zero-valent iron (Fe^0) (Lackovic et al. 1998; Nikolaidis et al. 2004; US Patent No. US 6,387,276 B1 "Immobilization of inorganic arsenic species using iron" - May 14, 2002). Fe^0 has already been shown to effectively remediate halogenated organic compounds (Johnson et al. 1996; Roberts et al. 1996; Chuang et al. 1995) and selected metals (Cantrell et al. 1995; Feidor et al. 1998; Powell et al. 1995; Blowes et al. 1997). The use of Fe^0 as a barrier for *in situ* remediation has been found to be very effective and arsenic removal represents a significant added benefit to the process (Blowes et al. 1997; Blowes et al. 2000).

There is an urgent need for an efficient, low maintenance and cost effective technology for arsenic removal that can be easily adaptable from household units to large-scale water treatment plants. This manuscript reviews the potential of using Fe^0 as filter media for the removal of arsenic from groundwater.

9.2 Mechanism of Removal and Competing Ion Effects

The mechanism of arsenic removal by Fe^0 has gained great attention in the scientific literature recently. Many research groups have focused their interest on the identification of the mechanism of arsenic removal by Fe^0, the exact products formed onto the surface of Fe^0 and the kinetics of arsenic removal. Different results were obtained depending on the experimental conditions and the overall reaction time.

The first step in the arsenic removal process by Fe^0 is the corrosion of iron and the formation of an electrochemical corrosion cell. When Fe^0 is immersed in an aqueous solution, electrons are released at local anodic sites (Stumm and Morgan 1981; Snoeyink and Jenkins 1980) and electrons are gained by oxidized species at local cathodic sites. Fe^0 can be oxidized to ferrous iron (Fe^{2+}) under aerobically conditions resulting in the consumption of oxygen in the solution:

$$2Fe^0 + 2H_2O + O_2 \rightarrow 2Fe^{2+} + 4OH^- \tag{9.1}$$

This reaction will utilize all oxygen. When all dissolved oxygen is consumed, the solution becomes anaerobic and iron oxidation is coupled with the hydrolysis of water (i.e., the production of hydrogen gas) and the reduction of sulphate in solution. (Lackovic et al. 2000; Nikolaidis et al. 2003). The main processes can be described by the following reactions:

Fe^0	=	$Fe^{2+} + 2e^-$	iron oxidation
Fe^{2+}	=	$Fe^{3+} + e^-$	iron oxidation
$8e^- + 9H^+ + SO_4^{2-}$	=	$HS^- + 4H_2O$	sulphate reduction
$2e^- + 2H^+$	=	$H_2(g)$	hydrolysis of water

Studies on the removal mechanisms of As(III) and As(V) by Fe^0 show that arsenic is removed due to adsorption and /or precipitation /coprecipitation with Fe(II) and Fe(III) oxides/hydroxides which are formed onto the surface of Fe^0 during its oxidation (Farrell et al. 2001; Lackovic et al. 2000; Nikolaidis et al. 2003; Melitas et. al. 2002a; Su and Puls 2001a; Manning et al. 2002; Su and Puls 2004a; Lien and Wilkin 2005; Kanel et al. 2005). Farrell et al. (2001) using Tafel analysis, indicated that arsenate did not directly oxidize Fe^0, instead water was the primary oxidant for its corrosion. In addition, it was suggested that in anaerobic solutions, Fe(II) was the initial oxidation product which reacted with hydroxide (OH^-) to form ferrous hydroxides [$Fe(OH)_2$]. After a period of hours to days, ferrous hydroxide was converted to magnetite (Fe_3O_4) or maghemite (γ-Fe_2O_3) and ferric hydroxide [$Fe(OH)_3$] (Farrell et al. 2001; Kanel et al. 2005). Similar corrosion products were identified with Fe^0 (Manning et al. 2002, Melitas et al. 2002a) and with nanoscale-Fe^0 (Kanel et al. 2005). These studies reveal that the corrosion products are a mixture of amorphous Fe(III) oxide/hydroxide, magnetite and/or maghemite and lepidocrocite (γ-FeOOH). The formation of various corrosion products on the surface of Fe^0 results in the creation of adsorption sites for both As(III) and As(V). The suggested mechanism of arsenic removal was the formation of inner-sphere bidentate As(III) and As(V) complexes on iron corrosion products (Manning et al. 2002,

Farrell et al. 2001). Arsenate complexation with iron (III) oxyhydroxides was consistent with the kinetic model proposed by Farrell et al. (2001) where arsenate removal kinetics in batch reactors was described by a dual-rate model according to the equation:

$$\frac{dC}{dt} = \frac{-k_0 C}{k_0 / k_1 + C} \tag{9.2}$$

where C is the arsenate concentration in solution, t is the time, k_0 is the zero-order rate constant and k_1 is the first-order rate constant. The reaction rates were first-order at low concentrations and zero-order for high concentrations. Other studies emphasized the importance of the formation of Fe(II)-Fe(III) hydroxycarbonate green rust (or carbonate green rust) on the surface of Fe^0 and its role on the arsenic removal mechanism (Su and Puls 2004a; Lien and Wilkin 2005). According to Lien and Wilkin (2005), the mechanism of arsenic removal by green rust is not certain. Arsenite removal due to green rust formation was related either to adsorption onto green rust or arsenite could replace carbonate in the interlayer position of green rust. Other studies (Lackovic et al. 2000; Nikolaidis et al. 2003) pointed out that surface precipitation is a predominant removal process. Nikolaidis et al. (2003) found a sulfur accumulation gradient in the spent-material indicating that sulphate was reduced and precipitated in the column. Spectroscopic results and laboratory leaching studies suggested that the arsenic surface precipitate was related to sulfur (possible formation of arsenopyrite) which was formed possibly due to sulphate reduction. They reported a strong sulfur smell when the columns were opened indicating sulfide formation. Sulphate reduction can also occur due to the presence of sulphate reducing bacteria but further studies should be conducted in order to elucidate the nature (biotic versus abiotic) of the mechanism. The Nikolaidis et al. (2003) study suggests that if a column is operated extended periods (months), the aging process makes the precipitation process irreversible. The media became more reactive and their surface area increased up to 40 times (the initial surface area of iron filings was about 1 m^2/g, and the final surface area increased up to 37.8 m^2/g). A schematic representation of the main mechanism of arsenic removal is presented in Fig. 9-1.

The reduction of arsenic in Fe^0 filters highly depends on the experimental and field conditions and different researchers have reported different results. Some researchers (Farrell et al. 2001; Melitas et al. 2002b; Lackovic et al. 2000) reported no or minimal reduction of arsenate to arsenite under potable water treatment conditions. Specifically, the work of Melitas et al. (2002b) showed that the electrochemical potential required for the As(V) reduction is lower than the potential produced at the corroding surface of Fe^0 in aqueous solution. Batch experiments by Su and Puls (2001a), with four different Fe^0 products (*Fisher*, *Peerless*, *Master Builders*, and *Aldrich*) showed that arsenic removal was affected by the reaction time, the iron type and to a lesser degree by the initial As species. X-ray Photoelectron Spectroscopy (XPS) on *Peerless* iron showed no reduction of As(V) to As(III) after 5 days. Partial reduction of As(V) to As(III) occurred on *Peerless* iron at 30 and 60 days, when As(V) was the initial arsenic species in the solution. The reduction of As(V) probably occurred by Fe^0 and ferrous iron produced on *Peerless* iron. An XPS study by Bang et al. (2005) showed that reduction of As(III) to As(0) by Fe^0 can

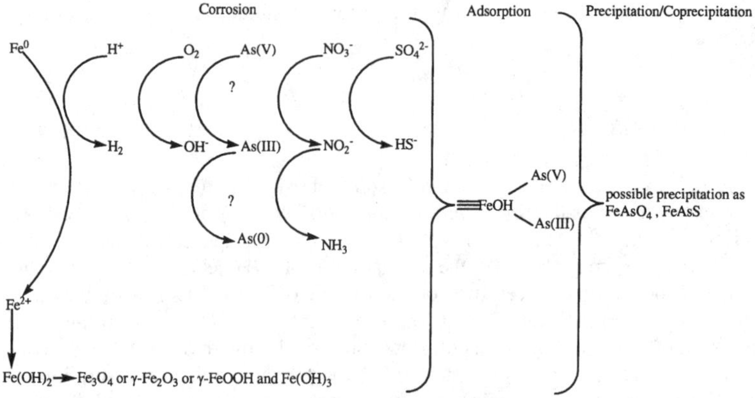

Fig. 9-1. Main mechanism of arsenic removal by Fe^0

occur under anoxic conditions within 5 day and soluble As(III) was formed when As(V) reacted with Fe^0.

Yet the conflict upon redox chemistry of arsenic on Fe^0 becomes more complicated because some research groups did not report reduction of arsenate, but oxidation of arsenite to arsenate. Even the same work described above by Su and Puls (2001a) reported oxidation of sorbed As(III) to As(V) which was accomplished when As(III) was the initial species in the solution. Su and Puls (2001a) suggested that the As(III) oxidation was probably accomplished by the presence of MnO_2 in Peerless iron. Oxidation of As(III) to As(V) and no reduction of As(V) by Fe^0 were also observed by other studies (Manning et al. 2002; Su and Puls 2004a; Lien and Wilkin 2005; Kanel et al. 2005). According to Manning et al. (2002) As(III) oxidation, under aerobic conditions, was caused by maghemite and hematite minerals formed during Fe^0 oxidation. In addition, As(III) oxidation is pH dependent since in pH values greater than the first pK_a for As(III) (i.e., 9.2), As(III) can be auto-oxidized (Manning and Goldberg 1997). High pH values were achieved due to the reduction of water molecules on the surface of corroding iron (Manning et al. 2002). XPS and combined microscopic and macroscopic wet chemistry indicated that oxidation of sorbed As(III) occurred at the surface of corroded iron where carbonate green rust was freshly accumulated (Su and Puls 2004a).

Even though Fe^0 can effectively remove both As(III) and As(V), removal rates of As(III) and As(V) differ from study to study. Batch experiments by Su and Puls (2001a), with four different iron filings (*Fisher*, *Peerless*, *Master Builders*, and *Aldrich*) showed that As(III) removal rates were greater than As(V) removal rates in all experiments. Batch studies (Bang et al. 2005) indicated that removal rates of As(III) and As(V) depended on the experimental conditions. Specifically, As(III) removal rate was faster than As(V) removal rate at pH 4-7 under anoxic conditions, whereas As(III) removal rate was slower than As(V) at pH 7 under oxic conditions. Laboratory column studies revealed greater removal rates for As(V) than As(III) (Lackovic et al. 2000). Additional batch experiments at two different temperatures (20 and 40 ^{0}C) demonstrated that As(V) removal rate by Fe^0 was greater than As(III) removal rate under the same conditions in both temperatures (Tyrovola et al. 2005). Temperature played a dominant role on the kinetics of arsenic removal. An increase in temperature resulted in an increase of the pseudo first-order rate constant of both As(III) and As(V). Specifically, for a 20 ^{0}C increase in temperature, the pseudo-first-order rate constant increased by 411% for As(III) and by 558% for As(V) rendering temperature as the dominant design parameter for water treatment units (Tyrovola et al. 2005).

The effect of competing ions on the kinetics of arsenic removal by Fe^0 was tested in batch and laboratory experiments (Su and Puls 2001b; Su and Puls 2003; Tyrovola et al. 2005). Su and Puls (2001b) evaluated the effect of phosphate, silicate, carbonate, borate, sulfate, chromate, molybdate and nitrate relative to chloride on arsenic removal by Fe^0. Phosphate caused the greatest decrease in arsenic removal, relative to chloride. Moreover, silicate, chromate, and molybdate caused strong inhibition of arsenic removal, followed by carbonate and nitrate. Borate and sulfate slightly inhibited As(III) removal. Other column studies (Su and Puls 2003) demonstrated the strong inhibition effect of phosphate and silicate on As(III) and As(V) removal. The presence of combined phosphate and silicate caused earlier breakthrough of arsenic. Consistent with earlier results (Su and Puls 2001b), the study of Tyrovola et al. (2005) showed that the presence of phosphate and nitrate decreased the removal rates of As(III) and As(V) at 20 and 40 ^{0}C. Both phosphate and nitrate ions were removed by Fe^0. Nitrate reduction was followed by the formation of nitrite and ammonia. Nitrate reduction by Fe^0 was also observed by other researchers in batch and column experiments. Nitrate reduction by Fe^0 can be accomplished by two different mechanisms. The first proposed mechanism involves direct reduction of nitrate by Fe^0 whereas the second mechanism involves indirect reduction of nitrate by hydrogen (generated by iron corrosion) (Huang and Zhang 2004; Choe et al. 2000; Huang et al. 1998; Westrhoff and James 2003; Su and Puls 2004b; Cheng et al. 1997). However, the effect of nitrate reduction on arsenic removal is not thoroughly clear. The study from Westerhoff and James (2003) suggested that nitrate could be adsorbed onto positively charged surfaces formed on the surface of Fe^0. Thus nitrate can possibly compete for reaction sites with arsenic.

9.3 Field/Pilot Studies and Modeling

A large number of field/pilot studies have been conducted to assess the efficacy of Fe^0 in removing arsenic from contaminated ground water (Lackovic et al. 2000; Nikolaidis et al. 2003; Nikolaidis et al. 2005; Tyrovola et al. 2005; Vlassopoulos et al. 2002; Meladiotis et al. 2002; Su and Puls 2003; Morrison et al. 2002; Lien and Wilkin 2005; Munir et al. 2001). Table 9-1 presents a comparison of the experimental conditions and the arsenic removal efficiencies of these studies.

The field pilot studies can be categorized according to the general categories of arsenic contaminated groundwater. Lackovic et al. (2000), Nikolaidis et al. (2003), Vlassopoulos et al. (2002), Su and Puls (2003), Morrison et al. (2002), and Lien and Wilkin (2005) have conducted pilot studies to deal with industrial contamination of mine runoff. Lackovic et al. (2000) has reported the first field pilot study on arsenic

Table 9-1. Comparison of Pilot Studies for Arsenic Removal Efficiency with Fe^0

Test Type and Field Location	Operating Parameters	Influent Arsenic Concentration	Treated PV (Duration)	Maximum Removal Efficiency	Reference
1. Field – Landfill Leachate (Maine, USA)	Flow rate: 12L/day Detention time: 110 - 150 min Filter volume: 2.4 L	300 μg/L	760 (2 months)	99.7% 1150 μg-As/g-Fe^0 (Fe^0 -*Connelly*) 470 μg-As/g-Fe^0 (Fe^0-J.T.Baker) Fe^0/sand ratio: 1:1 (w:w)	Lackovic et al. (2000)
2. Field – Landfill Leachate (Maine, USA)	Flow rate: 5444 L/d Detention time: 22 min Filter volume: 185.4 L	294 μg/L	48000 (20 months)	99.7% 4471 μg-As/g-Fe^0 (Fe^0 –*Connelly*) Fe^0/sand ratio: 1:1 (w:w)	Nikolaidis et al. (2003)
3. Field – Industrial Contamination (New Jersey, USA)	Flow rate: 9800 L/day Detention time: 100 min Filter volume: 1800 L	2800 μg/L	1500 (4 month)	91-99.8 % (Fe^0 –*Connelly*) Fe^0/sand ratio: 1:1 (w:w)	Vlassopoulos et al. (2002)
4. Field – Uranium Mill Tailings (Colorado, USA)	Flow rate: 445 L/day Detention time: 4200 min Filter volume: 1800 L	186 μg/L	380 (24 months)	99% (*Peerless* Fe^0) Fe^0: granular	Morrison et al. (2002)
5. Laboratory – Simulated Groundwater (US Coast Guard Support Center - North Carolina, USA)	Flow Rate: 0.67 L/day. Detention time: 215 min Filter volume: 0.1 L	2000 μg/L	1900 (4months)	99 % 357 μg-As/g-Fe^0 (*Peerless*- Fe^0) Fe^0/sand ratio: 1:1 (w:w)	Su and Puls (2003)
6. Laboratory – Simulated Groundwater (Smelting Facility-Montana, USA)	Flow rate: 0.4 L/d Detention time: 50 min Filter volume: 0.15 L	50000 μg/L	4500 (6.6 months)	7500 μg-As/g-Fe^0 100% Fe^0 *Peerless*	Lien and Wilkin (2005)
7. Field – Groundwater Drinking Water Supply Well (Chalastra, Greece)	Flow rate: 160 L/day Detention time: 7 min Filter volume: 5.4 L	160 μg/L	(0.7 month)	99 % (Fe^0 -*Connelly*) Fe^0/sand ratio: 1:1 (v:v)	Meladiotis et al. (2002)
8. Field – Groundwater Drinking Water Supply Well (Bangladesh)	Flow rate: 216 L/day Detention time: 9 min Filter volume: 5.4 L	560 μg/L	750 (8 days)	99 % (Fe^0 -*Connelly*) Fe^0/sand ratio: 1:1 (v:v)	Nikolaidis et al. (2005)
9. Field – Groundwater Drinking Water Supply Well (Geothermal Fields of Triglia, Greece)	Flow rate: 144 L/day Detention time: 6 min Filter volume: 5.4 L	137 μg/L	(1 month)	99 % (Fe^0 -*Connelly*) Fe^0/sand ratio: 1:1 (v:v)	Tyrovola et al. (2005)
10. Field – Industrial Inorganic and Methylated Arsenic Contamination (Superfund site, New Jersey, USA)	Flow rate: 144 L/day Detention time: 6 min Filter volume: 2.7 L	1000-1500 μg/L (30% organic As)	(4 month)	90 % (Fe^0 -*Connelly*) Fe^0/sand ratio: 1:1 (v:v)	Cheng et al. (2005)

removal by Fe^0. The study was conducted at a closed industrial landfill in Maine, USA. The groundwater was reducing and had a trivalent arsenic concentration of 300 µg/L. Two types of Fe^0 (*Connelly-GPM* and *JT Baker*) were tested for two months. The maximum removal efficiency was 1150 µg-As/g-Fe^0 (for Fe^0-*Connelly-GPM*) and 470 µg-As/g-Fe^0 (for Fe^0-*J.T.Baker*), and the effluent As concentration was less than 1 µg/L. The hydraulic retention time (HRT) of the water in the filter was 110 – 150 min. The column with *JT Baker* Fe^0 had an earlier breakthrough than that with *Connelly-GPM* Fe^0 and its effluent As concentration at initial breakthrough was less than the detection limit. This was attributed to the variability of initial reactivity of the two Fe^0. At the same location, a larger pilot study was conducted by Nikolaidis et al. (2003). The treated flow of the system was 5444 L/day, the HRT was 22 min and the maximum accumulation was 4471 µg-As/g-Fe^0. The pilot was run for a total of 20 months or 48000 pore volumes (PV). Breakthrough occurred at 19700 PV (8 months). This study illustrated that Fe^0 is potentially a viable treatment technology for arsenic removal. It removed arsenic to levels below the recommended drinking water standards and required no maintenance or a pre-oxidation step that is necessary for most of the other adsorption based methods. On the other hand, the Fe^0 filters exported iron making it necessary to be followed by an iron removal unit.

Vlassopoulos et al. (2002) constructed an *in situ* flow-through reactor for arsenic contaminated groundwater from a former pesticide manufacturing facility in New Jersey, USA. The experimental conditions were: average flow of 9800 L/day, HRT of 100 min, reactor volume of 1800 L and influent arsenic concentration of 2.8 mg/L. The experiment was operated for 1500 PV (about 4 months) and the efficiency of removal was 91-98%, which satisfied the criteria of the industrial groundwater treatment. The inflow groundwater was oxygenated and neutral in pH with low dissolved iron concentrations. Morrison et al. (2002) conducted a similar experiment in which three *in situ* passive treatment cells were created at a uranium mill tailings repository site in Colorado, USA. The objective was to evaluate the efficacy of three types of Fe^0 (foamed plates, steel wool and granular iron) on treating metals remained in the tailings. All types of Fe^0 were shown decreasing the concentrations of arsenic by more than 98%.

Two other studies have used simulated groundwater to evaluate the remedial performance of *Peerless* Fe^0 for arsenic removal at two Superfund sites in USA. Su and Puls (2003) conducted a laboratory pilot study to test the behavior of Fe^0 permeable reactive barrier (PRB) in removing arsenic by simulating the conditions of groundwater at the US Coast Guard Support Center in Elizabeth City, North Carolina, USA. The experimental conditions were: average groundwater flow of 0.67 L/day, HRT of 215 min, reactor volume of 0.1 L and influent arsenic concentration of 2.0 mg/L. The experiment was run for 1900 PV, the removal efficiency was 99%, and the average accumulation of arsenic on Fe^0 was 357 µg-As/g-Fe^0. Similarly, Lien and Wilkin (2005) conducted a column laboratory study by simulating the contaminated groundwater at a Superfund site in Helena, Montana, USA. The average flow of the experiment was 0.4 L/day, the HRT was 50 min, and the volume of the reactor was 0.15 L. The influent arsenic concentration was 50 mg/L, the experiment was run for 4500 PV and the maximum arsenic accumulation on the Fe^0 was 7500 µg-As/g-Fe^0. The authors also observed that there was a two-step reaction for the removal of arsenic: an initial fast reaction followed by a slow process. Nikolaidis et al. (2003) have also reported this slow process.

Small-scale pilot studies were conducted at two locations in Greece and one in Bangladesh to evaluate the efficacy of Fe^0 on removing arsenic for drinking water purposes. The experimental setup of these studies was comprised of 5.4 L Fe^0-sand media reactors, aeration vessels and iron removal filter units. Their flow rates and HRTs were similar, while their influent arsenic concentration was different. The origins of the contaminations were the Axios River sediments near the town of Chalastra, Greece (Meladiotis et al. 2002), the delta of the rivers in Bangladesh (Nikolaidis et al. 2005) and the geothermal field of the town of Triglia, Greece (Tyrovola et al. 2005). Later on, Cheng et al. (unpublished manuscript, 2005) conducted a similar field study to illustrate the removal of methylated arsenic species by Fe^0 at a Superfund site in New Jersey, USA. These studies have illustrated the ability of Fe^0 to remove both inorganic and methylated organic arsenic under different geochemical conditions and hydrogeochemical environments.

The above pilot studies have illustrated the applicability of using Fe^0 as the media for the removal of arsenic from

- industrial origin (i.e., Morrison et al. 2002),
- sedimentary deposits (i.e., Nikolaidis et al. 2005),
- geothermal fields (i.e., Tyrovola et al. 2005) and
- regions near lacustrine or marine deposits (i.e., Nikolaidis et al. 2003).

Fig. 9-2 presents the Fe/As molar ratio against the phosphate concentration in the groundwater conditions in field sites in Bangladesh, USA and Greece. Phosphate is an interfering anion with respect to many arsenic removal technologies. Even though phosphate interferes in the removal process of arsenic by Fe^0, the degree of interference is not sufficient to render the technology to be impracticable in high phosphate regions. The Fe^0 technology has been tested successfully against a wide range of Fe/As molar ratio and phosphate concentration as it is shown in Fig. 9-2.

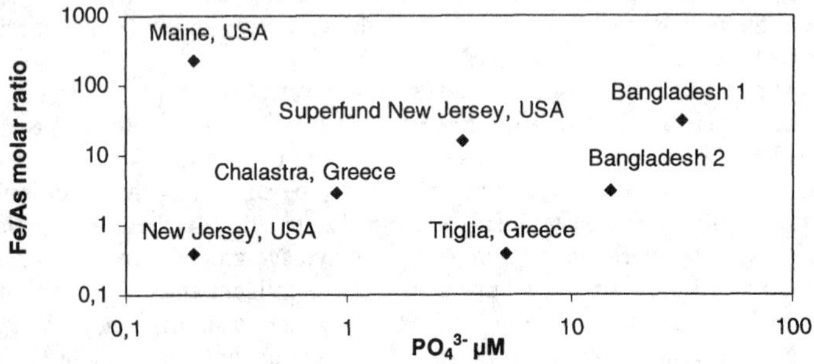

Fig. 9-2. Fe/As ratio versus PO_4^{3-} in the influent of the several pilot studies

Studies of removal of arsenic from acid rock drainage using Fe^0 have been conducted by Shokes and Moller (1999), and Wilkin and McNeil (2003), while Blowes et al. (2000) has proposed Fe^0-PRB for the removal of inorganics.

Three types of modeling approaches have been used in modeling the process of arsenic removal from aqueous solutions using Fe^0 or iron oxyhydroxides. One modeling approach is the 1-D analytical solution model (Nikolaidis et al. 2003; Lien and Wilkin 2005), the second approach is equilibrium modeling using the Double Layer Model (Cheng et al. 2004; Dixit and Hering 2003) or the Triple Layer Model (Goldberg and Johnston 2000) and the third approach is the reaction progress modeling (Morrison et al. 2002). Each approach is described briefly below.

Nikolaidis and coworkers used a 1-D analytical solution of the advective/dispersive equation with reactions to model the results from pilot experiments conducted in Maine, USA (Nikolaidis et al. 2003). They conceptualized two reactions as the main arsenic retardation mechanisms that occurred in the filters: surface adsorption and precipitation on iron and sand surfaces. Surface adsorption was modeled as a fast reaction (equilibrium) using a partitioning coefficient, K_d. Arsenic precipitation and co-precipitation in solution or diffusion through the iron oxides (corrosion products) and surface precipitation were modeled as a first-order loss reaction. The solution of the 1-D advective/dispersive equation with linear partitioning and a first-order loss term equation was shown as follows:

$$C = \frac{C_0}{2}\exp\frac{(U_x - V)X}{2D_x}\,erfc\left[\frac{RX - Vt}{2\sqrt{D_x Rt}}\right] + \frac{C_0}{2}\exp\frac{(U_x + V)X}{2D_x}\,erfc\left[\frac{RX + Vt}{2\sqrt{D_x Rt}}\right] \quad (9.3)$$

where C is arsenic concentration, C_0 is the initial concentration of arsenic, U_x is the velocity of water, $V = U_x\sqrt{(1 + 4kD_x / U_x^2)}$, K_d is linear equilibrium partitioning coefficient, ρ_s is the solid density of the particles, n is the effective porosity, k is the first-order mass transfer loss coefficient, D_x is dispersion coefficient in x direction, $D_x = \alpha\ U_x$, α is the apparent longitudinal dynamic dispersivity and L is the length of the flow path (calculated from $\alpha = 0.0175L^{1.46}$), R is retardation factor, $R = 1 + K_d\rho_s(1-n)/n$, x is the column length and t is the time of simulation. The linear equilibrium-partitioning coefficient, K_d, and the first-order mass transfer loss coefficient, k, were the two calibrated parameters. The model was used to simulate the breakthrough curves from three pilot columns. The influent concentrations, the experimental breakthrough curves and the simulation results for the two columns are presented in Fig. 9-3. Calibration of model parameters revealed that the equilibrium-partitioning coefficient was the same (4300 ml/g) for all three columns, while the first-order mass transfer loss coefficient was a function of the flow rate (1.08 to 1.44 hr^{-1}). The calibrated mass transfer loss coefficients were similar to those reported in the literature if they are normalized to the surface area of the media resulting in a half-life of 11 to 18 hours.

Lien and Wilkin (2005) used a similar approach to model flow-through columns packed with Fe^0 and for removing arsenic from simulated groundwater. The authors used the CFITIM program to curve fit the breakthrough data from the columns. The program uses 1-D analytical solution to solve non-equilibrium advective/dispersive

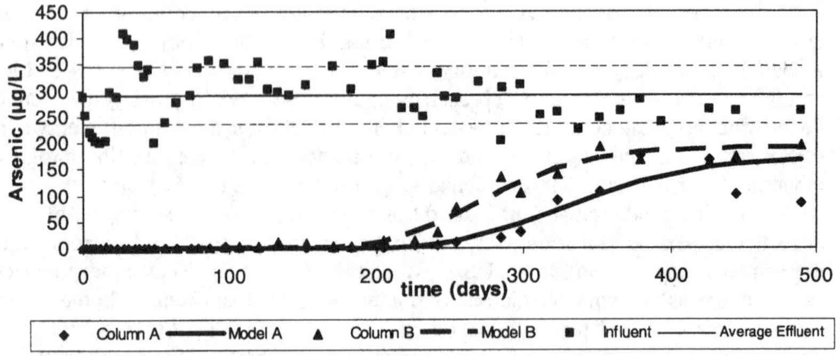

Fig. 9-3. Experimental and simulation data for the pilot columns

equation with arsenic uptake. The column capacity for removal of arsenic was estimated and the results were compared to other studies.

Cheng and coworkers used the windows version of MINTEQA2, Visual MINTEQ to model the results of a pilot experiment in Bangladesh (Cheng et al. 2004). The experiment used a household level arsenic removal system that was comprised of two buckets. The first bucket contained 20 L of arsenic-contaminated water and the reagents (ferric sulfate and calcium hypochlorite). The reagents were mixed with the water and were allowed to react for 5-10 min. Then the water containing iron oxyhydroxide flocs was poured into the next bucket which contained an 8 cm layer of sand. The sand retained the iron hydroxides that had adsorbed arsenic. The adsorption of arsenic on the iron hydroxide flocs was modeled using the Double Layer Model. Surface complexation model was used to predict the residual concentration for individual wells using the actual well water chemistry. The predictions were comparable to the observations. A similar modeling approach was used by Dixit and Hering (2003) with iron oxide minerals. The reactions and rate constants used in these modeling exercises are presented in Table 9-2. It is important to mention this modeling approach since sorption of arsenic onto the iron oxides produced from the corrosion of Fe^0 is a significant removal mechanism.

The last modeling approach was the reaction progress modeling used by Morrison et al. (2002). This approach was based on the concept of irreversible dissolution of mineral phases that were used to drive the chemical changes in a system. The geochemical speciation code, PHREEQC (Parkhurst 1995) was used in the modeling exercise. The approach was used to model arsenic removal from groundwater by Fe^0 in a series of treatment cells operated at a site near Durango, CO. In the modeling exercise, Fe^0 was allowed to dissolve slowly, while all other homogeneous and heterogeneous reactions were reacting fast. Equilibrium was assumed at each increment of iron dissolution. The modeling approach was useful in predicting the longevity of treatment cells.

Table 9-2. Reactions and the Corresponding Equilibrium Constants (data adapted from Cheng et al. 2004)

Reactions	
Surface Complexation Reactions	logK
(1) $SOH + H^+ + \exp(-F\psi_0/RT) = SOH_2^+$	7.29
(2) $SOH - H^+ - \exp(-F\psi_0/RT) = SO-$	-8.93
(3) $SOH + AsO_4^{3-} + 3H^+ = SH_2AsO_4 + H_2O$	29.31
(4) $SOH + AsO_4^{3-} + 2H^+ - \exp(-F\psi_0/RT) = SHAsO_4^- + H_2O$	23.51
(5) $SOH + AsO_4^{3-} - 3\exp(-F\psi_0/RT) = SOHAsO_4^{3-} + H_2O$	10.58
(6) $SOH + H_3AsO_3 = SH_2AsO_3 + H_2O$	5.41
(7) $SOH + PO_4^{3-} + 3H^+ = SH_2PO_4 + H_2O$	31.29
(8) $SOH + PO_4^{3-} + 2H^+ - \exp(-F\psi_0/RT) = SHPO_4^- + H_2O$	25.39
(9) $SOH + PO_4^{3-} + H^+ - 2\exp(-F\psi_0/RT) = SPO_4^{2-} + H_2O$	17.72
(10) $SOH + CO_3^{2-} + H^+ - \exp(-F\psi_0/RT) = SCO_3^- + H_2O$	12.78
(11) $SOH + CO_3^{2-} + 2H^+ = SCO_3H + H_2O$	20.37
(12) $SOH + H_4SiO_4 - 2H^+ - 2\exp(-F\psi_0/RT) = SOSiO_2OH^{2-} + H_2O$	-11.69
(13) $SOH + H_4SiO_4 - H^+ - \exp(-F\psi_0/RT) = SOSiO(OH)_2^- + H_2O$	-3.22
(14) $SOH + H_4SiO_4 = SOSi(OH) + H_2O$	4.28
Aqueous Reactions	logK
(15) $H_3AsO_4 = H_2AsO_4^- + H^+$	-2.24
(16) $H_3AsO_4 = HAsO_4^{2-} + 2H^+$	-9.20
(17) $H_3AsO_4 = AsO_4^{3-} + 3H^+$	-20.70
(18) $H_3AsO_3 = H_2AsO_3^- + H^+$	-9.29

9.4 Design Considerations

Fe^0 filters have been shown to remove effectively both inorganic and organic arsenic compounds. Field experiments have illustrated the ability of Fe^0 filters to operate under different geochemical environments in various parts of the world. The different types of Fe^0 have different performances and they should be examined prior to final full-scale deployment. In a few cases, there was an initial arsenic breakthrough followed by a complete arsenic removal (Lackovic et al. 2000; Nikolaidis et al. 2005). One disadvantage of the technology is the fact that the filters leach significant quantities of iron requiring therefore, the Fe^0 filter to be followed by an iron removal treatment unit if the water will be used for drinking water purposes. The following parameters should be considered in the design of Fe^0 filters for arsenic removal.

1. **Hydraulic Retention Time** – HRT is a critical design parameter of the Fe^0 filters. In the field experiments, the range of HRT used varied between 6 and 4200 min. It was shown in many experiments that a HRT less than 10 min is probably sufficient to remove inorganic arsenic to levels less than the drinking water standards. It has been shown in the literature that filters can be designed on a volumetric loading basis and not on a surface-loading basis.
2. **Degree of Corrossiveness** – Another important design parameters is the geochemistry of the water to be treated and its capacity to corrode iron. In the case of the field experiment in Chalastra, Greece (Meladiotis et al. 2002), the geochemistry of the water was not corrosive to iron (due to high alkalinity) and

that made the filter inoperable. To remedy the situation, the pH of the influent water had to be lowered to levels that made it corrosive. Empirically, it was found that if the water is supersaturated with respect to calcium carbonate (i.e., Langlier Index to be positive), then the water will not corrode iron. To activate the corrosiveness of the water, the pH had to lowered to a level where the Langlier Index is -0.5 (or half a pH unit below calcium carbonate equilibrium).

3. **Media Longevity** – A few long-term field studies showed that Fe^0 filters can remove arsenic for long periods of time (time to breakthrough is greater than 8 months) without requiring any maintenance. The highest arsenic accumulation measured in the field studies was 7500 μg-As/g-media. The longevity of the column and the fact that the system requires no maintenance (i.e., backwash) makes iron filings as a viable arsenic treatment alternative.
4. **Media Disposal** – Nikolaidis et al. (2003) analyzed the spent media for TCLP. The results indicated that the arsenic concentration in the leachate was two orders of magnitude lower than the 5 mg/L of TCLP for arsenic. These results suggest that the spent filter media are not hazardous and they can be disposed in regular landfills.

9.5 Conclusions

While the detailed mechanisms of arsenic removal by Fe^0 are not fully understood, the operative parameters are characterized sufficiently to permit design of removal units for groundwater remediation at industrial sites or as a treatment technology for drinking water. The technology is low maintenance and cost that can be used to treat water from small units to large treatment plants. Fe^0 is a promising technology that has the potential to solve, in a cost effective way, the arsenic problem around the world.

9.6 References

Andreae, M. O., and Klumpp, D. (1979). "Biosynthesis and release of organoarsenic compounds by marine algae." *Environ. Sci. & Technol.*, 13(6), 738-741.

Azcue, J. M., and Nriagu, J. O. (1994). "Arsenic: historical perspectives." In: *Arsenic in the Environment, Part I: Cycling and Characterization*. John Wiley & Sons, New York, 1-15.

Ball, J. W., Nordstrom, D. K., Jenne, E., and Vivit, D. V. (1998). "Chemical analyses of hot springs, pools, geysers, and surface waters from Yellowstone National Park, Wyoming, and vicinity 1974-1975." *U.S. Geological Survey Open-File Report*, 98-182.

Ballantyne, J. M., and Moore, J. N. (1988). "Arsenic geochemistry in geothermal systems." *Geochimica et Cosmochimica Acta*, 52(2), 475-483.

Bang, S., Johnson, M. D., Korfiatis, G. P., and Meng, X. (2005). "Chemical reactions between arsenic and zero-valent iron in water." *Wat. Res.*, 39(5), 763-770.

Bhumbla, D. K., and Keefer, R. F. (1994). "Arsenic mobilization and bioavailability in soils." In: *Arsenic in the Environment, Part I: Cycling and Characterization*, John Wiley & Sons, New York, 62-66.

Bissen, M., and Frimmel, F. H. (2003a). "Arsenic – A review. Part I: Occurrence, toxicity, speciation, mobility". *Acta hydrochim. hydrobiol.*, 31(1), 9–18.

Bissen, M., and Frimmel, F. H. (2003b). "Arsenic – A Review. Part II. Oxidation of arsenic and its removal in water treatment." *Acta hydrochim. hydrobiol.*, 31(2), 97-107.

Blowes, D. W., Ptacek, C. J., and Jambor, J. L. (1997). "In-situ remediation of Cr(VI)-contaminated ground water using permeable reactive walls: laboratory studies." *Environ. Sci. & Technol.*, 31(12), 3348-3357.

Blowes, D. W., Ptacek, C. J., Benner, S. G., McRae, C. W. T., Bennet, T. A., Puls, R. W. (2000). "Treatment of inorganic contaminants using permeable reactive barriers". *J. Contam. Hydrol.*, 45(1-2), 123-137.

Borgono, J. M., Vincent, P., Venturino, H., and Infante, A. (1977). "Arsenic in the drinking water of the city of Antofagasta: Epidemiological and clinical study before and after installation of a treatment plant." *Environ. Health Perspect.*, 19, 103-105.

Boudette, E. L,. Canney, F. C., Cotton, J. E., Davis, R. I., Ficklin, W. H., and Motooka, J. M. (1985). "High levels of arsenic in the ground waters of southeastern New Hampshire: A geochemical reconnaissance." *U.S. Geological Survey Open-File Report*, 85-202.

Boyle, R. W., and Jonasson, I. R. (1973). "The geochemistry of arsenic and its use as an indicator element for geochemical prospecting." *J. Geochem. Explor.*, 2(3), 251-296.

Cantrell, K. J., Kaplan, D. I., and Wietsma, T. W. (1995). "Zero-valent iron for the in situ remediation of selected metals in ground water." *J. Hazard. Mater.*, 42(2), 201-212.

Chatterjee, A., Das, D., Mandal, B. K., Roy Chowdhury, T., Samanta, G., and Charkraborti, D. (1995). "Arsenic in groundwater in six districts of West Bengal, India: The biggest arsenic calamity in the world. Part 1: Arsenic species in drinking water and urine of the affected people." *Analyst,* 120, 643-650.

Chen, S. L., Dzeng, S. R., Yang, M. H., Chiu, K. H., Shieh, G. M., and Wai, C. M. (1994). "Arsenic species in ground waters of the blackfoot disease area, Taiwan." *Environ. Sci. & Technol.*, 28(5), 877-881.

Cheng, I. F., Muftikian, R., Fernando, Q., and Korte, N. (1997). "Reduction of nitrate to ammonia by zero valent iron." *Chemosphere*, 35(11), 2689-2695.

Cheng, Z., van Geen, A., Jing, C., Meng, X., Seddique, A., and Ahmed, K. M. (2004). "Performance of a household-level arsenic removal system during 4-month deployments in Bangladesh." *Environ. Sci. & Technol.*, 38(12), 3442-3448.

Cheng, Z., Van Geen, A., Louis, R., Nikolaidis, N. P., and Bailey, R. (2005). "Removal of methylated arsenic species with iron filings." *Environ. Sci. & Technol.*, (unpublished manuscript).

Choe, S., Chang, Y., Hwang, K., and Khim, J. (2000). "Kinetics of reductive denitrification by nanoscale zero-valent iron." *Chemosphere*, 41(8), 1307-1311.

Chuang, F. W., Larson, R. A., and Wessman, M. S. (1995). "Zero-valent iron-promoted dechlorination of polychlorinated biphenyls." *Environ. Sci. & Technol.*, 29(9), 2460-2463.

Compeau, G., and Bartha, R. (1984) "Methylation and demethylation of mercury under controlled redox, pH, and salinity conditions." *Appl. Environ. Microbiol.*, 48, 1203-1207.

Crecelius, E. A., Johnson, C. J., and Hofer, G. C. (1974). "Contamination of soils near a copper smelter by arsenic, antimony, and lead." *Water, Air, and Soil Pollut'n.*, 3, 337-342.

Cullen, W. R., and Reimer, K. J. (1989). "Arsenic speciation in the environment." *Chem. Rev.,* 89(4), 713-764.

Das, D., Chatterjee, A., Mandal, B. K., Samanta, G., Chakraborti, D., and Chanda, B. (1995). "Arsenic in groundwater in six districts of West Bengal, India: The biggest arsenic calamity in the world. Part 2. Arsenic concentration in drinking water, hair, nails, urine, skin-scale and liver tissue (biopsy) of the affected people." *Analyst,* 120, 917-924.

Daus, B., Wennrich, R., and Weiss, H. (2004). "Sorption materials for arsenic removal from water: A comparison study." *Wat. Res.*, 38(12), 2948-2954.

Davis, J. A., and Kent, D. B. (1990). "Surface complexation modeling in aqueous geochemistry." In: *Mineral-Water Interface Geochemistry*, Mineral Society of America, 177-260.

Del Razo, L. M., Arellano, M. A., and Cebrian, M. E. (1990). "The oxidation states of arsenic in well water from a chronic arsenicism area of northern Mexico." *Environ. Pollut.,* 64(2), 143-153.

Dhar, R. K., Biswas, B. Kr., Samanta, G., Mandal, B. Kr., Chakraborti, D., Roy, S., Jafar, A., Islam, A., Ara, G., Kabir, S., Khan, A. W., Ahmed, S. A., and Hadi, S. A. (1997). "Groundwater arsenic calamity in Bangladesh." *Current Sci.,* 73(1), 48-58.

Dixit, S., and Hering, J. G. (2003). "Comparison of arsenic(V) and arsenic(III) sorption onto iron oxide minerals: Implications for arsenic mobility." *Environ. Sci. & Technol.*, 37(18), 4182-4189.

Duker, A. A., Carranza, E. J. M., and Hale, M. (2005). "Arsenic geochemistry and health." *Environment International*, 31(5), 631-641.

Edwards, M. A. (1994). "Chemistry of arsenic removal during coagulation and Fe-Mn oxidation." *J. Am. Water Works Assoc.*, 86 (9), 64-78.

Elkhatib, E. A., Bennett, O. L., and Wright, R. J. (1984). "Arsenite sorption and desorption in soils." *Soil Sci. Soc. of Am. J.*, 48, 1025-1030.

Farrell, J., Wang, J. P., O'Day, P., and Conklin, M. (2001). "Electrochemical and spectroscopic study of arsenate removal from water using zero-valent iron media." *Environ. Sci. & Technol.*, 35(10), 2026-2032.

Fiedor, J. N., Bostick, W. D., Jarabek, R. J., and Farrell, J. (1998). "Understanding the mechanism of uranium removal from ground water by zero-valent iron using x-ray photoelectron spectroscopy." *Environ. Sci. & Technol.*, 32(10), 1466-1473.

Feinglass, E. J. (1976). "Arsenic intoxication from well water in the United States." *New Eng. J. Medicine*, 288, (16), 828-830.

Frankenberger, W. T. (2002). *Environmental chemistry of arsenic*, Marker Dekker, Inc., New York.

Freeman, M. C., Aggett, J., and O'Brien, G. (1986). "Microbial transformations of arsenic in Lake Ohakuri, New Zealand." *Wat. Res.*, 20(3), 283-294.

Frey, M. M., Owen, D. M., Chowdhury, Z. K., Raucher, R. S., and Edwards, M. A. (1998). "Cost to utilities of a lower MCL for arsenic." *J. Am. Water Works Assoc.*, 90(3), 89-102.

Fuller, C. C., Davis, J. A., and Waychunas, G. A. (1993). "Surface chemistry of ferrihydrite: Part 2. kinetics of arsenate adsorption and coprecipitation." *Geochim. Cosmochim. Acta*, 57(10), 2271-2282.

Germani, M. S., and Zoller, W. H. (1988). "Vapor-phase concentrations of arsenic, selenium, bromine, iodine, and mercury in the stack of a coal-fired power plant." *Environ. Sci. & Technol.*, 22(9), 1079-1085.

Goldberg, S., and Johnston, C. T. (2000). "Mechanism of arsenic adsorption on amorphous oxides evaluated using macroscopic measurements, vibrational spectroscopy, and surface complexation modeling." *Journal of Colloid and Interface Science,* 234(1), 204-216.

Hering, J. G. (1997). "Chinatown revisited: Arsenic and the Los Angeles water supply." *Engineering & Science*, 3, 35-40.

Hering, J. G., Chen, P. Y., Wilkie, J. A., Elimelech, M., and Liang, S. (1996). "Arsenic removal by ferric chloride." *J. Am. Water Works Assoc.*, 88 (4), 155-167.

Huang, C. P., and Vane, L. M. (1989). "Enhanced As(+5) removal by a Fe(+2)-treated activated carbon." *J. Water Pollut. Control Fed.*, 61, 1596-1603.

Huang, C., Wang, H., and Chieu, P. (1998). "Nitrate reduction by metallic iron." *Wat. Res.*, 32(8), 2257-2264.

Huang, Y. H., and Zhang, T. C. (2004). "Effect of low pH on nitrate reduction by iron powder." *Wat. Res.*, 38(11), 2631-2642.

Jiang, J. Q. (2001). "Removing arsenic from groundwater for the developing world-a review." *Wat. Sci. & Technol.*, 44(6), 89-98.

Johnson, T. L., Scherer, M. M., and Tratnyek, P. G. (1996). "Kinetics of halogenated organic compound degradation by iron metal." *Environ. Sci. & Technol.* 30(8), 2634-40.

Joshi, A., and Chaudhuri, M. (1996). "Removal of arsenic from ground water by iron oxide-coated sand." *J. Environ. Eng., ASCE,* 122, 769-771.

Kanel, S. R., Manning, B., Charlet, L., and Choi, H. (2005). "Removal of arsenic(III) from groundwater by nanoscale zero-valent iron." *Envron. Sci. & Technol.*, 39(5), 1291-1298.

Kusnierova, M. (1994). "Prognosis of the stability of arsenic in the biotop of the brown-coal dump." *J. Hazard. Mater.*, 37(1), 27-31.

Lackovic, J. A., Nikolaidis, N. P., and Dobbs, G. M. (2000). "Inorganic arsenic removal by zero-valent iron." *Environmental Engineering Science*, 17(1), 29-39.

Lien, H. L., and Wilkin, R. T. (2005). "High-level arsenite removal from groundwater by zero-valent iron." *Chemosphere*, 59 (3), 377-386.

Mackie, D., and Zdepski, J. M. (1998). "Persistent pesticide residues in New Jersey soils: technical, historical, and regulatory perspectives." In *Proceedings of the 30th Mid-Atlantic Industrial and Hazardous Waste*, Villanova, PA, 177-184.

Maeda, S., Ohki, A., Tsurusaki, Y., and Takeshita, T. (1990). "Selective adsorption of arsenic(V) ion by use of iron(III) hydroxide-loaded coral limestone." *Separ. Sci. Technol.* 25, 547-555.

Mandal, B. K., Chowdhury, T. R., Samanta, G., Basu, G. K., Chowdhury, P. P., Chanda, C. R., Lodh, D., Karan, N. K., Dhar, R. K., Tamili, D. K., Das, D., Saha, K. C., and Chakraborti, D. (1996). "Arsenic in ground water in seven districts of West Bengal, India-the biggest arsenic calamity in the world." *Current Sci.,* 70(11), 976-986.

Manning, B. A., and Goldberg, S. (1997). "Adsorption and stability of arsenic(III) at the clay mineral-water interface." *Envron. Sci.& Technol.*, 31(7), 2005-2011.

Manning, B. A., Hunt, M. L., Amrhein, C., and Yarmoff, J. A. (2002). "Arsenic (III) and Arsenic (V) reactions with zerovalent iron corrosion products." *Envron. Sci. & Technol.*, 36(24), 5455-5461.

Marvinney, R. G, Loiselle, M. C., Hopeck J. T., Braley, D., and Krueger, J. A. (1986). "Arsenic in Maine ground water: an example from Buxton, Maine." In: *Maine Geological Survey Publication*, Augusta, Maine.

Matisoff, G., Khourey, C. J., Hall, J. F., Varnes, A. W., and Strain, W. H. (1982). "The nature and source of arsenic in northeastern Ohio ground water." *Ground Water*, 20, 446-455.

Meladiotis, I., Veranis, N. S., Nikolaidis, N. P. (2002). "Arsenic contamination in Central Macedonia, Northern Greece: Extent of the problem and potential solution." In *Proceedings of the Conference of Protection and Restoration of the Environment VI*, 1-5 July, Skiathos, Greece.

Melitas, N., Wang, J., Conklin, M., O' Day, P., Farrell, J. (2002a). "Understanding soluble arsenate removal kinetics by zerovalent iron media". *Environ. Sci. & Technol.,* 36(9), 2074-2081.

Melitas, N., Conklin, M., and Farrell, J. (2002b). "Electrochemical study of arsenate and water reduction on iron media used for arsenic removal from potable water." *Environ. Sci. & Technol.*, 36(14), 3188-3193.

Morrison, S. J., Metzer, D. R., and Dwyer, B. P. (2002). "Removal of As, Mn, Mo, Se, U, V and Zn from groundwater by zero-valent iron in a passive treatment cell: reaction progress modeling". *J. Contam. Hydrol.,* 56(1-2), 99-116.

Morton, W., Starr, G., Pohl, D., Stoner, J., Wagner, S., and Weswig, P. (1976). "Skin cancer and water arsenic in Lane County, Oregon." *Cancer,* 37, 2523-2532.

Munir, A. K. M., Rasul, S. B., Habibuddowla, M., Alauddin, M., Hussam, A., Khan, A. H. (2001). "Evaluation of performance of sono 3-kolshi filter for arsenic removal from groundwater using zero valent iron through laboratory and field studies." In *Technologies for Arsenic Removal from Drinking Water*, BUET and United Nations University, Dhaka.

Nikolaidis, N. P. (2002). "Treatment of arsenic contaminated groundwater using iron filings: Field Experiments." In: *Proceedings of Conference of Protection and Restoration of the Environment VI*, 1-5 July, Skiathos Island, Greece.

Nikolaidis, N. P., Dobbs, G. M., and Lackovic, J. A. (2003). "Arsenic removal by zero-valent iron: field, laboratory and modeling studies." *Wat. Res.*, 37(6), 1417-1425.

Nikolaidis, N. P., Cheng, Z., and van Geen, A. (2005). "Removal of arsenic from Bangladesh Groundwater with Zero-valent Iron." *Amer. Chem. Soc. Symp. Ser.*, 915, in press.

Nriagu, J. O. (1994a). *Arsenic in the environment, part II: Human health and ecosystem effects*, John Willey and Sons, New York.

Nriagu, J. O. (1994b). *Arsenic in the environment, part I: Cycling and characterization*, John Willey and Sons, New York.

Pontius, F. W., Brown, K. G., and Chen, C. J. (1994). "Health implications of arsenic in drinking water." *J. Am. Water Works Assoc.*, 86 (9), 52-63.

Powell, R. M., Puls, R. W., and Hightower, S. K. (1995). "Coupled iron corrosion and chromate reduction: Mechanisms for subsurface remediation." *Environ. Sci. & Technol.*, 29(8), 1913-22.

Ragaini, R. C., Ralston, H. R., and Roberts, N. (1977). "Environmental trace metal contamination in Kellogg, Idaho, near a lead smelting complex." *Environ. Sci. & Technol.*, 11(8), 773-781.

Rahman, M. (1999). *Nonmalignant health effects of arsenic exposure*, Medical Dissertation, No. 612, Linköping University.

Raven, K. P., Jain, A., and Loeppert, R. H. (1998). "Arsenite and arsenate adsorption on ferrihydrite: Kinetics, equilibrium, and adsorption envelopes." *Environ. Sci. & Technol.*, 32(3), 344-349.

Roberts, A. L., Totten, L. A., and Arnold, W. A. (1996). "Reductive elimination of chlorinated ethylenes by zero-valent metals." *Environ. Sci. & Technol.*, 30(8), 2654-2659.

Schlottmann, J. L., and Breit, G. N. (1992). "Mobilization of As and U in the central Oklahoma aquifer, USA." In: *Water-Rock Interaction*, Balkema, Rotterdam, 835-838.

Shokes, T. E., and Moller, G. (1999). "Removal of dissolved heavy metals from acid rock drainage using iron metal." *Environ. Sci. & Technol.*, 33(2), 282-287.

Smedley, P. L., and Kinniburgh, D. G. (2002). "A review of the source, behaviour and destribution of arsenic in natural waters." *Applied Geochemistry*, 17(5), 517-568.

Snoeyink, V. L., and Jenkins, D. (1980). *Water chemistry*, John Wiley & Sons, New York.

Stumm, W., and Morgan, J. J. (1981). *Aquatic chemistry*, John Wiley & Sons, New York.

Su, C., and Puls, R. W. (2001a). "Arsenate and arsenite removal by zero valent iron: Kinetics, redox, transformation, and implications for in situ ground water remediation." *Environ. Sci. & Technol.*, 35(7), 1487-1492.

Su, C., and Puls, R. W. (2001b). "Arsenate and arsenite removal by zerovalent iron: Effects of phosphate, silicate, carbonate, borate, sulfate, chromate, molybdate, nitrate relative to chloride." *Environ. Sci. & Technol.*, 35(22), 4562-4568.

Su, C., and Puls, R. W. (2003). "In situ remediation of arsenic in simulated groundwater using zerovalent iron: Laboratory column tests on combined effects of phosphate and silicate." *Environ. Sci. & Technol.*, 37(11), 2582-2587.

Su, C., and Puls, R. W. (2004a). "Significance of iron (II,III) hydroxycarbonate green rust in arsenic remediation using zerovalent iron in laboratory column tests." *Environ. Sci. & Technol.*, 38(19), 5224-5231.

Su, C., and Puls, R. W. (2004b). "Nitrate reduction by zerovalent iron: Effect of formate, oxalate, citrate, chloride, sulfate, borate, and phosphate." *Environ. Sci. & Technol.*, 38(9), 2715-2720.

Tyrovola, K., Nikolaidis, N. P., Veranis, N., Kallithrakas-Kontos, N., Koulouridakis, P. E. (2005). "Arsenic removal from geothermal waters with zero-valent iron - Effect of temperature, phosphate and nitrate." *Wat. Res.*, in review.

USEPA. (2000a). *Arsenic occurrence in public drinking water supplies*, EPA/815/R-00/023, Office of Water (4607), United States Environmental Protection Agency, Washington D.C. 20460 <http://www.epa.gov/safewater/ars/occurrence.pdf>.

USEPA (2000b). *Technologies and costs for removal of arsenic from drinking water*, EPA/ 815/R-00/028, Office of Water (4607), United States Environmental Protection Agency, Washington D.C. 20460 <http://www.epa.gov/safewater/ars/treatments_and_costs.pdf>.

USEPA. (2001). *Treatment of arsenic residuals from drinking water removal processes*, EPA/600/R-01/033, Office of Research and Development, United States Environmental Protection Agency, Washington D.C. 20460 <http://www.epa.gov/nrmrl/pubs/600r01033/600r01033.htm>.

USEPA. (2002a). *Arsenic treatment technologies for soil, waste, and water*, EPA/542/R-02/004, Solid Waste and Emergency Response (5102G), United States Environmental Protection Agency, Washington D.C. 20460 <http://www.epa.gov/tio/download/remed/542r02004/arsenic_report.pdf >

USEPA. (2002b). *Proven alternatives for aboveground treatment of arsenic in groundwater*, EPA/542/S-02/002, Solid Waste and Emergency Response (5102G), United States Environmental Protection Agency, Washington D.C. 20460 <http://www.epa.gov/tio/tsp/download/arsenic_issue_paper.pdf>

USEPA. (2003). *Arsenic treatment technology evaluation handbook for small systems*, EPA/816/R-03/014, Office of Water (4606M), Environmental Protection Agency, Washington D.C. 20460 <http://www.epa.gov/safewater/smallsys/arsenic_treatment_handbook_lo.pd f >

Vlassopoulos, D., Pochatila, J., Lundquist, A., Andrews, C. B., Rafferty, M. T., Chiang, K., Sorel, D., and Nikolaidis, N. P. (2002). "An elemental iron reactor for arsenic removal from groundwater." In *Proceedings of the 3rd International Conference on Remediation of Chlorinated and Recalcitrant Compounds*, May 20-23, Monterey, CA.

Waychunas, G. A., Rea, B. A., Fuller, C. C., and Davis, J. A. (1993). "Surface chemistry of ferrihydrite: Part 1. EXAFS studies of the geometry of coprecipitated and adsorbed arsenate." *Geochim. Cosmochim. Acta*, 57(10), 2251-2269.

Webster, J. G., and Nordstrom, D. K. (2003). "Geothermal arsenic. The source, transport and fate of arsenic in geothermal systems." In *Arsenic in ground water, geochemistry and occurrence*, Kluwer Academic Publishers, 101-125.

Welch, A. H., Lico, M. S., and Hughes, J. L. (1988). "Arsenic in ground water of the western United States." *Ground Water*, 26(3), 333-347.

Welch, A. H., Watkins, S. A., Helsel, D. R., and Focazio, M. F. (2000). "Arsenic in ground-water resources of the United States." In *U.S. Geological Survey Fact Sheet 063-00*, 4p.

Welch, A. H., and Stollenwerk, K. G. (2003). *Arsenic in ground water, geochemistry and occurrence*, Kluwer Academic Publishers.

Westerhoff, P., and James, J. (2003). "Nitrate removal in zero-valent iron packed columns." *Wat. Res.*, 37(8), 1818-1830.

WHO (1981). "Environmental health criteria 18 arsenic." *IPCS International Programme on Chemical Safety*, <http://www.inchem.org/documents/ehc/ehc/ehc018.htm>.

Wilkie, J. A., and Hering, J. G. (1996). "Adsorption of arsenic onto hydrous ferric oxide: Effects of adsorbate/adsorbent ratios and co-occurring solutes." *Colloids and Surfaces A: Physicochem. Eng. Aspects*, 107, 97-110.

Wilkie, J. A., and Hering, J. G. (1998). "Rapid oxidation of geothermal arsenic (III) in streamwaters of Eastern Sierra Nevada." *Environ. Sci. & Technol.*, 32(5), 657-662.

Wilkin, R., and McNeil, M. S. (2003). "Laboratory evaluation of zero-valent iron to treat water impacted by acid mine drainage." *Chemosphere*, 53, 715-725.

Yamauchi, H., and Fowler, B. A. (1994). "Toxicity and metabolism of inorganic and methylated arsenicals." *Adv. Environ. Sci. Technol.*, 27, 35-53.

Zuena, A. J., and Keane, N. W. (1985). "Arsenic contamination of private potable wells." In *Proceedings of ASCE Specialty Conference, Environmental Engineering*, Boston.

CHAPTER 10

The Performance of Palladized Granular Iron: Enhancement and Deactivation

Lai Gui[1], Robert W. Gillham[2], Molly M. McGuire[3], and D. Howard Fairbrother[4]

Abstract: Column tests were conducted to evaluate the performance of iron enhanced by the addition of palladium (Pd-Fe, 0.05 wt %) for reductive dechlorination of trichloroethylene (TCE). Although the displacement plating method was efficient on the surface of both commercial-grade granular iron (i.e. particles covered by a thick layer of iron oxide) and the acid-washed iron (i.e. iron oxide layer mostly removed) and resulted in fast initial TCE degradation rates ($t_{1/2} <$ 1 min), their deactivation rates differed dramatically. Palladium plated on acid-washed base materials (Pd-Fe-A) lost their enhanced reactivity within 10 hours (500 pore volumes). In contrast, Pd plated on oxide-covered iron particles (Pd-Fe-U) showed consistently enhanced reactivity during the 200-day experiment (about 40 000 pore volumes). Using various surface analytical methods including X-ray photoelectron and Auger electron spectroscopy, it was found that after plating, more palladium deposits were present on the surface of the Pd-Fe-U particles than on the Pd-Fe-A particles both before and after the column tests. Auger electron spectroscopy measurements carried out in conjunction with ion sputtering showed that palladium plated onto Pd-Fe-A particles was partially buried under an iron oxide layer in the near-surface region. These findings were attributed to the fact that acid-washed particles undergo a faster initial rate of corrosion during plating and during reactions in the column compared to the Pd-Fe-U particles due to the absence of a fully passive iron oxide overlayer. During the formation of this passivating iron oxide film on the Pd-Fe-A particles, a significant fraction of Pd is "buried" below the surface, rapidly reducing the reactivity of Pd-Fe-A particles to a level comparable to that observed in columns containing either acid-washed or unwashed granular iron without palladium. This study suggests that Pd-Fe-U may be used as an effective medium for rapid degradation of TCE over long periods of time.

CE Database subject headings: Degradation; Porous media; Iron; Ground-water pollution; Remedial action; Halogen organic compounds; Trichloroethylene.

---

[1]Research Assistant Professor, Dept. of Earth Sciences, University of Waterloo, 200 University Ave. W., Waterloo, Ontario N2L 3G1, Canada (corresponding author). E-mail:lgui@scimail.uwaterloo.ca
[2]Professor, Dept. of Earth Sciences, University of Waterloo, Waterloo, Ontario, N2L 3G1, Canada.
[3]Assistant Professor, Dept. of Chemistry, Bucknell University, Lewisburg, Pennsylvania 17837 USA.
[4]Dept. of Chemistry, The Johns Hopkins University, Baltimore, Maryland, 21218 USA.

10.1 Introduction

The use of granular iron to reductively degrade chlorinated solvents and other organic and inorganic groundwater contaminants has proven to be successful, cost-effective and environmentally benign (Gillham and O'Hannesin 1994; Agrawal and Tratnyek 1996; O'Hannesin and Gillham 1998; Blowes et al. 2000). Further research has demonstrated that plating a small amount of a second metal, such as palladium or nickel, onto the surface of granular iron can greatly enhance the efficiency of the reductant (Grittini et al. 1995; Muftikian et al. 1995; Wang and Zhang 1997; Gui et al. 2000). Compared with granular iron, bimetallic materials offer (1) enhanced reaction rates, commonly by one to two orders of magnitude (Grittini et al. 1995; Muftikian et al. 1995; Gui et al. 2000); (2) more complete destruction of contaminants (Liang et al. 1997; Zhang et al. 1998); and (3) effective in degrading a wider range of contaminants such as chlorinated aromatic compounds (Grittini et al. 1995; Wang and Zhang 1997; Xu et al. 2005).

Palladium, a known hydrogenation catalyst, is very effective in destroying halogenated contaminants (Grittini et al. 1995; Muftikian et al. 1995; Muftikian et al. 1996; Liang et al. 1997; Wang and Zhang 1997; Zhang et al. 1998; Lowry and Reinhard 1999; Alonso et al. 2002; Ukisu and Miyadera 2003; Xu et al. 2005). Conventionally, palladium (Pd) is deposited onto an inert catalyst support such as alumina or carbon (Muftikian et al. 1996; Lowry and Reinhard 1999; Ukisu and Miyadera 2003). Hydrogen gas is then supplied as an external source of hydrogen and electrons. In the case of palladized iron, hereafter referred to as Pd-Fe, palladium is deposited on a reactive base metal, iron, which produces hydrogen (H atoms and H_2), through iron corrosion, for the reductive hydrogenation reaction. The Pd-Fe is commonly prepared by a displacement plating method, where acid-washed iron is placed in a palladium salt (e.g. K_2PdCl_6) solution (Grittini et al. 1995; Muftikian et al. 1995; Xu et al. 2005). The purpose of the acid-washing step is to remove surface oxide films and thus to achieve a high plating efficiency. Raman spectroscopic measurements have shown that after acid washing, the surface oxide film commonly present on untreated iron material is largely removed to the extent that it is no longer detectable by normal Raman spectroscopy.

Though fresh Pd-Fe bimetals, prepared with the acid-wash procedure, have been shown to substantially outperform granular iron, the materials suffer rapid loss of the catalytic activity, despite the fact that most studies have been carried out in batch experiments that lasted for only a few hours. In an above-ground treatment study, Korte et al. (2000) reported that the half-life of TCE degradation increased by an order of magnitude within four weeks. Using X-ray photoelectron spectroscopy (XPS), Muftikian et al. (1996) attributed the observed deactivation to the formation of hydroxylated iron oxide film, effectively burying the Pd below the surface. The hydroxylated iron oxide film formed on the Pd-Fe materials can be removed by acid washing (Korte et al. 2000). However, iron corrosion as well as acid-washing may result in dissolution of iron in direct contact with the Pd deposits, resulting in losses of Pd content and permanent deactivation of Pd-Fe. Alternatively, there may be advantages in depositing the palladium on iron that has not been acid washed. Granular iron used in the commercial construction of permeable reactive barriers is generally produced from reprocessed iron cuttings and grindings that were passed through a rotary kiln, ground and sieved to size. As a consequence, the inner core of metallic iron is covered by a double layer of oxides, an inner layer of magnetite and an outer layer of hematite and maghemite (Odziemkowski et al. 2000). It is proposed that the passive nature of the outer layer of the surface film may reduce the deactivation rates of the Pd-Fe by reducing surface film growth. In this case, the thickness and nature of the oxide layer on the surface of the iron particles before plating may have a substantial influence on the longevity of the active material.

The objectives of the present study were (1) to compare the plating efficiency of Pd on oxide-covered and acid-cleaned iron surfaces and to evaluate the longevity of the resulting Pd-Fe materials and (2) to gain insight into the deactivation processes associated with the two preparation procedures and the use of Pd-Fe. In particular, we examined the influence of the initial surface composition of the iron on deactivation rates in order to derive a mechanistic explanation for the apparent differences in deactivation rates when acid-washed versus unwashed iron is used in the plating process. Column tests and various surface and near-surface analysis techniques including XPS and Auger electron spectroscopy (AES) were used in this study.

10.2 Experimental Section

10.2.1 Chemicals and Materials

Trichloroethylene (TCE, 97%) was purchased from *Fisher Scientific* (Fair Lawn, NJ); *cis*-1,2-dichloroethene (*c*-DCE, 97%) and 1,1-dichloroethene (1,1-DCE, 99%) were from *Aldrich* (Milwaukee, WI) and *trans*-1,2-dichloroethene (*t*-DCE, >95%) and vinyl chloride (VC, >99.5%) were from *Fluka* (Neu-Ilm, Switzerland). A gaseous hydrocarbon standard, consisting of ethene (0.503%) and ethane (0.500%) and nitrogen (balance), was obtained from *Praxair* (Mississauga, Ont). Potassium hexachloropalladate (99.9%) was purchased from *Alfa Aesar* (Ward Hill, MA). All chemicals were used without further purification.

Two sources of granular iron were used in this study, *Fisher* and *Connelly*. The *Fisher* iron (*Fisher Scientific*, Co.) contained > 85 % metallic iron with a relatively uniform particle size of 40 mesh. The average surface area, measured by the BET method, was 2.89 m^2g^{-1}. The *Connelly* iron was obtained from *Connelly-GPM, Inc.* (Chicago, IL) and contained 89.8% metallic iron. The particle size varied from –8 to +50 mesh with about 35% in the fraction of 40 mesh (Data provided by supplier). The average surface area was 1.75 m^2g^{-1}. Both sources of iron were covered by a thick layer of iron oxide. The iron material was used either as received (several rinses with distilled water only) or after acid washing. The acid washing procedure was similar to that of Muftikian et al. (1995). Iron particles were washed 6 times with 6 N HCl for 10 minutes each, then rinsed 5 times with distilled water. The acid-washing process removed the bulk iron oxide surface films as confirmed by Raman spectroscopy (data not shown).

Pd-Fe was prepared by a displacement plating method. Potassium hexachloropalladate was dissolved in 0.001 N HCl to produce a solution of 500 mg L^{-1} Pd. The solution was added to pre-weighed iron in sufficient quantity to produce 0.05 wt% Pd at 100% plating efficiency. After stirring for 10 minutes, the supernatant was decanted and the iron was rinsed 5 times with distilled water.

Plating efficiency was determined by analyzing residual Pd in the plating solutions and the rinse waters. For the remainder of this chapter, acid washed iron plated with Pd will be referred to as Pd-Fe-A and unwashed iron plated with Pd will be referred to as Pd-Fe-U.

For surface analysis, approximately 5 g of the freshly prepared Pd-Fe was dried under a stream of nitrogen gas (zero-oxygen grade) for 30 minutes. At the completion of the column tests, the columns were placed in an anaerobic glovebox (95% N_2 + 5% H_2) and Pd-Fe samples were extracted from the middle section of each column and dried under nitrogen gas. Once dry, samples remained exposed to the atmosphere for a period of days before surface analysis.

10.2.2 Column Tests

A series of column tests was conducted to evaluate the performance of the Pd-Fe materials for TCE degradation. The glass columns (3.5 cm long x 1.6 cm ID) were packed with the wet Pd-Fe immediately after plating. The porosity of the columns ranged between 0.53 and 0.71. Anaerobic TCE solutions, with initial concentrations (C_0) ranging between 11 and 16 mg L^{-1}, were stored in a collapsible double-layered Teflon bag with N_2 gas filling the space between the bags to reduce oxygen diffusion. The concentration of dissolved oxygen in the TCE solution was maintained below 0.4 mg/L. Solutions were pumped through the columns from the bottom using a multi-channel peristaltic pump. The flow rates in the columns was decreased from 3.5 to 0.4 mL/min over time in order to maintain measurable declines in TCE concentration across the columns. With the exception of the pump tubing, the column fittings and tubing were made of stainless steel to minimize oxygen diffusion. Control columns consisted of either acid-washed (Fe-A) or unwashed (Fe-U) iron only and were operated under the same conditions used in the Pd-Fe columns.

Chemical analysis of the effluent was carried out periodically to determine concentrations of TCE, chlorinated breakdown products (*t*-DCE, *c*-DCE, 1,1-DCE, and VC), hydrocarbons (ethane and ethene), chloride ion, and dissolved Pd. The first effluent samples were taken within 0.5 h after the Pd-Fe was prepared (5 min after starting the column flow).

10.2.3 Chemical Analyses

TCE was extracted with pentane (containing 500 μg/L dibromoethane as an internal standard) and analyzed using a HP 5890 gas chromatography (GC) system equipped with a Ni^{63} electron capture detector. Chlorinated breakdown products were analyzed by a GC (HP-5890) equipped with a photon ionization detector and a headspace autosampler. Hydrocarbons were analyzed by a GC (HP-5790) equipped with a flame ionization detector. The gas formed in the columns was collected and quantified by connecting a water-filled glass tube at the effluent end. Pd was analyzed in solution using Direct Current Plasma Optical Emission Spectroscopy (DCP).

10.2.4 Auger Electron Spectroscopy (AES)

Auger spectra and depth profiles were obtained with a Physical Electronics (Φ) 610 Scanning Auger Microprobe operating in an ultrahigh vacuum (UHV) chamber with a typical base pressure of 1 x 10^{-8} Torr. Palladized iron samples were probed using a 3 keV electron beam operating with an emission current of 100 μA at 30° from the sample normal. Spectra were acquired using a beam raster size of 100 μm x 100 μm.

A Φ 04-303 differentially pumped ion gun (3 keV, 25 mA) in conjunction with the AES spectrometer was used for the depth profiles using an Ar pressure within the gun of 15 mPa. The Ar pressure was held constant throughout all experiments in order to ensure equivalent sputtering rates. The gun was oriented approximately normal to the sample surface and focused to a 2 mm x 2 mm spot. Data were acquired at 10 minute intervals. The typical operating pressure within the UHV chamber during sputtering was 1 x 10^{-7} Torr.

10.2.5 X-Ray Photoelectron Spectroscopy (XPS)

All XPS analyses were carried out in a UHV chamber with a base pressure of 2 x 10^{-8} Torr. Spectra were obtained with a Physical Electronics (Φ) 10-360 Precision Energy Analyzer using the Mg Kα line (1253.6 eV) of a Φ 04–500 Dual Anode X-ray Source at 15 kV and 300 W. All spectra shown were obtained with a pass energy of 44.75 eV and a resolution of 0.125 eV/step. Binding energies were corrected using the adventitious carbon 1s XPS peak at 284.6 eV (Barr and Seal 1995).

10.3 Results and Discussion

10.3.1 Pd Plating Efficiency

Analysis of Pd residuals in the plating bath and rinse water showed that near complete removal of Pd from solutions was achieved in both acid-washed and unwashed plating baths in 10 minutes. A qualitative comparison of the XPS spectra obtained from *Connelly* based Pd-Fe-U and Pd-Fe-A showed that, although the Pd peak intensities were different, the peak shapes were comparable, suggesting that the distribution of Pd oxidation states on the two freshly prepared palladized iron samples were similar. In both cases, the Pd(3d) region was dominated by peaks at 335.1 and 340.1 eV (Pd $3d_{5/2}/3d_{3/2}$ doublet) (Wagner et al. 1979), consistent with the presence of Pd^0 (data not shown). These peak positions are also in good agreement with previous XPS measurements of palladized iron surfaces (335.0 and 340.3 eV) (Davis et al. 1976). Therefore, the displacement method is very efficient for plating Pd onto both unwashed and acid-washed iron particles. Based on solution analysis, the plating efficiency was estimated and summarized in Table 10-1. The reduction of hexachloropalladate to elemental palladium, as shown in Eq. (10.1), occurs rapidly on the existing iron surface films and the acid washing step does not increase plating efficiency.

$$PdCl_6^{2-} + 4\,e^- \rightarrow \quad Pd^0 + 6Cl^- \qquad (10.1)$$

Table 10-1. Pd Plating Efficiency Calculated from Solution Analysis

Base iron	Base iron condition	Plating efficiency	Pd-Fe surface area
Connelly	Acid-washed (Pd-Fe-A)	99.5 %	1.47 $m^2 g^{-1}$
	Unwashed (Pd-Fe-U)	99.1 %	1.75 $m^2 g^{-1}$
Fisher	Acid-washed (Pd-Fe-A)	100 %	2.88 $m^2 g^{-1}$
	Unwashed (Pd-Fe-U)	99.5 %	2.89 $m^2 g^{-1}$

10.3.2 Column Tests

Effluent concentrations of TCE and degradation products were monitored to evaluate the performance of the Pd-Fe materials in the columns over time. Initially, the columns were operated at a flow rate of 3.5 mL/min, thus the TCE residence time in all of the columns was less than 2 minutes. Complete conversion of TCE to ethane and ethene was observed in the columns containing the palladium plated onto unwashed *Fisher* and *Connelly* iron particles (Pd-Fe-U), while in the column containing palladium plated onto acid-washed *Fisher* and *Connelly* iron particles (Pd-Fe-A) and the control columns containing either unwashed iron (Fe-U) or acid-washed iron (Fe-A), the TCE removal curves indicated no measurable degradation of TCE. Fig. 10-1 shows the performance of the four columns containing *Connelly*-based iron materials during the two-day tests (3,000 PV), as an example. Similar results were obtained in the four columns containing *Fisher*-based iron materials.

Plotting the sum of the carbon-containing degradation products (ethane and ethene) as their TCE equivalent ($C/C_{0,\mathrm{TCE}}$) versus time, Fig. 10-1b shows that in the *Connelly* Pd-Fe-A column approximately 45%, 20%, and 5% of the influent TCE was converted to ethane and ethene after 0.5h, 3h, and 10h, respectively. After 10h (500 PV), more than 95% of the initial TCE broke through the Pd-Fe-A column, similar to the control columns containing only Fe, either acid-washed or unwashed. The results were confirmed by analyses of chlorinated breakdown products and chloride ions. Thus, the results indicated rapid deactivation of the acid-washed Pd-Fe-A material. Analyses of the products associated with TCE degradation in the Pd-Fe-A column also showed a rapid decrease in the ratio of ethane to ethene, from over 10,000 initially to 5,000 within 2 h of operation to less than 1 at the end of the column test. This final ratio was similar to that obtained in the column that contained iron alone. This observation indicates that the ability of the Pd-Fe-A column to produce hydrogenation products decreased over the time course of the experiment, consistent with the idea that the enhanced reactivity associated with the Pd-Fe particles is rapidly lost for the acid-washed iron particles. In contrast, the complete

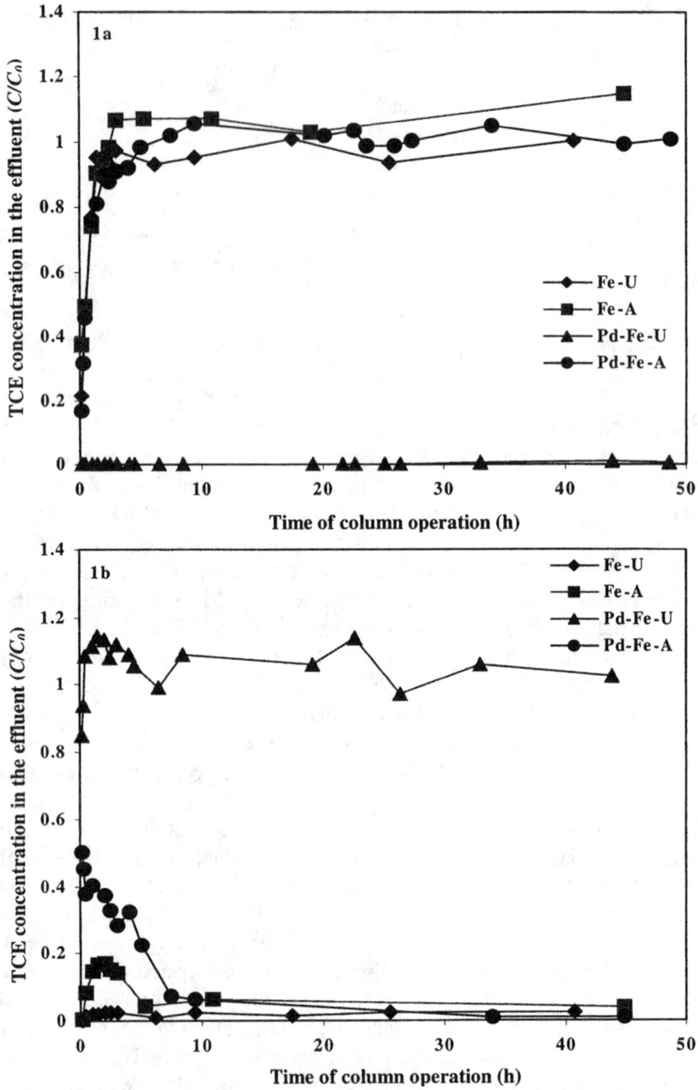

Fig. 10-1. Changes in TCE concentration during operation of columns consisting of *Connelly* based materials for the acid-washed (Pd-Fe-A) and unwashed (Pd-Fe-U) columns and their respective controls (Fe-A and Fe-U): (a) normalized TCE concentration (C/C_0) in the column effluents and (b) calculated TCE degradation based on the sum of ethane and ethene concentrations as TCE equivalent in the effluent samples. The initial TCE concentration was 15.8 mg/L.

reduction of TCE to ethene and ethane (Fig. 10-1b) was observed in the Pd-Fe-U column over the course of the experiment. Only a slight decrease in the ratio of ethane to ethene was observed during this period.

In the control columns that contained only iron, Fe-U degraded less than 5% of the initial TCE and the Fe-A degraded TCE to a maximum of 18% after 2 h of operation and rapidly decreased to about 5% for reaction times in excess of 3 h (Fig. 10-1b) due to the lower reactivity of Fe towards TCE, coupled with the fact that the TCE residence time in the columns was less than two minutes.

As a consequence of iron corrosion, formation of hydrogen gas and dissolution of iron occur concurrently with the reduction of TCE. Fig. 10-2 shows that the amount of hydrogen gas collected in the Pd-Fe-A columns was much greater than in the corresponding Pd-Fe-U columns for both types of iron, indicating that iron corrosion occurred much faster in the Pd-Fe-A columns than in the Pd-Fe-U columns. Considering the low rates of TCE degradation and high rates of hydrogen production (Figs. 10-1 and 10-2), it is clear that in the Pd-Fe-A columns iron corrosion was the dominant reaction. Depending on the water chemistry and the reduction potential, the ferrous iron may form various iron oxides on particle surfaces.

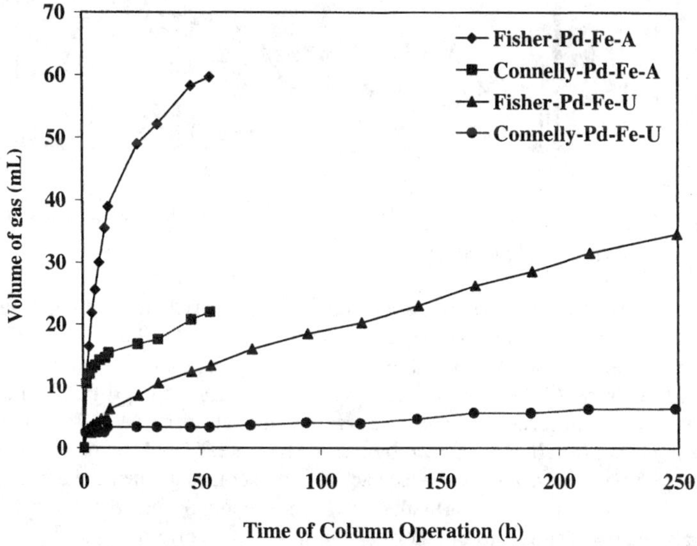

Fig. 10-2. Cumulative hydrogen gas collected during column operation

Fig. 10-3 shows that the Pd-Fe prepared by plating Pd onto oxide-covered iron particles, on both *Fisher* and *Connelly* iron materials, exhibited prolonged reactivity toward TCE reduction. After 40 000 PV (operated for 200 days) of TCE solution, less than 5 % and 30 % of the initial TCE was detected in the effluent of *Fisher* Pd-Fe-U and *Connelly* Pd-Fe-U, respectively (Fig. 10-3), at flow rates of 0.41 mL/min and 0.45 mL/min, respectively. Greater than 90% chlorine and carbon mass balances were obtained, with ethane as the primary product, confirming TCE degradation in the columns. If one assumes that TCE removal follows pseudo-first-order kinetics, the half-lives of TCE in the two columns were less than 2.5 min and 6.5 min, respectively, even at the end of the 200-day experiment (40 000 PV). Although the half-lives at the end of the 200-day operation were lower than those obtained at the initial time, they are still substantially shorter compared to those obtained with the un-plated iron materials which were not measurable at the residence times in the columns.

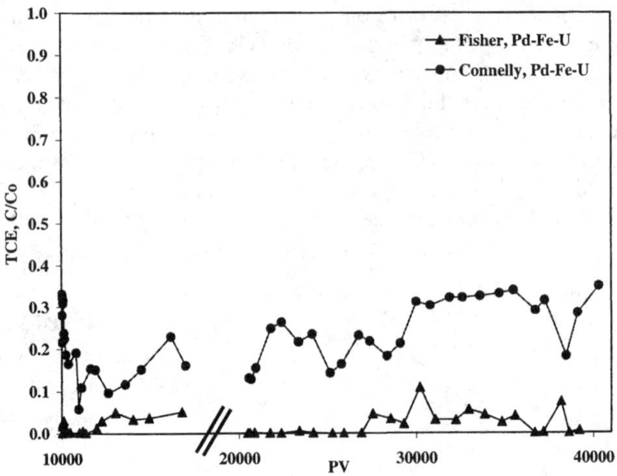

Fig. 10-3. Normalized effluent TCE concentration (C/C_0) obtained from the Pd-Fe-U columns during the 200-day operation

In this study, *Fisher* iron appeared to degrade TCE faster than *Connelly* iron. However, the apparent differences in reactivity may be due to the differences in surface area. As shown in Table 10-1, the surface areas of *Fisher* Pd-Fe-U and Pd-Fe-A were 65 % and 95 % greater than *Connelly* Pd-Fe-U and Pd-Fe-A, respectively. In order to compare reactivities of different base metals, kinetic constants are usually normalized by iron surface areas (Johnson et al. 1996; Lien and Zhang 2005). However, the surface area is not a constant when the corrosion process occurs continuously, in particular, when corrosion occurs rapidly due to the Pd catalytic effect, and therefore, normalization was not performed in this study.

10.3.3 Surface Analysis

To further determine the cause of the difference in performance between the acid-washed (Pd-Fe-A) and unwashed (Pd-Fe-U) materials, various surface analytic techniques were used to examine the characteristics of the Pd-Fe surfaces before and after the two-day column test. Only results from the Pd-Fe materials prepared with the *Connelly* base iron are presented here.

The Auger spectra (200 – 1000 eV) of palladized iron samples that were unwashed (a,b) and acid-washed (c,d) are shown in Fig. 10-4. Representative AES spectra of the palladized iron samples are shown both before (a,c) and after (b,d) reaction in the column. Fig. 10-4 exhibits AES peaks associated with carbon (268 eV), palladium (328 eV), oxygen (511 eV) and iron (595, 649, and 703 eV) (Davis et al. 1976).

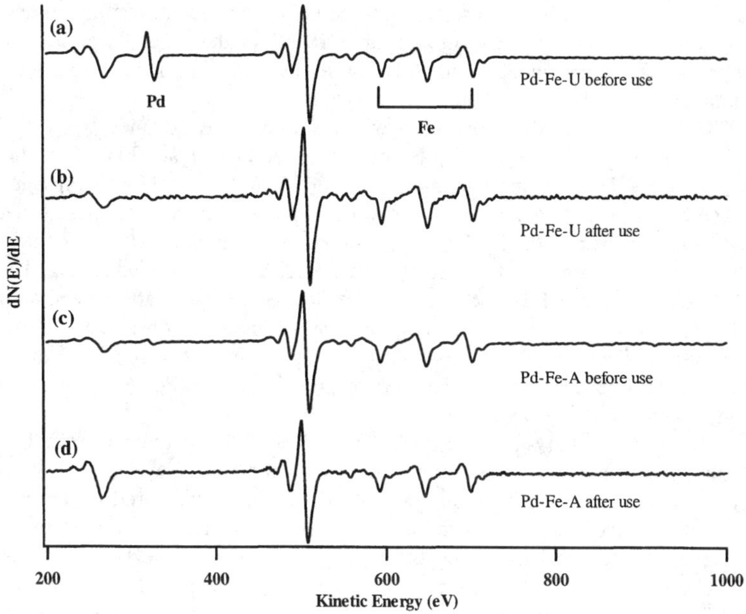

Fig. 10-4. Auger Electron Spectra (200-1000 eV) of unwashed palladized iron (Pd-Fe-U) samples before (a) and after (b) column reactions, and acid-washed palladized iron (Pd-Fe-A) samples before (c) and after (d) column reaction. Spectra have been normalized using the peak-to-peak height of the iron signal.

A comparison of Figs. 10-4a and 10-4c shows that the amount of palladium at the surface before reaction is significantly less on the acid-washed (Fig. 10-4c) compared to the unwashed (Fig. 10-4a) iron samples. This result is also reflected in Table 10-2, which indicates that the Pd:Fe ratio, calculated from the measured peak-to-peak heights and the relative AES atomic sensitivities of each element, is nearly an order of magnitude larger for the unwashed sample.

Table 10-2. Pd:Fe Ratios

	AES	XPS
Unwashed, before use	0.29	0.57
Unwashed, after use	0.04	0.07
Acid-washed, before use	0.04	0.10
Acid-washed, after use	-	0.03

Figs. 10-4b and 10-4d show the AES spectra of palladized iron surfaces after reaction in the column for unwashed (Fig. 10-4b) and acid-washed samples (Fig. 10-4d). With the exception of the Pd signals, these spectra are similar to the unwashed and acid-washed iron samples (Pd-Fe-U and Pd-Fe-A) before reaction in the columns (Figs. 10-4a and 10-4c, respectively). For the Pd-Fe-U material, the Pd signal intensity at 328 eV decreased substantially after reaction in the columns but remained above the detection limit (Table 10-2). In the case of the acid-washed sample, the Pd AES signal declined below the detection limit after reaction in the column (Fig. 10-4d).

XPS data provided confirmation of the Pd:Fe ratios obtained by AES (Table 10-2, the XPS spectra not shown). Similar to the AES analysis (Fig. 10-4), the XPS spectra of the Pd 3d region indicate that significantly more Pd is present at the surface of unwashed compared to acid-washed palladized iron samples. A decrease in Pd intensity for both the Pd-Fe-A and Pd-Fe-U samples after reaction in the columns is also observed. Consistent with the AES results shown in Fig. 10-4, the XPS intensity in the Pd 3d region for the Pd-Fe-A sample after reaction is now sufficiently low that the $3d_{5/2}$-$3d_{3/2}$ doublet can no longer be observed. Although the measured Pd:Fe ratios in Table 10-2 are different for AES and XPS due to the inherently different surface sensitivities of the techniques (Powell et al. 1999), the trends are in qualitative agreement.

Using XPS analysis, it was determined that the chemical oxidation state distributions of Pd were similar for both unwashed and acid-washed palladized iron samples after use in the columns, just as for the freshly prepared Pd-Fe materials. Therefore chemical differences in the oxidation state distribution are not responsible for the differences in observed reactivity.

The difference in surface coverage was further explored by examining the distribution of palladium as a function of depth below the surface, using Ar^+ ion sputtering in conjunction with AES. Fig. 10-5 contains plots of the Pd:Fe ratio obtained from the peak-to-peak heights of the differentiated AES spectra vs. sputtering time for the unwashed palladized iron sample before (a) and after (b) reaction in the column, as well as the acid-washed sample before (c) and after (d) reaction. In the case of the unwashed palladized iron sample before reaction (Fig. 10-5a), the Pd:Fe ratio falls off steeply during the first 200 minutes of sputtering. The unwashed palladized iron sample after reaction in the column (Fig. 10-5b) also shows a steady decrease in the Pd:Fe ratio but over a substantially longer sputtering time (1200 minutes).

In contrast, the acid-washed palladized iron samples exhibit different depth profiles. As seen in Fig. 10-5c, the Pd:Fe ratio in the acid-washed sample before reaction increases sharply during the first few sputtering cycles before decreasing. A similar trend is observed in the depth profile of the acid-washed sample after reaction (Fig. 10-5d), with the Pd:Fe ratio initially increasing before falling off over the next several hundred minutes of sputtering time.

The results suggest that the distribution of palladium through the near-surface region is dependent upon the sample preparation method. In the case of the Pd-Fe-U samples (Figs. 10-5a and 10-5b), the depth profiles indicate that the palladium concentration is highest at the surface. That is, the Pd:Fe ratio for these samples is

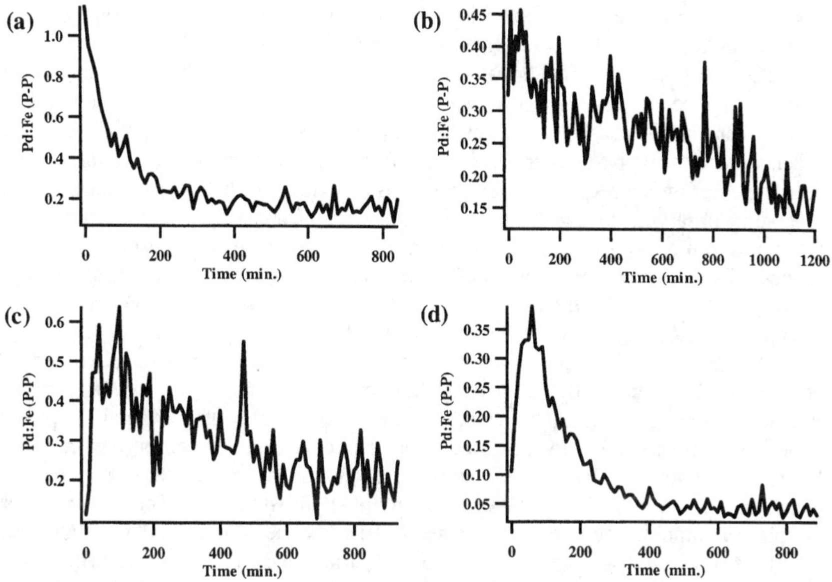

Fig. 10-5. AES depth profiles showing the change in Pd:Fe ratio through the near-surface region for unwashed palladized iron (Pd-Fe-U) samples before (a) and (b) after column reaction, and acid-washed palladized iron (Pd-Fe-A) samples before (c) and after (d) column reaction

initially high, and then declines as the palladium is removed by sputtering. In contrast, the Pd-Fe-A samples have a lower Pd:Fe ratio at the surface compared to the Pd-Fe-U samples, but the ratio increases during the early stages of the sputtering process. This initial increase in the Pd:Fe ratio is ascribed to the removal of an overlying iron oxide film. At longer sputtering times, the Pd:Fe ratio declines as palladium is removed by the sputtering process. Fitting a polynomial to the curves of Fig. 10-5 and integrating indicated that the amount of Pd deposited was similar on both the acid-washed and unwashed iron surfaces.

This difference in the near-surface distribution of palladium, rather than a difference in the total amount of palladium plated, explains the dissimilarity in reactivity between the Pd-Fe-A and Pd-Fe-U samples. Because the majority of the palladium on the Pd-Fe-A samples is concentrated below the surface, most of the Pd is not available to catalyse the TCE hydrodehalogenation reaction. This is further supported by the fact that the integrated areas under the various AES depth profile curves (Fig. 10-5) are comparable, indicating that the total amount of palladium plated onto each sample is approximately equal, consistent with the similar plating efficiencies determined by DCP on both the Pd-Fe-A and Pd-Fe-U samples. Although we cannot entirely rule out the possibility that Pd could be lost as particulates during column operation, Pd solution analysis and various surface analyses indicate that Pd loss during column operation was minimal.

The substantially smaller amount of exposed Pd on the surface of the acid-washed iron compared to the unwashed iron is believed to be a consequence of the fact that acid washing produces a non-passive surface layer. Specifically, the absence of a fully passive surface film after acid washing results in faster iron corrosion during subsequent Pd plating. This is evidenced by vigorous gas bubble formation observed during the plating process of the Pd-Fe-A samples, indicating that rapid iron corrosion occurred concurrently with Pd deposition during the plating process. The formation of the associated corrosion products can bury the Pd deposits as they form. Similarly, the changing composition of the surface of the iron during column operation can continue to affect the reactivity of the material.

Deposition of Pd onto the iron surfaces creates numerous micro-galvanic Pd-Fe cells as well as coupling between palladium and various iron oxides such as magnetite. The galvanic coupling results in positive shifts of corrosion potentials (Pourbaix 1973), leading to increased rates of anodic iron dissolution and cathodic hydrogen evolution. The Pd-Fe^0 coupling results in a greater shift in corrosion potential than in the case of Pd-oxide coupling (Pourbaix 1973). Thus, when Pd is coupled predominantly with iron oxides (Pd-Fe-U), electrochemical reaction rates will be slower than when Pd is coupled with iron (Pd-Fe-A), resulting in less iron corrosion and consequently less oxide film formation and hydrogen production. This may explain the more rapid rate of hydrogen production initially observed in the Pd-Fe-A columns compared to the Pd-Fe-U columns (Fig. 10-2).

Surface analyses of Pd-Fe-A and Pd-Fe-U samples after reaction in the column suggest that, for both types of sample, further oxide growth occurs over some of the surface palladium during reaction in the column, similar to the results of Muftikian (1996). AES (Fig. 10-4) and XPS data (Table 10-2) show a decrease in the Pd: Fe ratio after the palladized iron samples were used in the columns. In the case of the Pd-Fe-A samples, palladium concentrations fell below the detection limits of both electron spectroscopy techniques. Although both the unwashed and acid-washed palladized iron samples exhibit a decrease in the Pd:Fe ratio, the effect is more pronounced for the Pd-Fe-A samples due to the faster corrosion rate of the Pd-Fe galvanic couple. The gradual decrease in the reactivity of the Pd-Fe-U columns over the 200-day operation (Fig. 10-3) is consistent with the slower corrosion rates and gradual partial coverage of the surface Pd deposits. Furthermore, we speculate, that even with partial coverage of the palladium with iron oxides, at the TCE concentrations used in the experiments, the number of reactive surface sites was not the main rate limiting factor.

10.4 Conclusions

The present study clearly demonstrates that plating a small amount (0.05 wt %) of palladium onto oxide-covered granular iron, using the displacement plating method, is efficient and the resulting bimetals are effective for TCE degradation. Washing the base metal with acid before plating is not needed, thus making the plating procedure simpler and more practical. More importantly, the Pd-Fe-U material exhibited a much slower rate of deactivation than the Pd-Fe-A material. The nature of the iron surface preparation prior to Pd plating significantly affects the reactivity and deactivation rate of the resulting Pd-Fe materials. Surface and near-surface analyses using XPS and Auger as well as DCP measurements demonstrate that differences between Pd plated onto unwashed and acid-washed Fe particles are not due to variations in plating efficiency, nor to differences in Pd chemical oxidation state, but to the rate and extent of iron oxide growth during the plating process and hydrodechlorination reactions that, in the case of the acid-washed Fe particles, effectively bury the Pd deposits. Most studies of Pd-Fe bimetals for dehalogenation reactions have used acid-washed iron as the support material, and in the cases where longevity has been considered, have shown rapid deactivation of the catalyst. As a consequence, the practicability of using Pd-Fe materials has been in doubt. The results of this study indicate that the initial surface condition of the iron can have a major effect on rates of deactivation. By plating onto previously oxidized iron surfaces that have not been acid washed, the resulting product will be less prone to deactivation and may be suitable for prolonged practical applications.

Acknowledgements

This research was funded by the NSERC/DuPont/EnviroMetal Industrial Research Chair in Groundwater Remediation held by Dr. R. W. Gillham. The authors would also like to acknowledge the surface analysis laboratory of the Materials Research Science and Engineering Center at The Johns Hopkins University, funded through the National Science Foundation.

10.5 References

Alonso, F., Beletskaya, I. P., and Yus, M. (2002). "Metal-mediated reductive hydrodehalogenation of organic halides." *Chem. Rev.*, 102, 4009-4091.

Agrawal, A., and Tratnyek, P. G. (1996). "Reduction of nitro aromatic compounds by zero-valent iron metal." *Environ. Sci. & Technol.*, 30, 153-160.

Barr, T. L., and Seal, S. (1995). "Nature of the use of adventitious carbon as a binding energy standard." *J. Vac. Sci. Technol., A,* 13, 1239.

Blowes, D. W., Ptacek, C. J., Benner, S. G., McRae, C. W. T., Bennett, T. A., and Puls, R. W. (2000). "Treatment of inorganic contaminants using permeable reactive barriers." *J. Contam. Hydrol.*, 45, 123-137.

Davis, L. E., MacDonald, N. C., Palmberg, P. W., Riach, G. E., and Weber, R. E. (1976). *Handbook of Auger Electron Spectroscopy*, Physical Electronics Industries, Inc., Eden Prairie, Minnesota.

Gillham, R. W., and O'Hannesin, S. F. (1994). "Enhanced degradation of halogenated aliphatics by zero-valent iron." *Ground Water,* 32, 958-967.

Grittini, C., Fernando, Q., and Korte, N. (1995). "Rapid dechlorination of polychlorinated biphenyls on the surface of a Pd/Fe bimetallic system." *Environ. Sci. & Technol.*, 29, 2898-2900.

Gui, L., Gillham, R. W., and Odziemkowski, M. S. (2000). "Reduction of N-nitrosodimethylamine with granular iron and nickel-enhanced iron. 1. Pathways and kinetics." *Environ. Sci. & Technol.*, 34, 3489-3494.

Johnson, T. L., Scherer, M. M., and Tratnyek, P. G. (1996). "Kinetics of halogenated organic compound degradation by iron metal." *Environ. Sci. & Technol.*, 30, 2634-2640.

Korte, N., Zutman, J., Schlosser, R., Liang, L., Gu, B., and Fernando, Q. (2000). "Field application of palladized iron for the dehalogenation of trichloroethene." *Wast. Manag.*, 20, 687-694.

Liang, L., Korte, N., Goodlaxson, J. D., Clausen, J., Fernando, Q., and Muftikian, R. (1997). "Byproduct formation during the reduction of TCE by zero-valence iron and palladized iron." *Ground Water Monit. Rem.*, 17, 122-127.

Lien, H.-L., and Zhang, W.-X. (2005) "Hydrodechlorination of chlorinated ethanes by nanoscale Pd/Fe bimetallic particles." *J. Environ. Eng. ASCE*, 131(1), 4-10.

Lowry, G. V., and Reinhard, M. (1999). "Hydrodehalogenation of 1- to 3-carbon halogenated organic compounds in water using a palladium catalyst and hydrogen gas." *Environ. Sci. & Technol.*, 33, 1905-1910.

Muftikian, R., Fernando, Q., and Korte, N. E. (1995). "A method for the rapid dechlorination of low molecular weight chlorinated hydrocarbons in water." *Wat. Res.*, 29, 2434-2439.

Muftikian, R., Nebesny, K., Fernando, Q., and Korte, N. (1996). "X-ray photoelectron spectra of the palladium-iron bimetallic surface used for the rapid dechlorination of chlorinated organic environmental contaminants." *Environ. Sci. & Technol.*, 30, 3593-3596.

Odziemkowski, M. S., Gui, L., Gillham, R. W., and Irish, D. E. (2000). "The role of oxide films in the reduction of N-nitrosodimethylamine with reference to the iron groundwater remediation technology." In *Proceedings of the International Symposium*, The Electrochemical Society Proceedings Series, 4, 357-368.

O'Hannesin, S. F., and Gillham, R, W. (1998). "Long-term performance of an in situ "iron wall" for remediation of VOCs." *Ground Water*, 36, 164-170.

Pourbaix, M. (1973). *Lectures on electrochemical corrosion*, Plenum Press, New York.

Powell, C. J., Jablonski, A., Tilinin, I. S., Tanuma, S., and Penn, D. R. (1999). "Surface sensitivity of Auger-electron spectroscopy and x-ray photoelectron spectroscopy." *Elec. Spectrosc. Relat. Phenom.*, 98-99, 1-15.

Ukisu, Y., and Miyadera, T. (2003). "Hydrogen-transfer hydrodechlorination of polychlorinated dibenzo-*p*-dioxins and dibenzofurans catalyzed by supported palladium catalysts." *Appl. Catal. B: Environ.*, 40, 141-149.

Wagner, C. D., Riggs, W. M., Davis, L. E., Moulder, J. F., and Muilenberg, G. E. (1979). *The Handbook of X-ray Photoelectron Spectroscopy*, Perkin Elmer Corporation, Eden Prairie, Minnesota.

Wang, C., and Zhang, W. (1997). "Synthesizing nanoscale iron particles for rapid and complete dechlorination of TCE and PCBs." *Environ. Sci. & Technol.*, 31, 2154-2156.

Xu, X., Zhou, H., and Wang, D. (2005) "Structure relationship for catalytic dechlorination rate of dichlorobenzenes in water." *Chemosphere*, 58, 1497-1502.

Zhang, W., Wang, C.-B., and Lien, H.-L. (1998). "Treatment of chlorinated organic contaminants with nanoscale bimetallic particles." *Catal. Today,* 40, 387-395.

CHAPTER 11

Nanoscale Bimetallic Pd/Fe Particles for Remediation of Halogenated Methanes

Hsing-Lung Lien[1], Yuan-Pang Sun[2], and Wei-Xian Zhang[3]

Abstract: Iron nanoparticles, a great extension of conventional zero-valent iron technology, have been demonstrated as reactive reagents for remediation of a wide array of contaminants including both organics and inorganics. Because of their enormous feasibility and high reactivity, they could serve as cost-effective solutions to some of the most challenging environmental cleanup problems. As this technology is moving rapidly from laboratory research to real world implementation, a number of fundamental issues (e.g., mobility, environmental impact, reactivity and contaminant degradation mechanisms) related to its potential applications have been received a great attention. In this paper, halogenated methanes were selected to investigate contaminant degradation processes in reaction with nanoscale palladium/iron (Pd/Fe) particles using thermodynamic and kinetics techniques. It has been found that the reactivity of halogenated methanes is a function of bond strengths of carbon-halogen bond. This suggests that the bond strength of halogenated methanes is a significant factor in determining reaction mechanisms. The formation of methane and long-chain hydrocarbons (e.g., butane) in the transformation of halogenated methanes by nanoscale Pd/Fe particles has been observed. Kinetic simulations indicate that degradation processes of halogenated methanes involve complicated pathways including sequential, parallel, and multi-step surface reactions.

CE Database subject headings: Abatement and removal; Ground-water pollution; Halogen organic compounds; Iron; Particle size.

11.1 Introduction

Iron nanoparticle technology, utilizing zero-valent iron in the range of 1-100 nm as a reactive reagent for contaminant remediation, represents perhaps one of the first environmental nanotechnologies. Since 1995, work at Lehigh University has pioneered the research and development of iron nanoparticles for environmental remediation. The first synthesis of iron particles with the borohydride method was developed in 1997 (Wang and Zhang 1997). Work has established that nanoscale

[1] Assistant Professor, Dept. of Civil and Environmental Engineering, National University of Kaohsiung, 811 Kaohsiung, Taiwan, ROC.

[2] Dept. of Civil and Environmental Engineering, Lehigh University, Bethlehem, PA 18015, United States.

[3] Associate Professor, Dept. of Civil and Environmental Engineering, Lehigh University, Bethlehem, PA18015 United States (corresponding author). E-mail: wez3@lehigh.edu

iron particles were effective reductants and catalysts for a wide variety of common environmental contaminants (Table 11-1). Interests on the iron nanoparticle technology have been growing rapidly over last 3-4 years. Recent field tests have further demonstrated promising prospective for *in situ* remediation (Zhang 2003; Elliot and Zhang 2001).

Table 11-1. Common Environmental Contaminants that Can Be Treated with Bimetallic or Iron Nanoparticles

Type of Contaminants	Name	References
Halogenated aliphatic hydrocarbons	Carbon tetrachloride	Lien and Zhang (1999)
	Chloroform	Lien and Zhang (1999)
	Bromoform	This study
	Dibromomethane	This study
	Tetrachloroethylene	Lien and Zhang (2001)
	Trihchloroethlene	Lien and Zhang (2001)
	Dichloroethylenes	Lien and Zhang (2001)
	Vinyl chloride	Lien and Zhang (2001)
	Hexachloroethane	Lien and Zhang (2005)
	Pentachloroethane	Lien and Zhang (2005)
	Tetrachloroethanes	Lien and Zhang (2005)
	Trichloroethanes	Lien and Zhang (2005)
Halogenated aromatic hydrocarbons	Hexachlorobenzene	Xu and Zhang (2000)
	Pentachlorobenzene	Xu and Zhang (2000)
	Tetrachlorobenzenes	Xu and Zhang (2000)
	Trichlorobenzenes	Xu and Zhang (2000)
	PCBs	Wang and Zhang (1997); Yak et al. (1999); Lowry and Johnson (2004)
Pesticides	DDT	Engelmann et al. (2001)
	Carbothioate Herbicide Molinate	Joo et al. (2004)
		Joo et al. (2004)
Nitro aromatic compounds	Nitrotoluene/Dinitrotoluene	Choe et al. (2001)
	Nitrobenzene/Dinitrobenzene	Choe et al. (2001)
Heavy metals	Lead	Ponder et al. (2000)
	Arsenic	Kanel et al. (2005)
	Chromate	Ponder et al. (2000)
Other inorganic contaminants	Nitrate	Choe et al. (2000)
	Perchlorate	Masciangioli and Zhang (2003)

As this technology is moving rapidly from laboratory research to real world implementation, still, there are certainly a number of fundamental issues related to the potential applications of iron nanoparticles for environmental remediation. In particular, issues that are essential to be addressed include: (1) mobility, (2) cost-effectiveness, (3) environmental impact, and (4) reactivity and contaminant degradation mechanisms. An important attribute of nanoparticles is the potential high mobility in environmental media, especially in the subsurface environment. An iron particle in water is subject to various interactive forces, e.g., gravitational, electrostatic, magnetic, and random Brownian motion. For large particles (e.g., > 1 μm), the gravitational force dominates and could lead to rapid sedimentation of particles. For ultra-fine iron particles (< 100 nm), the random Brownian motion likely governs. Evidence observed suggests that iron nanoparticles have strong tendency to form much larger aggregates. A relatively stable dispersion of iron particles can be achieved by modifying the iron surface. The surface modified (i.e., charged) iron nanoparticles could remain in suspension for extended periods (Schick et al. 2004). Only a stable dispersion of iron particles offers the possibility of uncomplicated injection, mixing and rapid transport within the slow flowing groundwater.

An outstanding concern of cost-effectiveness of nanoscale iron particles is a question being asked constantly. In the early phase of our work (prior to 2000), there was no commercial supply of iron nanoparticles. Small quantities (e.g., < 10 kg) were prepared in laboratory for bench- and pilot-scale tests. The cost was rather high (e.g., > \$500/kg). Several vendors have since started limited production of iron nanoparticles. The price is coming down rapidly (e.g., << \$50/kg as of mid-2004). Even with the increased supply and reduced price, the general perception is that the nanoparticles are still too expensive for real world applications. A careful analysis suggests the contrary, that iron nanoparticles actually offer a much better economics due to its large surface area (Table 11-2). Research has demonstrated that the iron nanoparticles have much higher surface activity per unit of surface area. For example, the surface area normalized reaction rate for TCE and DCEs dechlorination for iron nanoparticles are 1-2 orders of magnitudes higher than bulk (~10 μm) iron powders. As shown in Table 11-2, iron nanoparticles have much lower cost per unit surface area and consequently should offer much higher return in term of performance per unit iron mass. The key hurdle for the application of iron nanoparticles is perhaps not the cost, but demonstrated field performance of the technology at large scale.

Table 11-2. Comparison of Unit Cost between Iron Powder and Nanoscale Iron

	Iron Powder	Nanoscale Iron
Diameter (nm)	500,000	50
Surface area (m^2/kg)	1.5	15,000
Cost (USD/kg)	~0.5	< 50
Iron surface (m^2) /USD	3	> 300

So far there is no report on the eco-toxicity of low-level iron in soil and water. However it is the authors' opinion that systematic research on the environmental transport, fate and ecotoxicity is needed to overcome increasing concerns in the environmental use of nanomaterials, and minimize any unintended impact. Iron nanoparticles could actually provide a valuable opportunity to demonstrate the positive effect on environmental quality. Iron is the fifth most used element; only hydrogen, carbon, oxygen and calcium were consumed in greater quantities. It has been found at the active center of many biological molecules and likely plays an important role in the chemistry of living organisms being. It is well documented that iron is an essential constituent of the blood and tissues of the animal body. Iron in the body is mostly present as iron porphyrin or heme proteins, which include hemoglobin in the blood, myoglobin and the heme enzymes (Ballou 1999). The heme enzymes permit the reversible combination with molecular oxygen. By this mechanism the red blood cells carry oxygen from one part of the body to another. Nature has evolved highly organized system of iron uptake (e.g., the siderophore-mediated iron uptake), transport (via transferrins) and storage of iron (e.g., in the form of ferritin). The intracellular concentration of free iron, i.e., iron not bound to organic ligands, is tightly controlled in animals, plants, and microbes. This is partly a result of the poor solubility of iron under aqueous and aerobic conditions: the solubility production of $Fe(OH)_3$ is 4×10^{-38}. Within the cell, free iron generally precipitates as polymeric hydroxides.

Three main areas where the amount of bioavailable iron affects human health are pointed out below. The one that is most important numerically is iron deficiency, with usually leads to low haemoglobin levels and anemia. A large proportion of the world's population are iron deficient and in most cases iron sufficiency has been achieved relatively easily through a greater dietary intake of bioavailable iron. Too much iron, or iron overload, is not so common. Primary haemochromatosis results from a genetic defect leading to enhanced iron uptake. Secondary haemochromatosis occurs because excess iron has accumulated in the body as a result of the extensive blood transfusions. The treatment is to removal of excess non-haem iron by chelation therapy, or the removal of both haem and non-haem iron by bleeding. The third area is the role of iron in infection. Because iron is essential for the growth of microbes, pathogenic microbes may replicate much faster so under iron-enriched conditions. The challenge for us is to determine the transport, aquatic and biochemistry, ultimate fate of manufactured iron nanoparticles in the environment.

Measurements conducted in our laboratory showed that in the presence of excessive (> 10 g/L) iron nanoparticles, concentrations of dissolved iron (mostly ferrous ion or Fe(II)) are typically in the range of 10-100 mg/L. The level of dissolved iron is naturally dependent on the solution chemistry and presence of ligands such as dissolved organic substances. Data from our several recent field tests suggest that the actual increase in the dissolved iron concentrations is less than 10 mg/L as a result of dilution, dispersion, and precipitation. The doses of iron nanoparticles were in the range of 0.1-4 g/L. In the U.S., iron is regulated by the National Secondary Drinking Water Regulations, non-enforceable guidelines regulating contaminants that may cause cosmetic effects or aesthetic effects in drinking water. The secondary maximum contaminant level of iron is set at 0.3 mg/L (Code of Federal Regulations 2002).

Nanoscale bimetallic iron particles (e.g., Pd/Fe, Ag/Fe) have been shown a high reactivity in degradation of a wide array of contaminants (Lien and Zhang 2001; Xu and Zhang 2000). This can be attributed to the presence of catalysts such as palladium or silver at the iron surface and large available surface areas of nanoscale iron particles that allow reactions to occur. Although degradation of contaminants with zero-valent iron is net results of surface-mediated reactions, the heterogeneous characteristics of metal-mediated reactions involving the interactions between contaminants and metals are still limited.

The objectives of this study were to gain a better understanding of metal-mediated dehalogenation reaction using thermodynamic and kinetics techniques, which are correlation analysis and kinetic simulation. Correlation of reaction rate constants with thermodynamic parameters (i.e., linear free energy relationships, LFERs) provides insights for the mechanistic investigations (Scherer et al. 1998). Kinetic simulation, breaking down complex processes into a sequence of simple reactions, is a powerful analytical tool to verify the reaction pathways that are deduced from the product information. Fundamental information required for analysis is product distributions and reaction rates that are provided in the study.

11.2 Experimental Section

11.2.1 Batch Experiments

Five halogenated methanes including carbon tetrachloride (CT), chloroform (CF), dichloromethane (DCM), bromoform (BF) and dibromomethane (DBM) were tested in the study. Batch experiments were conducted in 150 mL serum bottles. The metal loading of nanoscale Pd/Fe particles was 12.5 g/L in reaction with chlorinated methanes but it was reduced to 5 g/L in reaction with brominated methanes because of their high reactivity. For a typical experiment, 20 μL methanol solution of a halogenated methane was spiked into an aqueous solution to a desired concentration. The serum bottles were then capped with Teflon Mininert valves and mixed on a rotary shaker (30 rpm) at room temperature (22±1°C).

11.2.2 Headspace Analysis

At selected time intervals, 20 μL headspace gas aliquot was withdrawn by a gastight syringe for GC analysis. Concentrations of halogenated compounds were measured by a HP5890 GC equipped with a DB-624 capillary column (J&W, 30 m × 0.32 mm) and an electron capture detector (ECD). Temperature conditions were programmed as follows: oven temperature at 50°C for 5 minutes and then increased to 180°C for 5 minutes with a rate of 20°C/min; injection port temperature at 180°C; and detector temperature at 300°C. Carrier gas for GC was ultrapure nitrogen at a flow rate of 4.86 mL/min. Calibration curves for each model compound were made initially and the variability was checked daily before analysis (< 15%).

Hydrocarbon products in the headspace were qualitatively identified with a Shimadzu QP5000 GC-MS and further quantified with GC analysis. A quadrupole mass spectrometer was set to scan from 10 to 150 m/z with data collection every 0.1 seconds. Oven temperature was set at 30°C, injection port temperature at 150°C and detector temperature at 230°C. Identification of individual compounds was made by comparing the resulting mass spectral pattern with those of the National Testing and Information Service spectral library and verified with standards as well as the GC retention time. The GC was equipped with a flame ionization detector (FID) and an AT-Q column (*Alltech*, 30 m x 0.32 mm). Oven temperature was set at 30°C, injection port temperature at 250°C and detector temperature at 300°C. The GC retention time of the sample should be within 0.1 minute of the reference standard.

11.2.3 Kinetic Analysis

Reaction rates of dehalogenation in a batch reactor are described by pseudo-first order rate law (Johnson et al. 1996):

$$\frac{dC}{dt} = -k_a C = -k_{SA} a_s \rho_m C \tag{11.1}$$

where C is the concentration of organic compound in the aqueous phase (mg/L); k_a is the measured rate constant (h^{-1}); k_{SA} is the surface-area normalized rate constant ($L/h/m^2$); a_s is the specific surface area of metal (m^2/g); ρ_m is the mass concentration of metal (g/L); and t is time (h). For a specific system, k_{SA}, a_s and ρ_m are constants. BET analysis gave a specific surface area of approximately 33.5 m^2/g (Wang and Zhang 1997).

11.2.4 Materials and Chemicals

HPLC grade halogenated methanes were obtained from *Aldrich*. A standard gas mixture for GC analysis was obtained from *Supelco*. Single standard gases of 1.04% ethylene and 1.04% methane were acquired from *Aldrich*. Sodium borohydride ($NaBH_4$, 98%) and ferric chloride ($FeCl_3 \cdot 6H_2O$, 98%) were from *Aldrich*. Palladium acetate ($[Pd(C_2H_3O_2)_2]_3$, Pd 47.4%) was from *Alfa*.

11.3 Results

11.3.1 Product Distributions and Reaction Rates

Two analysis methods, correlation analysis and kinetic simulation, were used to explore the reaction mechanisms of halogenated methanes reacting with nanoscale Pd/Fe particles. Correlation of rate constants with other kinetic constants, with thermodynamic properties, and with other predictor variables is required to use a database of reaction rates of similar compounds. Accordingly, halogenated methanes including carbon tetrachloride (CT), chloroform (CF), bromoform (BF), dichloromethane (DCM), and dibromomethane (DBM) were selected for analysis. For the kinetic simulation, product information and rate constants are two essential quantities to verify the proposed reaction pathways.

Transformation of brominated methanes with nanoscale Pd/Fe particles is shown in Fig. 11-1. The initial concentration of BF and DBM were 30 mg/L and the metal particle loading was 5 g/L. As shown in Fig. 11-1a, complete degradation of BF was accomplished within 0.5 hour with a few end-products. Dibromomethane, bromomethane (BM), methane and ethane were detected within six hours. Trace amounts of long-chain hydrocarbons such as propene and butenes were also observed. At its peak concentration, BM accounted for about 30% of the BF disappearance at 30 minutes while a relatively small amount of DBM (less than 10%) was also detected. Methane was the major product, which accounted for approximately 70% of BF. Ethane was minor hydrocarbon product. Its concentration remained relatively constant with a yield of about 15%. Overall, the mass balance of carbon was between 85% and 100% except during the initial reaction period where the mass recovery was only 60%. This may be due to initial sorption of BF to the metal surface.

Dehalogenation of DBM with nanoscale Pd/Fe particles is shown in Fig. 11-1b. Similar to BF, degradation of DBM was completed in less than one hour. Methane was the major product accounting for about 80% of DBM disappearance. Only trace amounts of C_2 hydrocarbons including ethane and ethylene were detected (not shown

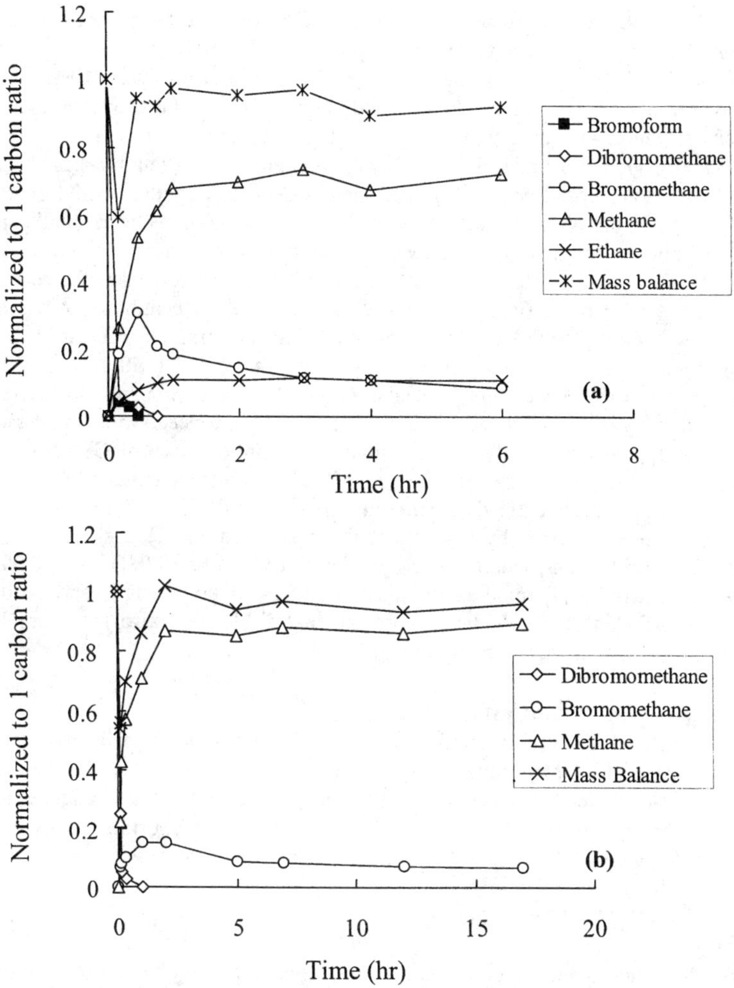

Fig. 11-1. Transformation of 30 mg/L of halogenated methanes using nanoscale Pd/Fe particles: (a) bromoform and (b) dibromomethane

in Fig. 11-1 b). Bromomethane was the only detectable brominated intermediate with a yield of about 15%. A similar carbon mass balance as described above was observed in this case where a deficiency of carbon in the aqueous phase was found during the initial phase.

Results of transformation of chlorinated methanes have been published in our previous work (Lien and Zhang 1999). The order of degradation rate was CT>CF>DCM. It should be noted that only little reaction was found in the transformation of DCM. In the transformation of CT, methane (~55%), CF (max. 25%) and DCM (~25%) accounted for approximately 80% of the CT lost. Tetrachloroethylene (PCE), trichloroethylene (TCE) and chloromethane (CM) were detected in a minor amount. Currently, our study further indicated that trace amounts of hexachloroethane and acetylene were also produced. Formation of acetylene involved both hydrogenolysis and reductive-β-elimination reactions has been reported in the transformation of TCE with zero-valent iron (Arnold and Roberts 2000). Quantification of hydrocarbons under low concentrations is a challenge because the rate of hydrocarbon production decreases significantly as the chain length grows. For example, it is very difficult to detect hydrocarbons with a chain length more then 3 carbons (e.g., propane) when the initial concentration of parent compounds was only 30 mg/L. In order to overcome this problem caused by the detection limit, experiments using a high concentration of CT (150 mg/L) were conducted. Hydrocarbons including methane, ethylene, ethane, propane, propylene, propyne, and n-butane were detected. Similar to the low concentration, methane, accounted for approximately 50% of the initial mole concentration of CT, was the major product whereas concentrations of C_2 and C_3 compounds were much lower.

Based on the experimental data presented in Fig. 11-1, specific reaction rate constant, k_{SA}, can be obtained by best fits of the data to Eq. (11.1), and results are summarized in Table 11-3, where the k_{SA} of BF and DBM were 0.079 and 0.044 ($L/h/m^2$), respectively. Thermodynamic data including bond strength and redox potential are included. It should be pointed out that Table 11-3 also incorporated results of our previous work (Lien and Zhang 1999).

11.3.2 Correlation Analysis

The correlation analysis was accomplished with the two-electron transfer reduction potential via rates of hydrogenolysis as given in Fig. 11-2a. The standard reduction potentials of halogenated methanes were estimated based on Eq. (11.2) (Vogel et al. 1987) where the conditions were carried out at 25°C, pH 7 and chloride activity and bromide activity were 0.001 and 0.00001, respectively.

$$RX + H^+ + 2e^- \rightarrow RH + X^- \qquad (11.2)$$

Attempts were not made to examine the one-electron reduction potential for halogenated methanes because no sufficient thermodynamic data are available. The lack of systematic trends (R^2 = 0.025) was observed in the correlation between k_{SA} values and reduction potentials. However, for the correlation analysis using bond strength as the descriptor variable, a better correlation between the k_{SA} values and the bond strength was determined (R^2 = 0.5681, Fig. 11-2b). These results suggest that breaking of the carbon-halogen bond in the dehalogenation of halogenated methanes is an important factor in reaction mechanisms, although only limited thermodynamic data are available at this stage.

Table 11-3. Summary of the Intermediates, End-Products, and Reaction Rates for Dehalogenation of Halogenated Methanes by Nanoscale Pd/Fe Particles

Halogenated Methanes	Products	k_a (h^{-1})	k_{SA} ($L/h/m^2$)	Metal loading (g/L)	Redox potential (V)[a]	Bond strength (kJ/mol)[b]
Carbon Tetrachloride (CT)[1]	CF (max. 24%), DCM (25%), CH_4 (55%) C_2H_6 (3.3%) Unaccounted-for mass (11.7%)	3.81	0.009	12.5	0.67	305.9
Chloroform (CF)[1]	DCM (12.6%), CH_4 (70%) Unaccounted-for mass (17.4%)	2.84	0.0065	12.5	0.56	338.5
Dichloromethane (DCM)[1]	No reaction			12.5	0.49	350.2
Bromoform (BF)[2]	DBM (max. 9%), BM (8.2%), CH_4 (71%), C_2H_6 (11%) Unaccounted-for mass (9.8%)	13.8	0.079	5	0.61	292
Dibromomethane (DBM)[2]	BM (6.2%), CH_4 (90%), C_2H_6 (1%) Unaccounted-for mass (2.8%)	7.77	0.044	5	0.5	296.7

[1]Data obtained for Lien and Zhang (1999).
[2]This study.
[a]Vogel et al. 1987.
[b]Lide 1993.

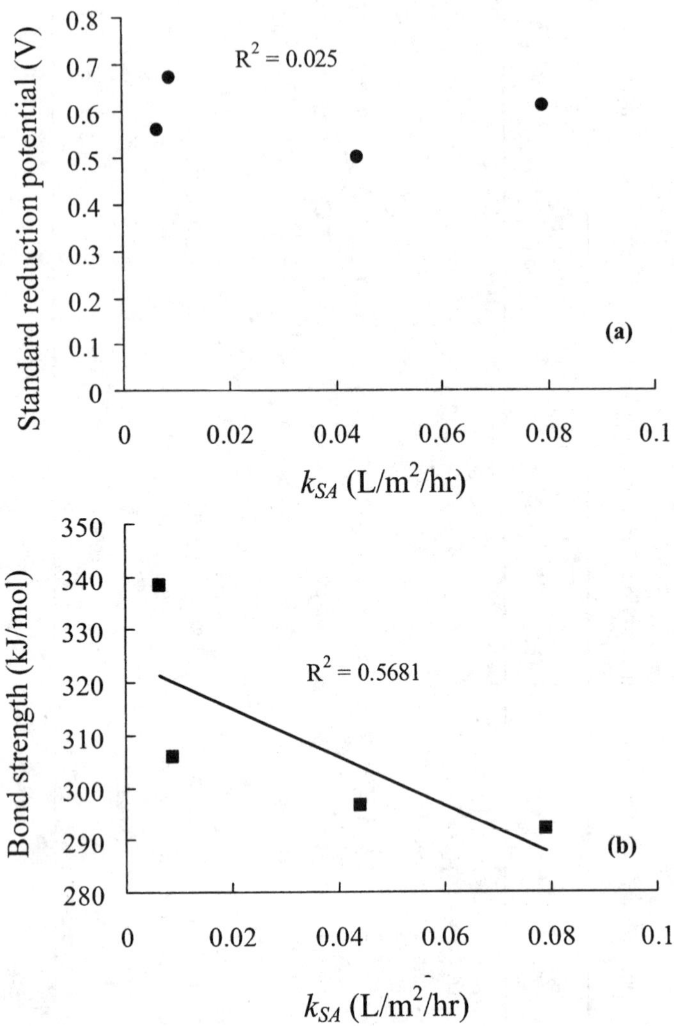

Fig. 11-2. Regression analysis between surface-area normalized rate constants (k_{SA}) and thermodynamic parameters of halogenated methanes: (a) reduction potentials and (b) bond strength of carbon-halogen bonds

11.3.3 Kinetic Simulation

Kinetic simulation was conducted to verify possible reaction pathways using a set of experimental data collected from the reaction of chlorinated methanes with nanoscale Pd/Fe particles. Bromated methanes were not taken into account because of insufficient experimental data determined. Nevertheless, the simulation results should be applicable to bromated methanes under identical conditions.

Sequential Pathways. Sequential dehalogenation is apparently reasonable to account for the transformation of halogenated methanes based on the observed products. For example, intermediates such as CF and DCM and CH_4 were observed in the transformation of CT by nanoscale Pd/Fe particles. One might be explained on the basis of the following set of sequential reactions, which are hydrogenolysis reactions as described in Eq. (11.3):

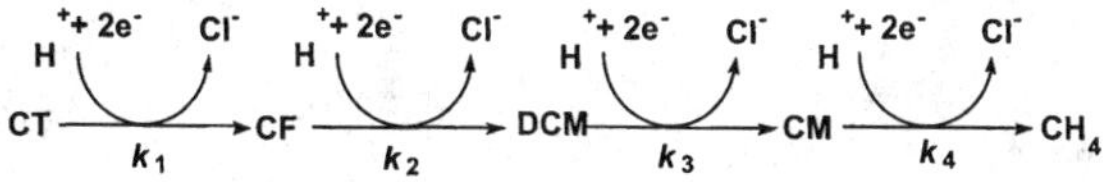

(11.3)

Accordingly, using the measured rates constants, which have been shown in Table 11-3, the sequential dehalogenation of CT can be simulated by following equations:

$$\frac{d(CT)}{dt} = -k_1(CT) \tag{11.4}$$

$$\frac{d(CF)}{dt} = +k_1(CT) - k_2(CF) \tag{11.5}$$

$$\frac{d(DCM)}{dt} = +k_2(CF) - k_3(DCM) \tag{11.6}$$

$$\frac{d(CM)}{dt} = +k_3(DCM) - k_4(CM) \tag{11.7}$$

$$\frac{d(CH_4)}{dt} = +k_4(CM) \tag{11.8}$$

where k_1, k_2, and k_3 are the measured rate constants. It should be pointed out that because no measurable reaction of DCM to CM was observed, Eqs. (11.7) and (11.8) can be neglected for the simulation.

The values of k_1, and k_2 were determined to be 3.81, and 2.84 (h^{-1}), respectively, based on the experimental data of first-order rate constants showing in Table 11-3. Because of lack reactivity of DCM, the reaction rate constant of DCM (k_3) was assigned to be zero. The simulation results of sequential reactions along with the experiment data are shown in Fig. 11-3. The yield of DCM was approximately 80% of the initial concentration of CT and minor amounts of methane (< 20%) were produced during the sequential dehalogenation of CT. However, in comparison to the experiment results, the yield of DCM accounted for only 25% of CT disappearance whereas the yield of methane was up to about 50%. Consequently, the simulation results indicated that sequential dehalogenation alone is unlikely to constitute the main pathway of CT in the presence of nanoscale Pd/Fe particles.

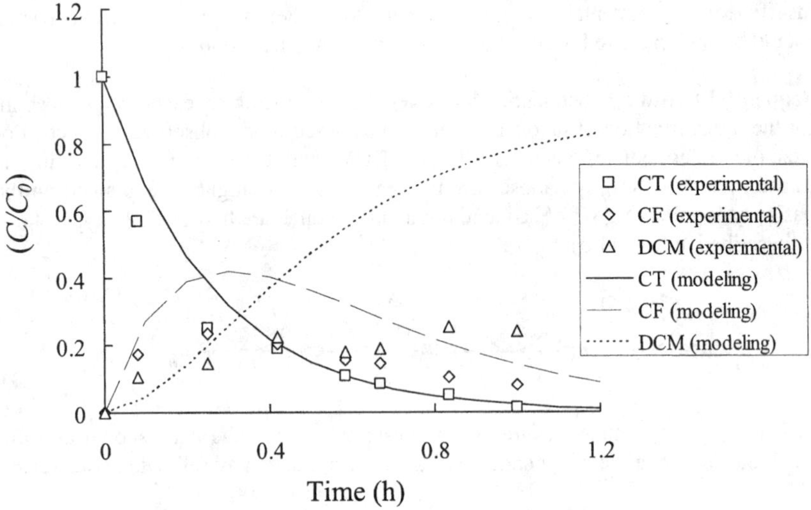

Fig. 11-3. Simulation of sequential dehalogenation reactions of CT with experimental data points entered

Parallel and Multi-Step Pathways. In considering the importance of both parallel and sequential reactions in the degradation of CT, conceptual reaction pathways are proposed in Scheme 1.

Scheme 1

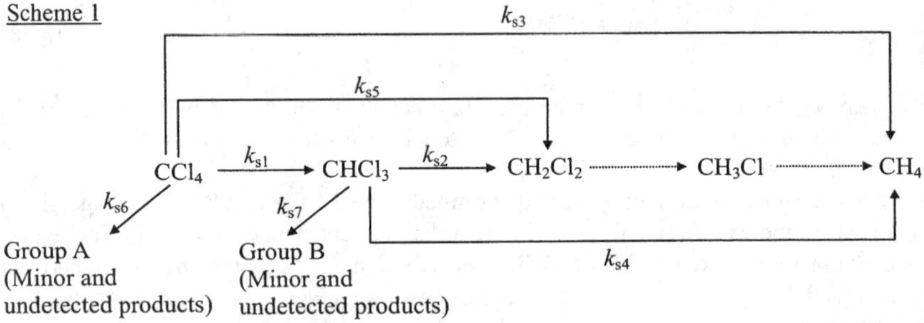

Parallel pathways including the transformation of CT (k_{s3}) and CF (k_{s4}) via a direct pathway to CH_4, and the transformation of CT to DCM (k_{s5}) are incorporated. Sequential hydrogenolysis reactions, indicating the degradation of CT to CF (k_{s1}) and from CF to DCM (k_{s2}), are also considered. As mentioned above, no rate constant of hydrogenolysis of DCM to CM was assigned because the degradation rate of DCM is negligible. Further, the bond strength of carbon-halogen bond of CM is about 352 kJ/mol, which is similar to that of DCM. Therefore, it is unlikely to be degraded according to the correction analysis as shown in Fig. 11-2b. To satisfy the carbon mass balance, rate constants, k_{s6} and k_{s7}, corresponding to the pathway of CT to group A and that of CF to group B, respectively, were therefore incorporated into the scheme. Both groups A and B accounted for minor (~3%) and undetected (~10-17%) products during reactions. Accordingly, the pathway of CT to group A and that of CF to group B are not a single-step stoichiometric reaction. Nevertheless, the validity of simulation result was not affected without taking into account of the stoichiometric ratio in Eqs. (11.13) and (11.17) as listed below. It should be noted that direct transform of CT to CM or CF to CM is not possible because the production of CM was negligible.

A set of rate equations can be written for this reaction scheme:

$$\frac{d(CT)}{dt} = -(k_{s1} + k_{s3} + k_{s5} + k_{s6})(CT) \tag{11.9}$$

$$\frac{d(CF)}{dt} = -(k_{s2} + k_{s4} + k_{s7})(CF) + k_{s1}(CT) \tag{11.10}$$

$$\frac{d(DCM)}{dt} = +k_{s5}(CT) + k_{s2}(CF) \tag{11.11}$$

$$\frac{d(CH_4)}{dt} = k_{s3}(CT) + k_{s4}(CF) \tag{11.12}$$

$$\frac{d(A)}{dt} = k_{s6}(CT) \tag{11.13}$$

The sum of k_{s1}, k_{s3}, k_{s5}, and k_{s6} should be equal to the measured rate constant of CT disappearance (k_1) while the sum of k_{s2}, k_{s4}, and k_{s7} should account for the measured rate constant of CF disappearance (k_2). Rate constants, k_{s2}, k_{s4}, and k_{s7} can be independently determined by directly fitting with the experimental data of the CF degradation. The governing equations used to simulate the reaction process of the only transformation of CF alone as shown in Scheme 1 are expressed as follows:

$$\frac{d(CF)}{dt} = -(k_{s2} + k_{s4} + k_{s7})(CF) \tag{11.14}$$

$$\frac{d(DCM)}{dt} = +k_{s2}(CF) \tag{11.15}$$

$$\frac{d(CH_4)}{dt} = +k_{s4}(CF) \tag{11.16}$$

$$\frac{d(B)}{dt} = +k_{s7}(CF) \tag{11.17}$$

Fig. 11-4 shows the simulated degradation curve of CF and formation curves of DCM and CH_4 with the experimental data points entered. The agreement between the simulation curves and experimental data is good with a small portion of unknown reactions which account for the deficiency of the carbon mass balance. The best fit data of k_{s2}, k_{s4}, and k_{s7} are 0.5, 2.0, and 0.34 (h^{-1}), respectively.

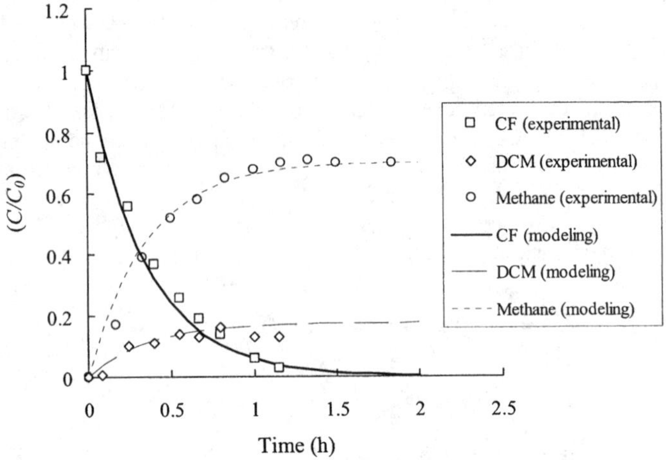

Fig. 11-4. Simulation results for the degradation of CF with the experimental data points entered

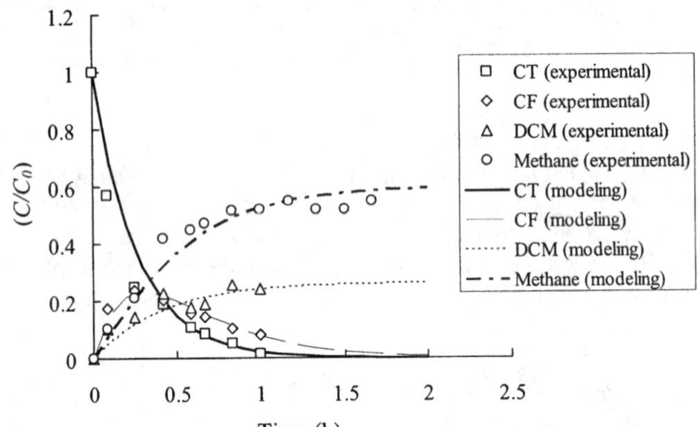

Fig. 11-5. Simulation results for the degradation of CT with the experimental data points entered

Using the best-fit rate constants, k_{s2}, k_{s4}, and k_{s7}, the rest of rate constants can be determined by simulating the degradation of CT. Fig. 11-5 shows the simulation results for the degradation of CT with the experimental data points entered. The best-fit results of k_{s1}, k_{s3}, k_{s5}, and k_{s6} are 2.0, 0.9, 0.72, and 0.3 (h^{-1}), respectively. Again, a small portion of unknown reactions may account for C_2H_6, PCE and TCE produced and the deficiency of carbon mass balance for CT. A consistence between the experimental data and simulated curves indicated that the reaction pathways proposed in Scheme 1 offer a better explanation of degradation processes than sequential hydrogenolysis reactions in the transformation of chlorinated methanes by nanoscale Pd/Fe particles.

11.4 Discussion

Reduction potentials and bond strengths are the thermodynamic parameters that may have significant impact on reaction mechanisms in the chemical processes. The use of these descriptor variables is to examine the validity of the following assumptions: (1) dehalogenation reactions are limited by the electron transfer process (e.g., $RX + H^+ + 2e^- \rightarrow RH + X^-$); or (2) dehalogenation reactions are a surface reaction in which the breaking of carbon-halogen bonds is key step. For halogenated methanes, correlation analysis suggested that the bond strength of carbon-halogen bonds in halogenated methanes, rather than the reduction potentials, is a significant factor influencing reaction mechanisms.

The lack of reactivity of DCM further supports the significance of the bond strength of carbon-halogen bonds for degradation of halogenated methanes. The bond strength of C-Cl bond for DCM (350.2±0.8 kJ/mol) is notably higher than that of either CT (305.9±7.5 kJ/mol) or CF (338.5±4.2 kJ/mol) (Lide 1993). On the contrary, DBM, which has same molecular structure as DCM, possesses a relatively low bond strength (296.7 kJ/mol). It showed a high kinetic rate in reaction with nanoscale Pd/Fe particles.

Kinetic simulation suggested that the conventional sequential reaction is not a major reaction pathway. The use of sequential hydrogenolysis reactions to explain the formation of methane failed because of the lack of reactivity for the transformation of DCM. In contrast, a combination of sequential, parallel, and multi-step reactions as proposed in Scheme 1 provides a better explanation of experimental data. Scheme 1 also revealed that the direct transformation of CT or CF to methane is a possible pathway. This suggested a surface reaction has to occur in the metal-mediated dehalogenation reaction.

Further, in this study, the formation of methane and long-chain hydrocarbons in the transformation of halogenated methanes by nanoscale Pd/Fe particles was observed. Long-chain hydrocarbons (longer than parent compounds) such as C_3 to C_5 alkanes (e.g., propane, hexane) in the dehalogenation of halogenated aliphatic compounds (HACs) using zero-valent iron and palladium have been widely reported (Liang et al. 1997; Campbell et al. 1997; Fennelly and Roberts 1998). The production of many hydrocarbons in the presence of iron was also observed even in the absence of HACs (Deng et al. 1997; Hardy and Gillham 1996). Possible carbon sources for those hydrocarbons appearing in the absence of HACs may include carbide carbon in the iron (Deng et al. 1997) and aqueous CO_2 (Hardy and Gillham 1996). Although the sources of carbon are still an open question, it is believed that hydrocarbon formation is likely the result of surface-mediated reactions similar to the well-known Fischer-Tropsch synthesis.

Evidence has indicated a surface reaction characteristic of zero-valent iron reaction with many contaminants (Matheson and Tratnyek 1994; Weber 1996). Consequentially, the importance of surface intermediates in the influence of reaction mechanisms should be taken into account. Many surface intermediates have been proposed for chlorinated methanes reaction with active metals as indicated in Table 11-4. The surface intermediates of chlorinated methanes included a variety of moieties such as CH_3 and CH_2 species that indicated the possibility of conversion of CT to methane and other chlorinated intermediates on the metal surface. However, in homogeneous systems, the formation of methane from CT transformation has been shown to be very difficult. For example, in the corrinoid-mediated dechlorination of CT, the production of methane was observed rarely (Lewis et al. 1996).

Table 11-4. Surface Intermediates of Halogenated Methanes on Reactive Surfaces under Various Conditions

Chemicals	Surface	Environment	Surface	References
$CH_{4-n}Cl_n$ (n=0~4) species	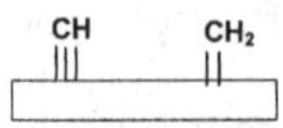	Aqueous	Pt, Pd	Müller et al. 1997
CCl_4	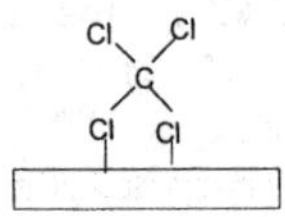	Gas phase	Fe, Pt	Smentkowski et al. 1990
CCl_4, CCl_2H_2	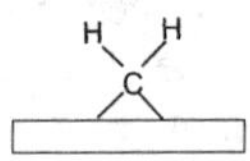	Gas phase	Co, Ni	Steinbach et al. 1985
CH_3Cl, CH_3I	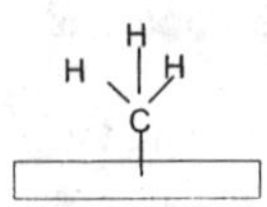	Gas phase	Cu, Co, Ni	Steinbach et al. 1985; Park et al. 1996

Similar to Fischer-Tropsch synthesis, it is possible that the transformation of CT at Pd/Fe surface results in the carbon-chlorine bond dissociation to surface intermediates such as carbyne (≡CH) or carbene (=CH_2) and eventually to form either methane or long-chain hydrocarbons. Through surface reactions, iron is capable of directly transferring CT to methane. The observation of dimerization products (i.e., hexachloroethane) in the transformation of CT suggests that trichloromethyl radicals were generated during the reaction. Trichloromethyl radicals are typical intermediates occur in homogeneous systems of CT reductive degradation (Choi and Hoffmann 1996). In heterogeneous system, the trichloromethyl radical is likely a product desorbed from the metal surface. It has been identified in the reaction of CT with iron in the gas phase (Smentkowski et al. 1990). The trichloromethyl radical undergoes either a one-electron transfer reaction to CF (major route) or a dimerization reaction to hexachloroethane (minor route).

Hexachloroethane further undergoes elimination to PCE followed by hydrogenolysis to TCE. Taken together, our data, along with evidence from the literature, suggest that the transformation of halogenated methanes involved complicated processes including metal-medicated surface reaction, radical reaction and elimination reaction.

11.5 Conclusions

Transformation of halogenated methanes by nanoscale Pd/Fe particles was examined in the study. The results of the study indicate that nanoscale Pd/Fe particles can serve as effective remedial reagents for environmental remediation. The reactivity of halogenated methanes is a function of bond strengths of carbon-halogen bond. The rank order of reactivity was bromoform > dibromomethane >> carbon tetrachloride > chloroform >> dichloromethane in reaction with nanoscale Pd/Fe particles. Methane was the major product while minor amounts of long-chain hydrocarbons were also observed. Formation of methane, ethane, ethylene and long-chain hydrocarbons is likely the result of surface-mediated reactions similar to the well-known Fischer-Tropsch synthesis. Kinetic simulations suggest that sequential hydrogenolysis reaction is unlikely the major reaction pathways in the dehalogenation of halogenated methanes. It has been demonstrated that the reaction processes involve complicated pathways including sequential, parallel and multi-step surface reactions.

11.6 References

Arnold, W. A., and Roberts, A. L. (2000). "Pathways and kinetics of chlorinated ethylene and chlorinated acetylene reaction with Fe(0) particles." *Environ. Sci. & Technol.,* 34, 1794-1805.

Ballou, D. P., Ed. (1999). *Essays in biochmemistry: Metalloproteins.* Princeton University Press, Princeton, NJ, United States.

Campbell, T. J., Burris, D. R., Roberts, A. L., and Wells, J. R. (1997). "Trichloroethylene and tetrachloroethylene reduction in a metallic iron-water-vapor batch system." *Environmental Toxicology and Chemistry.,* 16, 625-630.

Choe, S., Chang, Y. Y., Hwang, K. Y., and Khim, J. (2000). "Kinetics of reductive denitrification by nanoscale zero-valent iron." *Chemosphere*, 41, 1307-1311.

Choe, S., Lee, S.-H., Chang, Y.-Y. Hwang, K.-Y., and Khim, J. (2001). "Rapid reductive destruction of hazardous organic compounds by nanoscale Fe0." *Chemosphere*, 42, 367-372.

Choi, W., and Hoffmann, M. R. (1996). "Kinetics and mechanism of CCl_4 photoreductive degradation on TiO_2: The role of trichloromethyl radical and dichlorocarbene." *J. Phys. Chem.*, 100, 2161-2170.

Code of Federal Regulations (2002) *National secondary drinking water regulations*, Title 40, Part 143.

Deng, B. L., Campbell, T. J., and Burris, D. R. (1997). "Hydrocarbon formation in metallic iron/water systems." *Environ. Sci. & Technol.*, 31, 1185-1190.

Elliot, D. W., and Zhang, W. (2001). "Field assessment of nanoscale bimetallic particles for groundwater treatment." *Environ. Sci. & Technol.*, 35, 4922-4926.

Engelmann, M. D., Doyle, J. G., and Cheng, I. F. (2001). "The complete dechlorination of DDT by magnesium/palladium bimetallic particles." *Chemosphere*, 43, 195-198.

Fennelly, J. P., and Roberts, A. L. (1998). "Reaction of 1,1,1-trichloroethane with zero-valent metals and bimetallic reductants." *Environ. Sci. & Technol.*, 32, 1980-1988.

Hardy, L. I., and Gillham, R. W. (1996). "Formation of hydrocarbons from the reduction of aqueous CO_2 by zero-valent iron." *Environ. Sci. & Technol.*, 30, 57-65.

Johson, T. L., Scherer, M. M., and Tratnyek, P.G. (1996). "Kinetics of halogenated organic compound degradation by iron metal." *Environ. Sci. & Technol.*, 30, 2634-2640.

Joo, S. H., Feitz, A. J., and Waite, T. D. (2004). "Oxidative degradation of the carbothioate herbicide, molinate, using nanoscale zero-valent iron." *Environ. Sci. & Technol.*, 38, 2242-2248.

Kanel, S. R., Manning, B., Charlet, L., and Choi, H. (2005). "Removal of Arsenic (III) from groundwater by nanoscale zero-valent iron." *Environ. Sci. & Technol.*, 39, 1291-1298.

Lewis, T. A., Morra, M. J., and Brown, P. D. (1996). "Comparative product analysis of carbon tetrachloride dehalogenation catalyzed by cobalt corrins in the presence of thiol or titanium (III) reducing agents." *Environ. Sci. & Technol.*, 30, 292-300.

Lide, D. R., (1993). *Handbook of chem. & phys.* 74th Ed., CRC Press.

Liang, L., Korte, N., Goodlaxson, J. D., Clausen, J., Fernando, Q., and Muftikian, R. (1997). "Byproduct formation during the reduction of TCE by zero-valence iron and palladized iron. Ground water monitoring and remediation." *Ground Water Monitoring and Remediation,* Winter, 122-127.

Lien, H.-L., and Zhang, W-X. (2001). "Nanoscale iron particles for complete reduction of chlorinated ethenes." *Colloids and Surfaces A: Physicochemical & Eng. Aspects,* 191, 97-106.

Lien, H.-L., and Zhang, W-X. (1999). "Dechlorination of chlorinated methanes in aqueous solutions using nanoscale bimetallic particles." *J. Environ. Eng. ASCE,* 125, 1042-1047.

Lien, H.-L., and Zhang, W-X. (2004). "Hydrodechlorination of chlorinated ethanes by nanoscale Pd/Fe bimetallic particles," *J. Environ. Eng. ASCE,* 131, 4-10.

Lowry, G. V., and Johnson, K. M. (2004). "Congener-specific dechlorination of dissolved PCBs by microscale and nanoscale zerovalent Iron in a water/methanol solution." *Environ. Sci. & Technol.,* 38, 5208-5212.

Masciangioli, T., and Zhang, W.-X. (2003). "Environmental technologies at the nanoscale." *Environ. Sci. & Technol.,* 102A-108A.

Matheson, L. J., and Tratnyek, P. G. (1994). "Reductive dehalogenation of chlorinated methanes by iron metal." *Environ. Sci. & Technol.*, 28, 2045-2053.

Müller, U., Dülberg, A., Stoyanova, A., and Baltruschat, H. (1997). "Reactions of halogenated hydrocarbons at Pt-group — II. On the adsorption rate at Pt and Pd electrodes." *Electrochimica Acta.*, 42, 2499-2509.

Park, Y.-S., Kim, J.-Y., and Lee, J. (1996). "Abstraction of CH_3 group from the CH_3I molecule adsorbed on Cu(100) by gas-phase hydrogen atoms." *Surface Science,* 363, 62-67.

Ponder, S. M., Darab, J. G., and Mallouk, T. E. (2000). "Remediation of Cr(VI) and Pb(II) aqueous solutions using supported, nanoscale zero-valent Iron." *Environ. Sci. & Technol.*, 34, 2564-2569.

Satterfield, C. N. (1991). *Heterogeneous catalysis in industrial practice.* 2nd Ed., McGraw-Hill, Inc, New York, NY, United States.

Scherer, M. M., Balko, B. A., Gallagher, D. A., and Tratnyek, P. G. (1998). "Correlation analysis of rate constants for dechlorination by zero-valent iron." *Environ. Sci. & Technol.,* 32, 3026-3033.

Schrick, B., Hydutsky, B. W., Blough, J. L., and Mallouk, T. E. (2004). "Delivery vehicles for zerovalent metal nanoparticles in soil and groundwater," *Chem. Mater.*, 16, 2187-2193.

Smentkowski, V. S., Cheng, C. C., and Yates, J. T., Jr. (1990). "The Interaction of carbon tetrachloride with Fe (110): A system of tribological importance." *Langmuir*, 6, 147-158.

Steinbach, F., Kiss, J., and Krall, R. (1985). "Identification and stability of CH_3, CH_2, and CH species on Co and Ni surfaces, A PES investigation." *Surface Science*, 157, 401-412.

Vogel, T. M., Criddle, C. S., and McCarty, P. L. (1987). "Transformation of halogenated aliphatic compounds." *Environ. Sci. & Technol.*, 21, 722-736.

Wang, C.-B., and Zhang, W.-X. (1997). "Synthesizing nanoscale iron particles for rapid and complete dechlorination of TCE and PCBs." *Environ. Sci. & Technol.*, 31, 2154-2156.

Weber, E. J. (1996). "Iron-mediated reductive transformations: Investigation of reaction mechanism." *Environ. Sci. & Technol.*, 30, 716-719.

Xu, Y., and Zhang, W.-X. (2000) "Subcolloidal Fe/Ag particles for reductive dehalogenation of chlorinated benzenes." *Industrial & engineering chemistry research*, 39, 2238-2244.

Yak, H. K., Wenclawiak, B. W., Cheng, I. F., Doyle, J. G., and Wai, C. M. (1999). "Reductive dechlorination of polychlorinated biphenyls by zerovalent iron in subcritical water." *Environ. Sci. & Technol.*, 33, 1307-1310.

Zhang, W.-X. (2003). "Nanoscale iron particles for environmental remediation: an Overview." *J. Nanoparticle Research*, 5, 323-332.

CHAPTER 12

Reduction by Bimetallic Reactive Materials Containing Zero-valent Iron

Hocheol Song[1], YoungHun Kim[2], and Elizabeth R. Carraway[3]

Abstract: For nearly a decade the use of bimetallic reactive materials containing zero-valent iron has been investigated because of the substantial reaction rate enhancements and applicability to a broader range of contaminants compared to those materials in which iron is the only reactive metal. However, despite the knowledge gained regarding bimetallic treatment approaches, details of reaction pathways remain elusive. This paper presents a review of some fundamental processes and underlying reactions of bimetallic materials in order to provide a backdrop for interpreting reports of bimetallic material reactivity as well as understanding current approaches to improving, controlling, and utilizing their characteristics.

CE Database subject headings: Chemical reactions; Copper; Halogen organic compounds; Hydrogen; Iron; Metals.

12.1 Introduction

Reductive dechlorination has been shown to be a major degradation pathway for chlorinated hydrocarbons in reducing environments. Of the numerous reports in this field, a large portion of the research has dealt with biological processes and has focused on understanding biochemical pathways and developing effective strains of microorganisms in metabolizing chlorinated compounds. Recognition of zero-valent iron as a possible treatment medium to accomplish reductive dechlorination only dates back a decade, but there has been an explosion of research activities in the exploration of this new approach. As a reducing agent, iron is a powerful reductant, inexpensive, and able to degrade a wide range of contaminants. It is also readily available and nontoxic in nature. For such reasons, iron has been used in permeable reactive barriers and has proved to be a successful degradation strategy for the remediation of groundwater contaminated with chlorinated hydrocarbons (O'Hannesin and Gillham 1998; Vogan et al. 1999; Puls et al. 1999).

Several studies have noted shortcomings of iron-based treatment system, including accumulation of undesirable byproducts and a decrease in reactivity over

[1] Postdoctoral Research Associate, Dept. of Environmental Engineering and Science, Clemson University, 342 Computer Court, Anderson, SC 29625, United States.

[2] Assistant Professor, Dept. of Environmental Engineering, Andong National University, Andong, KyungPook 760-469, Korea.

[3] Associate Professor, Dept. of Environmental Engineering and Science, Clemson University, 342 Computer Court, Anderson, SC 29625, United States (corresponding author). E-mail:ecarraw@clemson.edu

long time periods (Orth and Gillham 1996; Vikesland et al. 2003; Klausen et al. 2003). These factors have led many researchers to investigate ways to improve metallic iron-based treatment systems. Among the numerous efforts, the use of bimetallic reductants has drawn much attention in the past few years as a possible alternative to iron metal alone. The main advantage of using bimetallic reductants lies in their substantial enhancement of reaction rates compared to unmodified irons. In many cases, reaction rates are orders of magnitude more rapid with bimetallics than unmodified irons under comparable conditions. Such an enhancement is largely due to catalytic effects of the secondary metals deposited on the iron metal surface. The metals used for deposition onto iron are Group VIII noble metals such as palladium (Pd), platinum (Pt), nickel (Ni), silver (Ag), and copper (Cu). These metals can serve as catalysts and route dechlorination reactions through more rapid catalytic pathways that also yield fewer or no byproducts of concern. Reactions of some classes of compounds that are impracticably slow with iron only become usable with bimetallic materials. Examples of these include chlorinated phenols, benzenes, and biphenyls.

The first use of a bimetallic reductant to treat organic contaminants was reported by Muftikian et al. (1995), who prepared a palladium amended iron (Pd/Fe) and used it for reduction of chlorinated methanes and ethylenes. The Pd/Fe was prepared by a procedure that is typical of many bimetallic material studies, in which a small amount (a few percent by mass) of palladium was plated onto the surface of iron metal via the spontaneous redox reaction between a palladium metal salt precursor and iron in aqueous solution. Pd/Fe rapidly transformed all the compounds tested to fully dechlorinated products without yielding chlorinated byproducts. Thereafter, a number of studies followed using other secondary or plating metals to treat various classes of compounds. Among the noble metals, palladium is the most frequently utilized for its catalytic ability and relatively higher stability against catalyst poisoning (Fung and Sinfelt 1987; Kulkarni et al. 1999). A summary of the literature reports to date of reactions of iron-containing bimetallic materials, organized by the class of chlorinated contaminant tested, is presented in Table 12-1.

Despite burgeoning research activities and the knowledge gained using reactive bimetallic materials, many of the details of important reaction pathways remain elusive, largely due to the rapidity of the reactions and sensitivity to reaction conditions. In addition, there is still considerable uncertainty regarding the long-term effectiveness of bimetallic reductants in permeable reactive barriers. To date, the principles underlying reactions of iron-containing bimetallic materials as applied to water and groundwater remediation have yet to be critically reviewed. Therefore, this paper is intended to review fundamental principles and characteristics of these reactions and to report progress in the application of these materials for the reduction of environmental organic contaminants in water and groundwater. It is hoped this review provides a framework for better understanding and applications of iron-containing bimetallic materials.

12.2 Noble Metals as Reduction Catalysts

Several noble metals have been noted for their catalytic activity in reduction reactions of halogenated organic compounds (March 1985). In most applications,

Table 12-1. Literature on Reactions of Iron Containing Bimetallic Reductants, Organized by the Types of Contaminants

Contaminants treated	Bimetallic reductants	References
Chlorinated methanes	Pd/Fe	Li and Farrell (2000)
		Lien and Zhang (1999)
		Muftikian et al. (1995)
		Wan et al. (1999)
Chlorinated ethylenes	Pd/Fe	Kim and Carraway (2003a)
		Korte et al. (2000)
		Li and Farrell (2000)
		Liang et al. (1997)
		Lien and Zhang (2001)
		Muftikian et al (1995)
		Muftikian et al (1996)
		Wang and Zhang (1997)
	Ni/Fe	Grenier et al. (2004)
		Schrick et al. (2002)
	Pd/Fe, Ni/Fe	Cheng and Wu (2000)
	Pd/Fe, Ni/Fe, Cu/Fe, Pd/Fe,	Kim and Carraway (2003b)
	Ni/Fe, Pt/Fe	Zhang et al. (1998)
	Pd/Fe, Ru/Fe, Pt/Fe, Au/Fe	Lin et al. (2004)
Chlorinated ethanes	Pd/Fe	Lien and Zhang (2005)
Trihalomethanes	Pd/Fe	Guasp and Wei (2003)
Methylene blue	Cu/Fe	Ma et al. (2004)
N-nitrosodimethyl amine	Ni/Fe	Gui et al. (2000)
		Odziemkowski et al. (2000)
Chlorobenzenes	Pd/Fe	Liu et al. (2003)
		Xu et al. (2005a)
		Xu et al. (2005b)
		Zhang et al. (1998)
	Ag/Fe	Xu and Zhang (2000)
Chlorophenols	Pd/Fe	Kim and Carraway (2000)
		Liu et al. (2001)
		Morales et al. (2002)
		Neurath et al. (1997)
	Pd/Fe	Xu and Zhang (2000)
Polychlorinated biphenyls	Pd/Fe	Doyle et al. (1998)
		Filipek et al. (2003)
		Grittini et al. (1995)
		Kim et al. (2004)
		Korte et al. (2002)
		Wang and Zhang (1997)

noble metal catalysts are used in combination with support materials such as silica, alumina, and activated carbon (Coq et al. 1993; Rajagopal et al. 1995; Mukhopadhyay et al. 1999; Navarro et al. 2000). The noble metals are anchored or dispersed over the supports to increase the available catalytic surface area and thereby enhance reaction efficiency. The catalysis of dechlorination reactions by noble metals is generally believed to occur on the surface of the catalytic metal and, for a given metal, the reaction mechanism has been observed to be independent of the support material type. However, supports commonly used have been selected in part for their inert characteristics, in contrast to zero-valent iron based bimetallics. Studies of noble metal catalysis have also shown that the primary dechlorination mechanism is hydrodehalogenation in which the replacement of chlorine by hydrogen occurs at catalytic sites that concentrate and co-locate reactive atomic hydrogen and surface-adsorbed compounds (Schreier et al. 1995; Rajagopal et al. 1995; Mukhpadhyay et al. 2000). In this process, metal-hydride (M-H) complexes formed from the noble metal and dissociation of molecular hydrogen (or other hydrogen sources) on the catalyst surface serve as the direct reductant for contaminants. More detailed discussion on the formation of metal-hydrides is provided in the section on reactions of bimetallic reductants.

Various types of hydrogen sources have been reported for use with catalysts. These sources include hydrogen gas (Coq et al. 1993; Mukhopadhyay et al. 1999; Navarro et al. 2000), alcohols (Kopinke et al. 2004), electrochemically generated hydrogen (McNab et al. 1998), hydrazine (Miyabe et al. 2003; Kopinke et al. 2004), formic acid (Rajagopal et al. 1995; Kopinke et al. 2004), and bimetallic materials (Wan et al. 1999; Mukhopadhyay et al. 2000; Liu et al. 2001; Lien et al. 2001; Xu et al. 2005a; Xu et al. 2005b).

12.3 Preparation of Bimetallic Reductants

In most of the previous studies, bimetallic reductants were prepared by mixing the secondary metal precursors (soluble metal salts) with iron metal in aqueous solution. Since the reduction potentials of noble metals are greater than that of iron (-0.44 V vs. SHE), a redox reaction between the precursors and iron results in reduction of noble metals, oxidation of a small amount of iron, and deposition of the noble metal on the surface of iron. For example, reduction of Pd(IV) (2.2 V vs. SHE) by iron can be expressed by following reaction.

$$Pd^{4+} + 2\,Fe^{0} \rightarrow Pd^{0} + 2\,Fe^{2+} \tag{12.1}$$

This redox reaction is rapid, usually completed in a few minutes, and is reported to proceed by stepwise reduction involving initial reduction of Pd(IV) to Pd(II), formation of Pd(II)-O-Fe complex, and further reduction of Pd(II) to Pd^0 on the iron surface (Muftikian et al. 1996). Similar stepwise reduction patterns were also observed for Au(III), Ru(III), and Pt(IV) (Lin et al. 2004).

The deposition of small amounts of a secondary metal onto iron results in an increase in surface area compared to that of untreated iron, which has been attributed to formation of islands of the noble metal on the iron surface (Liu et al. 2001; Lin et al. 2004; Xu et al. 2005a). However, the increase in surface area is limited to low loadings, indicating excess loading leads to the formation of larger islands rather than a larger number of small islands (Lin et al. 2004). Spectroscopic evidence has confirmed that palladium was unevenly distributed on the iron surface and reactivity indicated reduced specific palladium surface area at higher loadings (Kim and Carraway 2003a). A SEM image of Pd/Fe is shown in Fig. 12-1, along with the corresponding energy dispersive X-ray spectroscopic pattern. The figure reveals several Pd islands (bright spots) of differing size on a partially oxidized zero-valent iron particle, confirmed by EDX spectra.

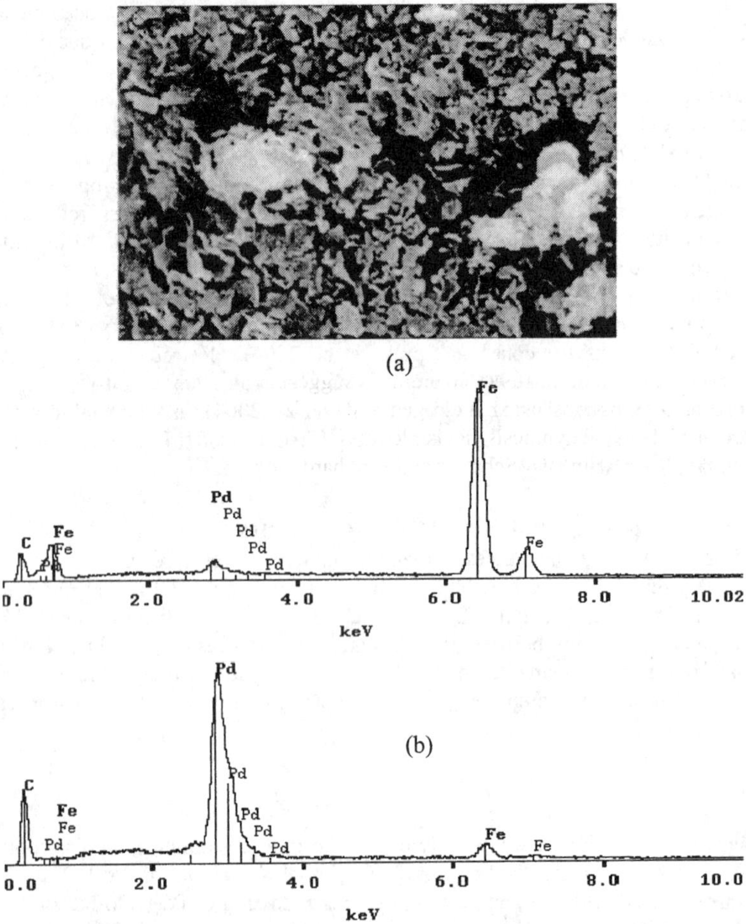

Fig.12-1. (a) Scanning electron microscope (SEM) image of Pd/Fe particle (0.8% Pd by mass, 5000x magnification), and (b) energy dispersive X-ray spectra of the Pd/Fe particle obtained at 2 locations. Upper spectrum corresponds to dark spot and lower refers to bright spot of the SEM image (Kim 1999).

12.4 Reduction Reactions of Bimetallic Materials

Most of the applications of bimetallic reductants reported to date have focused on reductive dechlorination transforming chlorinated organic compounds into their nonchlorinated analogs - that is, the carbon backbone or structure of the compound is not changed, but chlorine is replaced by hydrogen. Reductive dechlorination by metallic reductants is a complex reaction that may involve a variety of mechanisms (Matheson and Tratnyek 1994). For example, results of a study using Pd/Fe indicated that dechlorination of trichloroethylene proceeded via a catalytic reaction involving hydrogen, whereas carbon tetrachloride was reduced mainly by direct electron transfer (Li and Farrell 2000). However, for bimetallic reductants, more emphasis has generally been given to catalytic aspects of the reaction, due to the strong positive correlation between reaction rate and hydrogen concentration observed in many studies (Schreier and Reinhard 1995; Cheng et al. 1997; Perrone et al. 1998; McNab and Ruiz 1998; Prati and Rossi 1999; McNab et al. 2000; Lowry and Reinhard 2001).

The catalytic properties of the noble metals stem in part from their ability to adsorb hydrogen and intercalate it within the metal lattice, forming metal hydrides (Cheng et al. 1997; Korte et al. 2000; Gui et al. 2000). It is important to note that in this context the term hydride denotes a simple model of neutral hydrogen atoms dispersed in the metal, not anionic atomic hydrogen (Swaddle 1997). Metal hydrides are powerful reducing agents capable of transforming the majority of halogenated environmental contaminants of concern, as suggested by Table 12-1. Dehalogenation of organic compounds using hydrogen and noble metal catalysts is a widely used process in chemical synthesis (Rylander 1979), and has been used for treating many chlorinated contaminants (Schreier and Reinhard 1995; Cheng et al. 1997; Perrone et al. 1998; McNab and Ruiz 1998; Prati and Rossi 1999 McNab et al. 2000; Lowry and Reinhard 2000, 2001). Catalytic hydrodehalogenation using zero-valent iron containing bimetallic reductants differs from the reactions in the aforementioned studies in that hydrogen is supplied by the catalyst support itself (zero-valent iron). Noble metal catalysts on inert supports require a separate hydrogen supply (typically hydrogen gas, H_2), whereas in iron metal-based bimetallics, hydrogen is produced through anaerobic corrosion of the zero-valent iron.

In the reaction of bimetallic reductants, the formation of the direct reducing agent (e.g. Pd-H) results from two distinctive pathways by which hydrogen is introduced into the noble metal: (1) direct adsorption of atomic hydrogen (H·) onto the noble metal, and (2) formation of diatomic or molecular hydrogen (H_2), followed by dissociation into H· on the noble metal. These routes are illustrated in Fig. 12-2 and further described below.

12.4.1 Direct Adsorption of Atomic Hydrogen

As described by Bokris et al. (1974), the galvanic corrosion of a metal such as iron is accelerated in the presence of a noble metal such as palladium, because the electrochemical potential gap between the metals is greater than that of the iron-water pair. The electrochemical reaction of a material such as Pd/Fe may be written as

$$Fe \rightarrow Fe^{2+} + 2e^- \qquad \text{anode (Fe)} \qquad (12.2)$$

$$H^+ + e^- \rightarrow H\cdot \qquad \text{cathode (Pd)} \qquad (12.3)$$

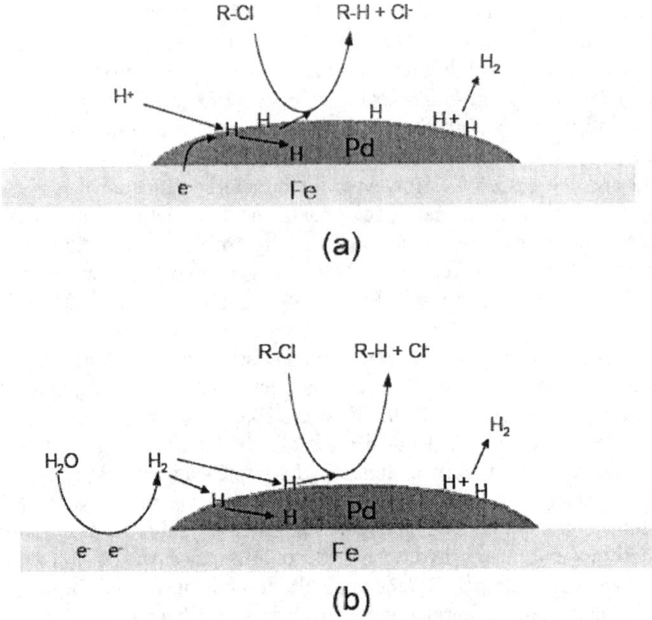

Fig. 12-2. Schematic representation of proposed reaction mechanisms of catalytic hydrodehalogenation by zero-valent iron containing bimetallic materials: (a) direct adsorption of atomic hydrogen, and (b) dissociative adsorption of diatomic hydrogen

Thus, in this model, zero-valent iron serves as the source of electrons for the one-electron reduction of protons at the cathodic Pd surface. Once formed, the fate of H· may be: 1) adsorption near the Pd surface forming a metal hydride complex, 2) diffusion or absorption into the metal, ultimately causing embrittlement, and 3) reaction with another H· to evolve H_2 (Perng and Wu 2003). These processes are shown in Fig. 12-2a and Eqs. (12.4) to (12.6) below.

$$H\cdot + Pd \rightarrow Pd\text{-}H_{(ads)} \quad (12.4)$$

$$Pd\text{-}H_{(ads)} \rightarrow Pd\text{-}H_{(abs)} \quad (12.5)$$

$$Pd\text{-}H_{(ads)} + Pd\text{-}H_{(ads)} \rightarrow H_2 + 2Pd \quad (12.6)$$

Here, both adsorbed and absorbed hydrogen are formed via a direct interaction with Pd. This pathway for the formation of reactive hydrogen in bimetallic materials has been suggested by several investigators (Gui et al. 2000; Odziemkowski et al. 2000; Li and Farrell 2000; Lin et al. 2004; Wang and Farrell 2004). The extent of hydrogen adsorption and absorption depends on the hydrogen storage capability of the metal (Perng and Wu 2003). Due to their low hydrogen overpotentials, noble metals tend to exhibit high hydrogen adsorption capacity as well as high diatomic hydrogen evolution rates (Bockris et al. 1974). Among the noble metals that have been used in combination with iron, palladium has been shown to bear the highest hydrogen adsorption capacity and to maintain a high surface concentration of hydrogen (Lawson et al. 1991; Lin et al. 2004). For example, Wu (1992) reported hydrogen absorption efficiency of palladium close to 100%. Cheng et al. (1997) reported much faster dechlorination of 4-chlorophenol using palladium compared to platinum despite the same flux of hydrogen generation on both surfaces. The authors concluded that hydrogen escaped more easily from the platinum surface. Lin et al. (2004) observed the highest hydrogen evolution and dechlorination rate of TCE by Pd/Fe. Taken together, these works indicate that the hydrogen evolution rate is not alone a good predictor of dechlorination rate. Rather, the balance between adsorption, absorption, and diatomic hydrogen evolution affects the concentration of adsorbed hydrogen on the surface of the noble metal and consequently the dechlorination reaction rate.

12.4.2 Dissociative Adsorption of Diatomic Hydrogen

Since the surfaces of zero-valent iron containing bimetallic reductants are only partially covered by a noble metal, iron is inevitably exposed to solution during the reaction and reacts with water to produce hydrogen. Iron metal has a much smaller hydrogen adsorption capacity than noble metals (Wu 1992). Thus, the concentration of Fe-H would be relatively low, and much of the H· will be evolved into solution as diatomic hydrogen. The presence of H_2 in the vicinity of the noble metal catalytic surface creates an environment similar to that used for catalysts on inert supports in which H_2 is supplied externally, typically as a gas. In both cases, when aqueous diatomic hydrogen comes into contact with the noble metal surface, it is easily adsorbed and dissociated to hydrogen or the metal hydride. The zero and small energy barriers to physical and chemical adsorption, respectively, of diatomic hydrogen on a nickel metal surface are illustrated in Fig. 12-3 (Swaddle 1997). In other words, on a catalytic surface (such as Ni), the H-H bond is easily broken with a small input of energy to form the powerful metal-hydride reductant for contaminant dechlorination. In this reaction scheme, the noble metal acts as a collector of hydrogen (produced by iron corrosion) and the dechlorination reaction rate is predicted to be proportional to the aqueous H_2 concentration (Lowry and Reinhard 2001). This pathway of reactive hydrogen generation has been hypothesized for bimetals by a number of researchers (Grittini et al. 1995; Cheng et al. 1997; Korte et al. 2000; Liu et al. 2001; Xu et al. 2005a) and for noble metals (Pd) supported on Al_2O_3 or carbon (Schreier and Reinhard 1995; Lowry and Reinhard 2001). In addition, hydrogen generation through iron corrosion followed by catalytic reaction with TCE on an iron-boron surface has been described by Liu et al. (2005) and Carraway and Song (2004).

12.4.3 Roles of Iron as the Primary Metal

Based on the two pathways for the catalysis of dechlorination reactions through the noble metal described above, two primary roles or pathways for zero-valent iron may be proposed. If direct adsorption of atomic hydrogen on the noble metal is the better model of bimetallic material reactions, role of zero-valent iron is to supply electrons

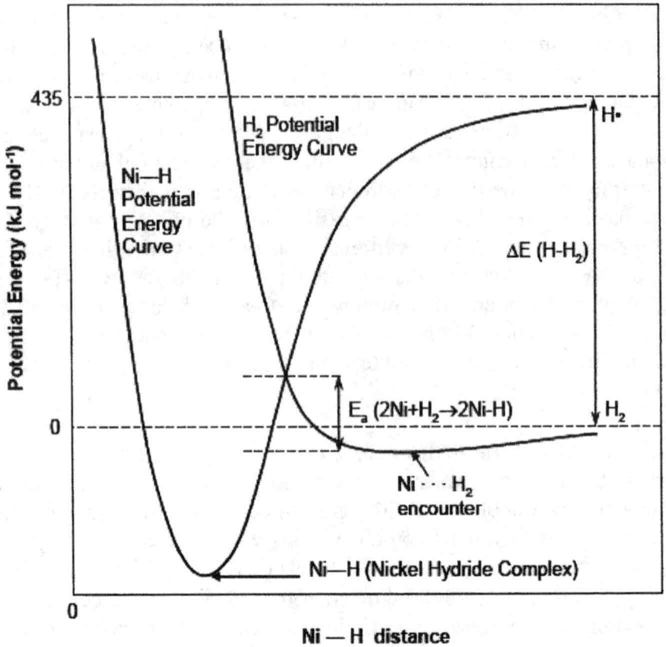

Fig. 12-3. Activation energies in adsorption of hydrogen on nickel (Adapted from Swaddle 1997)

to the catalytic metal through galvanic corrosion. In the bimetallic material, two metals of differing reduction potentials are in electrical contact. The difference in potential between the metals serves as the driving force for electron flow from the anodic base metal to the cathodic noble metal. Due to this electron flow, the base metal corrodes while the catalyst remains reduced and acts as a conduit for electrons to surface reactions or other ultimate fates. Factors that can affect the transfer of electrons between the primary and catalytic metals include: 1) the reduction potential of the noble metal, 2) the nature and kinetics of the cathodic and anodic reactions of the bimetallic material, 3) the relative surface areas of the primary and catalytic metals, and 4) the nature of the solution surrounding the bimetal. Although electron transfer to the noble metal from iron is more thermodynamically favored than to contaminants or water adsorbed at the zero-valent iron surface, kinetic considerations dictate that direct reactions between iron and such compounds will occur, resulting in products which may include partially dechlorinated contaminants or diatomic hydrogen.

If dissociative adsorption of diatomic hydrogen is the better description, then the primary role of zero-valent iron is to produce hydrogen by anaerobic corrosion. As described above and shown in Fig. 12-2b, iron can reduce water and supply hydrogen for hydrodechlorination reaction taking place at catalyst surface. It should be noted that iron itself can act as a catalyst, although it is a much weaker hydrogen collector than other noble metals (Wu 1992). For example, based on reaction products obtained from zero-valent iron reactions vs. electrochemical conversions employing a mercury electrode, Odziemkowski et al. (2000) concluded that reduction of N-nitrosodimethylamine by zero-valent iron occurs via catalytic hydrogenation on the iron surface. Liu et al. (2005) also argued that trichloroethylene was primarily dechlorinated via catalytic hydrogenolysis by laboratory synthesized nano-sized iron. Further, Carraway and Song (2004) showed that catalytic hydrogenolysis is the dominant reaction pathway for the reduction of all chlorinated ethylenes by similar nano-sized iron, providing evidence of strong correlation between reaction rate and hydrogen concentration in the solution. However, in the latter two studies, the presence of boron in the zero-valent iron at approximately 5% contributes to the catalytic activity of the iron surface and is not typical of commercial irons currently used in treatment systems (Liu et al. 2005).

12.5 Factors Affecting Reaction of Bimetallic Reductants

There are number of factors that can affect reaction of bimetallic reductants. These include the type of noble metal used, noble metal content, surface area of noble metal, and solution chemistry factors such as pH and dissolved oxygen.

12.5.1 Effect of the Type of Noble Metal

The extent of reactivity for bimetals varies dramatically depending on which noble metal is present on iron surface. Kim and Carraway (2003b) investigated copper, nickel, and palladium plated zero-valent iron and reported a reactivity order of Pd/Fe >> Ni/Fe > Cu/Fe for the reduction of trichloroethylene under similar loadings of the noble metals ($\approx$ 0.05 wt%). Compared to the reaction rate for the base iron metal, the enhancement of reaction rate achieved by Pd was more than an order of magnitude greater than that of Ni and Cu. A high reaction efficiency using Pd was also observed by Lin et al. (2004). At constant loading of noble metal (0.25 wt%), Pd reacted with trichloethylene 2-3 orders of magnitude faster than other metals, giving the reactivity order of Pd/Fe >> Ru/Fe > Pt/Fe > Au/Fe. However, the authors recommend Ru as a potential alternative to Pd for use in practical applications because: 1) Ru/Fe was more effective in maintaining its reactivity in long term operation, 2) Ru/Fe resulted in no accumulation of chlorinated reaction by products (similar to Pd/Fe reactions), and 3) Ru costs only 10% of the price of Pd.

12.5.2 Effect of Noble Metal Content and Surface Area

The relationship between reaction rate and noble metal content has been demonstrated by several investigators (Grittini et al. 1995; Korte et al. 2000; Liu et al. 2001; Kim and Carraway 2003a; Lin et al. 2004; Xu et al. 2005a). In general, the reaction rate of bimetals is proportional to the noble metal content. However, in some cases, this proportionality has been shown to hold only to a certain point, above which the reaction rate plateaus (Kim and Carraway 2003a; Lin et al. 2004). This behavior implies that additional noble metal is not used as efficiently for iron surface coverage or reactive surface generation as smaller amounts of noble metal, as evinced by the decrease in surface area at higher loadings of the noble metal (Lin et al. 2004). For a given noble metal, the surface area of noble metal is directly proportional to the reaction rate (Lin et al. 2004). No exceptions to this observation have been reported to date.

12.5.3 Effect of Solution Chemistry

In the reaction of bimetallic materials in aqueous solution, pH may have significant impact in the contaminant reaction rate. The reaction rate of contaminants is expected to increase with decreasing solution pH as there is greater likelihood of anaerobic corrosion of iron in acidic conditions, which produces hydrogen for catalytic reaction. On the contrary, elevated solution pH may increase the possibility of surface passivation of iron or catalyst surface by precipitation of iron (hydr)oxide minerals. The precipitation of iron (hydr)oxides may serve as a physical barrier at the iron-/catalyst- water interface, reducing the generation of hydrogen that may drive the catalytic reduction of some contaminants (Scherer et al. 1998). The increased reaction rate with decreasing pH was reported for reduction trichloroethylene and carbon tetrachloride by Pd/Fe (Li and Farrell 2000). Xu et al. (2005a) also observed the effect of pH on the Pd/Fe catalyzed reduction of *para*-dichlorobenzene that is consistent with the expected trend.

The presence of dissolved oxygen would be expected to decrease dechlorination reaction rates since oxygen consumes hydrogen in a catalyzed reaction to form water on Pd surfaces (Fogelberg and Petersson 1996). Also, oxygen is a stronger oxidant than most of chlorinated organic contaminants, thus would outcompete chlorinated contaminants for reaction sites. However, within the range of typical oxygen level in ground and surface waters (less than 10 mg/L), oxygen did not affect the reaction rates of trichloroethylene and carbon tetrachloride by Pd/Fe (Li and Farrell 2000). Such a minimal effect of oxygen was also observed for the reductions using Pd on alumina and H_2 gas (Lowry and Reinhard 2001).

12.6 Deactivation of Bimetallic Reductants

A significant problem with bimetallic reductants is the loss of reactivity over time (Muftikian et al. 1996; Korte et al. 2000; Korte et al. 2002; Gui et al. 2003; Lin et al. 2004; Grenier et al. 2004). The main reasons for loss of reactivity of bimetals over time may include buildup of iron oxide layers on the surface, catalyst poisoning by various inorganic species, and loss of the noble metal particle from iron surface. Catalyst deactivation through the growth of oxide layers results from long exposures of catalysts to aqueous solutions as surface iron oxides becomes more oxidized and progressively converted to passivating materials such as hematite (α-Fe_2O_3) or maghemite (γ-Fe_2O_3) (Davenport et al. 1995; Muftikian et al. 1996), and ongoing dissolution and precipitation of these minerals can gradually lead to coverage of noble metal surfaces. As noted earlier, the development of passivating mineral phases on bimetallic surface can result in surface deactivation by acting as a physical barrier that limits the electron transfer reaction with water or contaminants. Catalysts deactivation incurred by buildup of oxide layers may be recovered by washing the particles with dilute acid solutions (Muftikian et al. 1996; Lowry and Reinhard 2000; Grenier et al. 2004), however, it is sometimes difficult to achieve full recovery of the catalyst reactivity (Gui et al. 2000; Korte et al. 2000). Further, acid washing of the bimetallic materials containing nickel and copper may cause leaching of these transition metals raising environmental concerns for toxicity to biota. Although there is no reported serious toxic effect of Pd to humans and it is not currently regulated by drinking water standards, potential toxic effects of Pd warrant concern.

Catalyst poisoning by inorganic species has been demonstrated in several studies (Schreier and Reinhard 1995; McNab and Ruiz 1998; Lowry and Reinhard 2000). Deactivation of Pd by reduced sulfur species (HS^-, SO_3^{2-}) has been attributed to chemisorption of S onto Pd that results in inhibition of adsorption of contaminants and modification of electronic properties of Pd (Hughes 1984). In addition, carbonate species and carbon monoxide (which can be formed from Pd-catalyzed reaction of bicarbonate) have been reported as potential deactivators of catalysts (Kramer et al 1995). Li and Farrell (2000) reported reduced reactivity of Pd/Fe in the presence of reduced sulfur species (HS^- and H_2S), although the catalysts recovered reactivity after removing sulfides from the solution. Similar Pd poisoning by sulfur species was reported by Lowry and Reinhard (2000). Also, sulfate reducing bacteria adversely affected the performance of *in situ* treatment Pd/Fe column reactor by taking up hydrogen and converting sulfate to reduced sulfur species (Korte et al. 2000). On the contrary, carbonate species did not affect catalyst reactivity (Lowry and Reinhard 2000; Li and Farrell 2000).

12.7 Nano-sized Bimetallic Reductants

As a chemical reactant, nano-sized metals exhibit unique and advantageous characteristics of small-sized particle. These include very rapid mass transfer, facile *in situ* delivery of particle, and potentially favorable changes in reaction mechanism with very small particle size. For such reasons, the use of nano-sized metals in environmental applications to treat organic contaminants has received significant attention in recent years. A direct benefit of using nano-sized bimetallic particles is that the reactivity of bimetals can be further enhanced from reduction of particle size. Wang and Zhang (1997) synthesized Pd/Fe by borohydride ($NaBH_4$) reduction of a iron salt ($FeCl_36H_2O$) in an aqueous phase, followed by Pd deposition on the iron surface, and used the resulting particles as reductants for TCE and a PCB mixture (Aroclor 1254). The results indicated that nano-sized Pd/Fe and unamended nano-sized Fe showed substantially higher reactivity compared to commercial grade micro-sized iron.

Following the work by Wang and Zhang, many reports have appeared on the use of nano-sized bimetals to treat various types of contaminants including chlorinated ethylenes (Zhang et al. 1998; Lien and Zhang 2001; Schrick et al. 2002), chlorinated methanes (Lien and Zhang 1999), chlorinated ethanes (Lien and Zhang 2005), and chlorinated benzenes (Zhang et al. 1998; Xu and Zhang 2000). In addition, a field assessment of nanoscale Pd/Fe particles exhibited nearly complete removal of trichloroethylene over a 4 week period (Elliott and Zhang 2001). These studies demonstrate that this new approach not only substantially improves contaminant degradation, but also curtails production of undesirable byproducts from the reduction. This is especially for chlorinated ethylenes, in which reduction of PCE and TCE by commercial grade iron has been observed to produce *cis*-DCE and vinyl chloride.

12.8 Conclusions

Bimetallic reductants have been shown to substantially enhance reductive dehalogenation of many environmental organic contaminants by routing reactions through catalytic pathways. The catalytic nature of reaction renders the bimetals effective reductants for some of the more recalcitrant contaminants such as chlorinated biphenyl and chlorinated phenols (Grittini et al. 1995; Neurath et al. 1997; Doyle et al. 1998; Kim and Carraway 2000; Morales et al. 2002; Kim et al. 2004). Results obtained so far for the performance of bimetallic systems look promising. However, there are some problems that need to be resolved in order for bimetallic treatment systems to be developed into a sound and viable remedial option. First, the high cost of Pd/Fe prohibits its application in large scale operations such as permeable reactive barriers for groundwater treatment. At present, the most practical applications are relatively small scale aboveground treatment systems. For example, it has been recently demonstrated that bimetallic materials can form effective treatment systems for organohalides in landfill gas and soil vapor extraction offgas (Grenier et al. 2004). It may also be effectively used for remediation of groundwater by the "pump and treat" method (Korte et al. 2000). Second, although bimetallic materials provide markedly faster reaction rates, it has been shown that bimetals are susceptible to surface deactivation, requiring frequent acid washing to restore the reactivity. This factor poses additional requirements on placing bimetals in underground permeable reactive barriers since regeneration of the media would present challenges in applying and removing the acid wash solutions. Finally, the risk of toxic effects of noble metals emplaced in the subsurface should be carefully considered. Treatment systems such as permeable reactive barriers employing reactive bimetallic materials will likely use minimum amounts and may be designed for removal or immobilization of the bimetallic components upon conclusion of the treatment.

The complexity of reaction pathways that may occur with zero-valent iron based bimetallic materials provide sequential, concurrent, and competing routes for contaminant dechlorination. While our understanding of these pathways and how they are affected in the short and long term by various conditions and parameters is incomplete, it is clear that a broad range of environmental contaminants can be effectively treated using these types of materials. Continued laboratory and field research in this area should result in sound and achievable approaches for the remediation of chemicals that currently present difficulties due to slow reaction rates and formation of problematic intermediate products, as well as for situations in which rapid and complete contaminant removal is crucial.

12.9 References

Bockris, J. O'M., Bonciocat, N., and Gutmann, F. (1974). *An introduction to electrochemical science*, Springer-Verlag, New York.

Carraway, E., and Song, H. (2004). "Reduction of chlorinated ethenes by nano-sized zero-valent iron." In *Extended Abstract of National Meeting, Div. Environ. Chem*, ACS, Philadelphia, PA, 44(2), 397-401.

Cheng, I. F., Fernando, Q., and Korte, N. (1997). "Electrochemical dechlorination of 4-chlorophenol to phenol." *Environ. Sci. & Technol.*, 31(4), 1074-1078.

Cheng, S. F., and Wu, S. C. F. (2000). "The enhancement methods for the degradation of TCE by zero-valent metals." *Chemosphere*, 41(8), 1263-1270.

Coq, B., Hub, S., Figueras, F., and Tournigant, D. (1993). "Conversion under hydrogen of dichlorodifluoromethane over bimetallic palladium catalysts." *App. Catal. A.*, 101(1), 41-50.

Davenport, A. J., and Sansone, M. (1995). "High-resolution in-situ XANES investigation of the nature of the passive film on iron in a pH 8.4 borate buffer." *J. Electrochem. Soc.*, 142(3), 725-730.

Doyle, J. G., Mile, T.-A., Parker, E., and Cheng, I. F. (1998). "Quantification of total polychlorinated biphenyl by dechlorination to biphenyl by Pd/Fe and Pd/Mg bimetallic particles." *Microchem. J.*, 60, 290-295.

Elliott, D. W., and Zhang, W. X. (2001). "Field assessment of nanoscale bimetallic particles for groundwater treatment." *Environ. Sci. & Technol.*, 35(24), 4922-4926.

Filipek, L. B., Coon, C. C., Geiger, C. L., Clausen, C. A., Quinn, J., and DeVor, R. (2003). "Dechlorination of polychlorinated biphenyls in solution by Pd/Fe bimetallic emulsions." In *Extended Abstract of National Meeting, Div. Environ. Chem*, ACS, New Orleans, LA.

Fogelberg, J., and Petersson, L. G. (1996). "Kinetic modelling of the H_2O_2 reaction on Pd and of its influence on the hydrogen response of a hydrogen sensitive Pd metal-oxide-semiconductor device." *Surf. Sci.*, 350(1-3), 91-102.

Fung, S. C., and Sinfelt, J. H. (1987). "Hydrogenolysis of methyl chloride on metals." *J. Catal.*, 103(1), 220-223.

Grenier, A. C., McGuire, M. M., Fairbrother, D. H., and Roberts, A. L. (2004). "Treatment of vapor-phase organohalides with zero-valent iron and Ni/Fe reductants." *Environ. Eng. Sci.*, 21(4), 421-435.

Grittini, C., Malcomson, M., Fernando, Q., and Korte, N. (1995). "Rapid dechlorination of polychlorinated-biphenyls on the surface of a Pd/Fe bimetallic system." *Environ. Sci. & Technol.*, 29(11), 2898-2900.

Guasp, E., and Wei, R. (2003). "Dehalogenation of trihalomethanes in drinking water on Pd-Fe bimetallic surface." *J. Chem. Tech. Biotech.*, 78(6), 654-658.

Gui, L., and Gillham, R. W. (2003). "Preparation and regeneration of nickel-iron for reduction of organic contaminants." In *Chlorinated Solvent and DNAPL Remediation. ACS Symposium Series*, 837, 206-216.

Gui, L., Gillham, R. W., and Odziemkowski, M. S. (2000). "Reduction of N-nitrosodimethylamine with granular iron and nickel enhanced iron. 1. Pathways and kinetics." *Environ. Sci. & Technol.*, 34(16), 3489-3494.

Hughes, R. (1984). *Deactivation of catalysts*, Academic Press, London.

Kim, Y. H. (1999). *Dechlorination of chlorinated aliphatic and aromatic compounds using zero valent metals: Modified metals and electron mediators*. Ph.D. dissertation, Texas A&M University, College Station, TX.

Kim, Y. H., and Carraway, E. R. (2000). “Dechlorination of pentachlorophenol by zero valent iron and modified zero valent irons.” *Environ. Sci. & Technol.*, 34(10), 2014-2017.

Kim, Y. H., and Carraway, E. R. (2003a). “Dechlorination of chlorinated ethenes and acetylenes by palladized iron.” *Environ. Technol.*, 24(7), 809-819.

Kim, Y. H., and Carraway, E. R. (2003b). "Reductive dechlorination of TCE by zero valent bimetals." *Environ. Technol.*, 24(1), 69-75.

Kim, Y. H., Shin, W. S., and Ko, S. O. (2004). “Reductive dechlorination of chlorinated biphenyls by palladized zero-valent metals.” *J. Environ. Sci. Heal. A*, 39(5), 1177-1188.

Klausen, J., Vikesland, P. J., Kohn, T., Burris, D. R., Ball, W. P., and Roberts, A. L. (2003). “Longevity of granular iron in groundwater treatment processes: Solution composition effects on reduction of organohalides and nitroaromatic compounds.” *Environ. Sci. & Technol.*, 37(6), 1208-1218.

Kopinke, F-D., Mackenzie, K., Koehler, R., and Georgi, A. (2004). “Alternative sources of hydrogen for hydrodechlorination of chlorinated organic compounds in water on Pd catalysts.” *Appl. Catal. A-Gen.*, 271, 119-128.

Korte, N. E., West, O. R., Liang, L., Gu, B., Zutman, J. L., and Fernando, Q. (2002). “The effect of solvent concentration on the use of palladized-iron for the step-wise dechlorination of polychlorinated biphenyls in soil extracts.” *Waste Manage.*, 22(3), 343-349.

Korte, N. E., Zutman, J. L., Schlosser, R. M., Liang, L., Gu, B., and Fernando, Q. (2000). “Field application of palladized iron for the dechlorination of trichloroethene.” *Waste Manage.*, 20(8), 687-694.

Kramer, H., Levy, M., and Warshawsky, A. (1995). “Hydrogen storage by the bicarbonate formate reaction - Studies on the activity of Pd catalysts.” *Int. J. Hydrogen Energ.*, 20(3), 229-233.

Kulkarni, P. P., Deshmukh, S. S., Kovalchuk, V. I., and D'Itri, J. L. (1999). “Hydrodechlorination of dichlorodifluoromethane on carbon-supported Group VIII noble metal catalysts.” *Catal. Lett.*, 61(3-4), 161-166.

Lawson, D. R., Tierney, M. J., Cheng, I. F., Vandyke, L. S., Espenscheid, M. W., and Martin, C. R. (1991). “Use of a coulometric assay technique to study the variables affecting deuterium loading levels within palladium electrodes.” *Electroch. Acta.*, 36(9), 1515-1522.

Li, T., and Farrell, J. (2000). “Reductive dechlorination of trichloroethene and carbon tetrachloride using iron and palladized-iron cathodes.” *Environ. Sci. & Technol.* 34(1), 173-179.

Liang, L. Y., Korte, N., Goodlaxson, J. D., Clausen, J., Fernando, Q., and Muftikian, R. (1997). “Byproduct formation during the reduction of TCE by zero-valence iron and palladized iron.” *Ground Water Monit. R.*, 17(1), 122-127.

Lien, H. L., and Zhang, W. X. (1999). “Transformation of chlorinated methanes by nanoscale iron particles.” *J. Environ. Eng. ASCE*, 125(11), 1042-1047.

Lien, H. L., and Zhang, W. X. (2001). “Nanoscale iron particles for complete reduction of chlorinated ethenes.” *Colloid. Surface. A.*, 191(1-2), 97-105.

Lien, H. L., and Zhang, W. X. (2005). “Hydrodechlorination of chlorinated ethanes by nanoscale Pd/Fe bimetallic particles.” *J. Environ. Eng. ASCE*, 131(1), 4-10.

Lin, C. J., Lo, S. L., and Liou, Y. H. (2004). "Dechlorination of trichloroethylene in aqueous solution by noble metal-modified iron." *J. Hazard. Mater.*, 116(3), 219-228.

Liu, Y. H., Yang, F. L., Chen, J. W., Gao, L. N., and Chen, G. H. (2003). "Linear free energy relationships for dechlorination of aromatic chlorides by Pd/Fe." *Chemosphere*, 50(10), 1275-1279.

Liu, Y. H., Yang, F. L., Yue, P. L., and Chen, G. H. (2001). "Catalytic dechlorination of chlorophenols in water by palladium/iron." *Water Res.*, 35(8), 1887-1890.

Liu, Y. Q., Majetich, S. A., Tilton, R. D., Sholl, D. S., and Lowry, G. V. (2005). "TCE dechlorination rates, pathways, and efficiency of nanoscale iron particles with different properties." *Environ. Sci. & Technol.*, 39(5), 1338-1345.

Lowry, G. V., and Reinhard, M. (2000). "Pd-catalyzed TCE dechlorination in groundwater: Solute effects, biological control, and oxidative catalyst regeneration." *Environ. Sci. & Technol.*, 34(15), 3217-3223.

Lowry, G. V., and Reinhard, M. (2001). "Pd-catalyzed TCE dechlorination in water: Effect of H_2 (aq) and H_2-utilizing competitive solutes on the TCE dechlorination rate and product distribution." *Environ. Sci. & Technol.*, 35(4), 696-702.

Ma, L. M., Ding, Z. G., Gao, T. Y., Zhou, R. F., Xu, W. Y., and Liu, J. (2004). "Discoloration of methylene blue and wastewater from a plant by a Fe/Cu bimetallic system." *Chemosphere*, 55(9), 1207-1212.

March, J. (1985). *Advanced organic chemistry*, McGraw-Hill, New York.

Matheson, L. J., and Tratnyek, P. G. (1994). "Reductive dehalogenation of chlorinated methanes by iron metal." *Environ. Sci. & Technol.*, 28(12), 2045-2053.

McNab, W. W., and Ruiz, R. (1998). "Palladium-catalyzed reductive dehalogenation of dissolved chlorinated aliphatics using electrolytically-generated hydrogen." *Chemosphere*, 37(5), 925-936.

McNab, W. W., Ruiz, R., and Pray, L. (2000). "In-situ destruction of chlorinated hydrocarbons in groundwater using catalytic reductive dehalogenation in a reactive well: Testing and operational experiences." *Environ. Sci. & Technol.*, 34(1), 149-153.

Miyabe, K., Taniguchi, N., Imura, A., and Tezuka, T. (2003). "Kinetic study of the hydrodechlorination of trichloroethene in water using a platinum catalyst and hydrazine." *Water Environ. Res.*, 75(5), 472-477.

Morales, J., Hutcheson, R., and Cheng, I. F. (2002). "Dechlorination of chlorinated phenols by catalyzed and uncatalyzed Fe(0) and Mg(0) particles." *J. Hazard. Mater.*, 90(1), 97-108.

Muftikian, R., Fernando, Q., and Korte, N. (1995). "A Method for the rapid dechlorination of low-molecular-weight chlorinated hydrocarbons in water." *Wat. Res.*, 29(10), 2434-2439.

Muftikian, R., Nebesny, K., Fernando, Q., and Korte, N. (1996). "X-ray photoelectron spectra of the palladium-iron bimetallic surface used for the rapid dechlorination of chlorinated organic environmental contaminants." *Environ. Sci. & Technol.*, 30(12), 3593-3596.

Mukhopadhyay, S., Rothenberg, G., Wiener, H., and Sasson, Y. (1999). "Palladium-catalyzed aryl-aryl coupling in water using molecular hydrogen: Kinetics and process optimization of a solid-liquid-gas system." *Tetrahedron*, 55, 14763-14768.

Mukhopadhyay, S., Rothenberg, G., Gitis, D., and Sasson, Y. (2000). "On the mechanism of palladium-catalyzed coupling of haloaryls to biaryls in water with zinc." *Org. Lett.*, 2(2), 211-214.

Navarro, R. M., Pawelec, B., Trejo, J. M., Mariscal, R., and Fierro J. L (2000). "Hydrogenation of aromatics on sulfur-resistant PtPd bimetallic catalysts." *J. Catal.,* 189, 184-194.

Neurath, S. A., Ferguson, W. J., Dean, S. B., Foose, D., and Agrawal, A. (1997). "Rapid and complete dehalogenation of chlorinated phenols by Fe-Pd bimetallic reductants in bench-scale reactors: Implications for soil and ground water remediation." In *Extended Abstract of National Meeting, Div. Environ. Chem*, ACS, San Fransico, CA.

Odziemkowski, M. S., Gui, L, and Gillham, R. W. (2000). "Reduction of N-nitrosodimethylamine with granular iron and nickel-enhanced iron. 2. Mechanistic studies." *Environ. Sci. & Technol.,* 34(16), 3495-3500.

O'Hannesin, S. F., and Gillham, R. W. (1998). "Long-term performance of an in situ 'iron wall' for remediation of VOCs." *Ground Water*, 36(1), 164-170.

Orth, W. S., and Gillham, R. W. (1996). "Dechlorination of trichloroethene in aqueous solution using Fe^0." *Environ. Sci. & Technol.*, 30(1), 66-71.

Perng, T. P., and Wu, J. K. (2003). "A brief review note on mechanisms of hydrogen entry into metals." *Mater. Lett.*, 57(22-23), 3437-3438.

Perrone, L., Prati, L., and Rossi, M. (1998). "Removal of chlorinated organic compounds from water by catalytic dehydrohalogenation." *Appl. Catal. B-Environ.*, 15(3-4), 241-246.

Prati, L., and Rossi, M. (1999). "Reductive catalytic dehalogenation of light chlorocarbons." *Appl. Catal. B-Environ.*, 23(2-3), 135-142.

Puls, R. W., Blowes, D. W., and Gillham, R. W. (1999). "Long-term performance monitoring for a permeable reactive barrier at the US Coast Guard Support Center, Elizabeth City, North Carolina." *J. Hazard. Mater.*, 68(1-2), 109-124.

Rajagopal, S., and Spatola, A. F. (1995). "Mechanism of palladium-catalyzed transfer hydrogenolysis of aryl chlorides by formate salts." *J. Org. Chem.*, 60, 1347-1355.

Rylander, P. N. (1979). *Catalytic hydrogenation in organic synthesis*, Academic Press, New York.

Scherer, M. M., Balko, A. D., and Tratnyek, P. G. (1998). "The role of oxides in the reduction reaction at the metal-water interface." In *Mineral-Water Interfacial Reactions: Kinetics and Mechanisms*. American Chemical Society, Washington, DC, 301-322.

Schreier, C. G., and Reinhard, M. (1995). "Catalytic hydrodehalogenation of chlorinated ethylenes using palladium and hydrogen for the treatment of contaminated water." *Chemosphere*, 31(6), 3475-3487.

Schrick, B., Blough, J. L., Jones, A. D., and Mallouk, T. E. (2002). "Hydrodechlorination of trichloroethylene to hydrocarbons using bimetallic nickel-iron nanoparticles." *Chem. Mater.*, 14(12), 5140-5147.

Swaddle, T. W. (1997). *Inorganic Chemistry: An industrial and environmental perspective*, Academic Press, San Diego, California.

Vikesland, P. J., Klausen, J., Zimmermann, H., Roberts, A. L., Ball, W. P. (2003). "Longevity of granular iron in groundwater treatment processes: Changes in solute transport properties over time." *J. Contam. Hydrol.*, 64(1-2), 3-33.

Vogan, J. L., Focht, R. M., Clark, D. K., and Graham, S. L. (1999). "Performance evaluation of a permeable reactive barrier for remediation of dissolved chlorinated solvents in groundwater." *J. Hazard. Mater.*, 68(1-2), 97-108.

Wan, C. H., Chen, Y. H., and Wei, R. (1999). "Dechlorination of chloromethanes on iron and palladium-iron bimetallic surface in aqueous systems." *Environ. Toxicol. Chem.*, 18(6), 1091-1096.

Wang, J. K., Blowers, P., and Farrell, J. (2004). "Understanding reduction of carbon tetrachloride at nickel surfaces." *Environ. Sci. & Technol.*, 38(5), 1576-1581.

Wang, C. B., and Zhang, W. X. (1997) "Synthesizing nanoscale iron particles for rapid and complete dechlorination of TCE and PCBs." *Environ. Sci. & Technol.*, 31(7), 2154-2156.

Wu, J. K. (1992). "Electrochemical method for studying hydrogen in iron, nickel and palladium." *Int. J. Hydrogen Energ.*, 17(12), 917-921.

Xu, X. H., Zhou, H. Y., He, P., and Wang, D. H. (2005a). "Catalytic dechlorination kinetics of p-dichlorobenzene over Pd/Fe catalysts." *Chemosphere*, 58(8), 1135-1140.

Xu, X. H., Zhou, H. Y., and Wang, D. H. (2005b). "Structure relationship for catalytic dechlorination rate of dichlorobenzenes in water." *Chemosphere*, 58(11), 1497-1502.

Xu, Y., and Zhang, W. X. (2000). "Subcolloidal Fe/Ag particles for reductive dehalogenation of chlorinated benzenes." *Ind. Eng. Chem. Res.*, 39(7), 2238-2244.

Zhang, W. X., Wang, C. B., and Lien, H. L. (1998). "Treatment of chlorinated organic contaminants with nanoscale bimetallic particles." *Catal. Today*, 40(4), 387-395.

CHAPTER 13

Configuration and Construction of Zero-valent Iron Reactive Barriers

Keith C. K. Lai[1], Irene M. C. Lo[2], and Rao Surampalli[3]

Abstract: Configuration of zero-valent iron (Fe^0)-based permeable reactive barriers (PRBs) mainly refers to the configuration of the reactive materials and impermeable funnels emplaced in subsurface. Continuous, funnel-and-gate, trench and caisson PRBs, and GeoSiphon™/GeoFlow are the configurations currently used in PRB sites. Installation of Fe^0 PRBs mainly involves emplacement of a permeable treatment zone containing Fe^0 and impermeable funnels. Emplacement of the permeable treatment zone into subsurface currently can be achieved through the techniques of trench excavation, caisson- and mandrel-based emplacements, and continuous trenching. In addition, steel sheet piling barrier and slurry walls are the barriers conventionally used as impermeable funnels in PRB technology. Innovative techniques, such as jetting, hydraulic fracturing and deep soil mixing, provide alternatives to the conventional emplacement technique and all of them do not involve an excavation.

CE Database subject headings: Barriers; Configuration; Construction; Ground-water pollution; Iron; Remedial action.

13.1 Introduction

The earliest known report of using zero-valent metals to reductively degrade chlorinated organic compounds was in the patent literature (Sweeney 1972) in which reductive degradation of pesticide was concerned. More recently, Senzaki and Kumagai (1989) examined the reductive degradation of aqueous trichloroethylene (TCE) in contact with electrolytic-iron powder. The experiment conducted by Gillham and O'Hannesin (1992) provided further evidence supporting the abiotic degradation of chlorinated aliphatic hydrocarbons (CAHs) by iron filings.

Granular zero-valent iron (Fe^0) is a strong reducing agent and it can abiotically reduce many types of contaminants in groundwater, such as CAHs, chromium, uranium, nitrate, chlorate and bromate etc (Gu et al. 1998; Lai and Lo 2002; Westerhoff 2003; Lo et al. 2005). The idea of Fe^0-based permeable reactive barriers (PRBs) is first prompted by O'Hannesin and Gillham (1992) and afterwards the first Fe^0 PRB was installed at the Canadian Forces Base (CFB) Borden site at

[1] Research Associate, Dept. of Civil Engineering, The Hong Kong University of Science and Technology, Clear Water Bay, Kowloon, Hong Kong.
[2] Associate Professor, Dept. of Civil Engineering, The Hong Kong University of Science and Technology, Clear Water Bay, Kowloon, Hong Kong (corresponding author). E-mail: cemclo@ust.hk
[3] Engineer Director, United States Environmental Protection Agency, Kanasas City, Kansas 66117, United States.

Ontario, Canada for the remediation of CAH contaminated groundwater (O'Hannesin and Gillham 1998). Powell and Puls (1997) defined that reactive barriers materials in subsurface to intercept a contaminant plume (see Fig. 13-1). As contaminated groundwater passes through the reactive zone, contaminants are either immobilized or transformed into more desirable states, thereby attaining remediation concentration goals downgradient of the barrier (Powell et al. 1998).

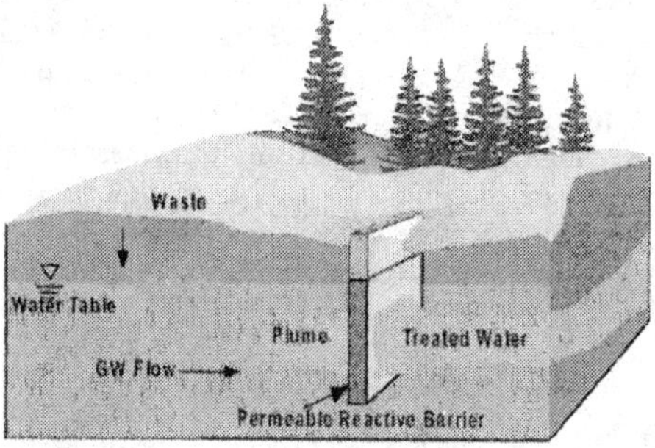

Fig. 13-1. Schematic diagram of a permeable reactive barrier in subsurface environment (adapted from Powell 1998, EPA/600/R-98/125)

The driving force behind leading wide spread of interest of PRB technology in the past decade is due to its advantages over conventional remediation approaches, such as pump-and-treat system, air sparging and bioremediation (Bedient et al. 1999). For example, groundwater remediation using PRB technology requires no pumping or aboveground treatment and reactive barriers act passively after installation, thereby resulting in hardly any annual operating costs other than site monitoring. Moreover, unlike air sparging limiting to volatile organics and bioremediation focusing on biodegradable contaminants, PRB technology can remediate a wide range of contaminants by using various reactive media (EnviroMetal Technology Inc. 1998; Gavaskar et al. 2000; Lai and Lo 2002; Lo et al. 2005; Lai et al. 2006a).

Reactive materials and impermeable funnels emplaced in subsurface are the key components of the PRB technology and various configurations of the emplaced reactive materials and funnels lead to various PRB configurations. In this chapter, the major PRB configurations currently being used in hazardous waste sites are first introduced. Then the conventional and innovative emplacement techniques for PRB installations are described.

13.2 Permeable Reactive Barrier Configurations

Nowadays, there are five main types of PRB configuration including continuous, funnel-and-gate, caisson and Trench PRBs, and GeoSiphon™ or GeoFlow (Powell et al. 1998; Ott 2000). Up to 2002, 34 out of 48 PRB sites are of either continuous or funnel-and-gate configuration (USEPA 2002). Virtually, most recent PRB applications are in continuous configuration since it is easier to install and generates less complex flow pattern compared to funnel-and-gate configuration (Gavaskar et al. 2000).

13.2.1 Continuous Permeable Reactive Barriers

A continuous PRB is a vertically oriented wall of reactive medium installed perpendicular to groundwater flow and downgradient of a contaminant plume or source (see Fig. 13-2a). It incorporates no funnel and thereby allows contaminated groundwater to pass through the barrier under its natural gradient and at its natural flow velocity (Powell et al. 1998). Further, the cross-sectional area of a continuous PRB should be comparable to the area of contaminant plume. Ideally, the bottom of a continuous PRB should be keyed into an impermeable aquitard to prevent the underflow of contaminated groundwater. If an aquifer is too deep, hanging barrier configuration should be used in which a continuous PRB is built to a depth which can over-encompass the vertical dimension of contaminant plume.

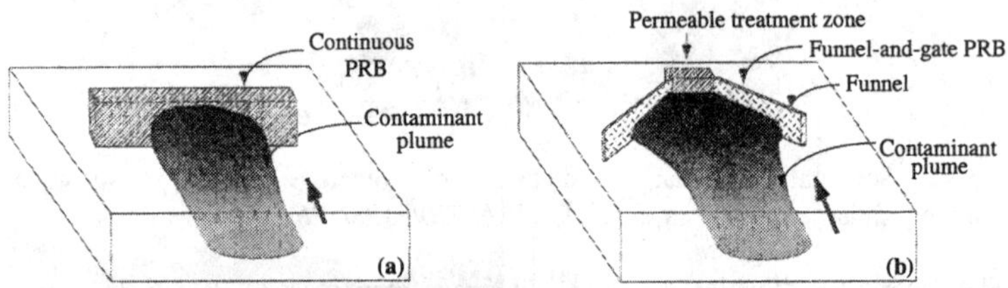

Fig. 13-2. Schematic diagrams of a (a) continuous permeable reactive barrier and (b) funnel-and-gate permeable reactive barrier (adapted from Powell et al. 1998, EPA/600/R-98/125)

13.2.2 Funnel-and-gate Permeable Reactive Barriers

As illustrated in Fig. 13-2b, funnel-and-gate PRB relies on a low permeability funnel composed of sheet piling, slurry wall or grout curtain to channel contaminated groundwater into a high permeability treatment zone containing reactive materials (Powell et al. 1998; Bedient et al. 1999). Typically, the ratio of the length of the funnel to the length of the permeable treatment zone is less than 6. In case of extremely large contaminant plume or highly heterogeneous aquifer, the funnel-and-gate system can be modified to contain multiple reactive zones (Ott 2000; ESTCP 2003). Due to directing large cross-sectional area of contaminated groundwater through the much smaller cross-sectional area of a permeable treatment zone, groundwater velocity within the treatment zone is usually 2 to 5 times higher than that resulting from natural gradient, depending on the funnel to treatment zone ratio (Day et al. 1999). To avoid the contaminant underflow, the bottom of the impermeable funnel and permeable treatment zone is also preferably keyed into an impermeable aquitard in which an impermeable funnel usually is keyed about 5 ft into the aquitard.

13.2.3 Caisson Permeable Reactive Barriers

In analogy to a funnel-and-gate PRB, caisson PRB or sometimes called *in situ* reactive vessel also consists of low permeability funnels to direct contaminated groundwater to the caisson containing reactive materials (Fig. 13-3a). However, as seen in Fig. 13-3b, contaminated groundwater flows through a caisson PRB in vertically upward direction. The advantages of a caisson PRB are that the upward flow of groundwater through the reactive medium is believed to be more uniform and closer to plug-flow pattern than the groundwater flows through continuous and funnel-and-gate PRBs. Moreover, the rigidity of a caisson allows for easier construction, particularly into existing slurry wall. Besides, the reactive medium within a caisson can be readily rejuvenated or replaced without excavation if necessary (Day et al. 1999; Elder 2000).

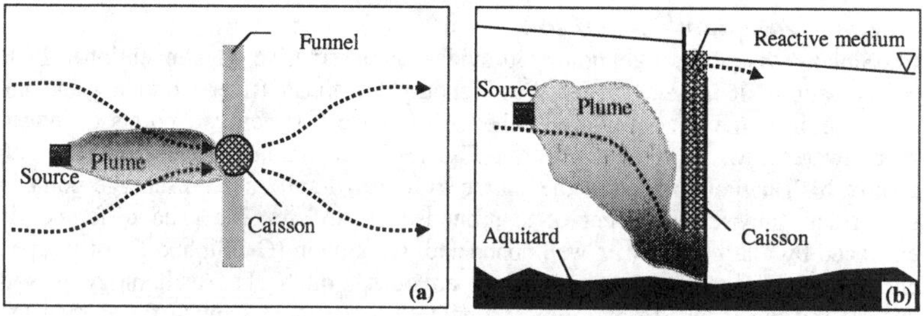

Fig. 13-3. (a) Plan view and (b) side view of a caisson permeable reactive barrier (dotted arrows represent the groundwater flow) (redrawn from Elder 2000)

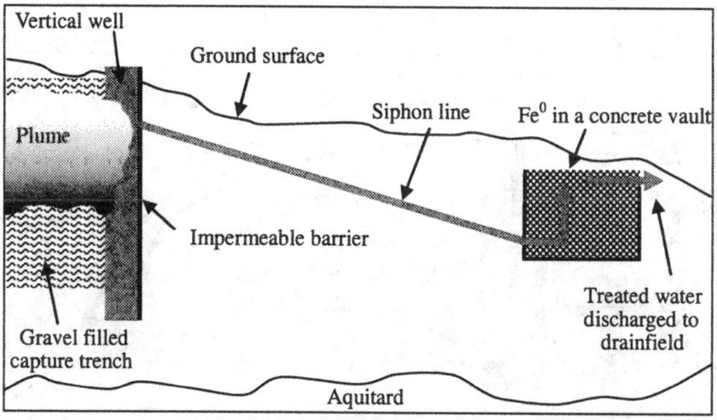

Fig. 13-4. Schematic diagram showing the trench permeable reactive barrier (redrawn from USDOE 2002, DOE/EM-0623)

13.2.4 Trench Permeable Reactive Barriers

Trench PRB mainly involves a groundwater capture trench that is installed subparallel to the direction of groundwater flow with an impermeable barrier located on the downgradient side of the trench, and a treatment zone containing reactive materials (e.g., Fe^0). As shown in Fig. 13-4, contaminated groundwater is collected through the capture zone portion of the trench and then is directed through the treatment zone using a siphon line. Finally, treated groundwater is discharged from the treatment zone back to the aquifer. The vertical well illustrated in Fig. 13-4 serves as a passive connection point for a siphon line and does not include any pumping system. Virtually, the flow of contaminated groundwater from the capture zone to the treatment zone is driven by the hydraulic head difference between the two zones (USDOE 2002).

13.2.5 GeoSiphon™/GeoFlow

GeoSiphon™/GeoFlow technology provides an alternative to conventional PRB configurations. It is patented by Westinghouse Savannah River Company and the system utilizes the natural hydraulic head difference between two points to induce groundwater flow. As you can see from Fig. 13-5, contaminant plume flows from an area of high hydraulic head at the source to low hydraulic head discharge point in subsurface, into surface water or at groundwater surface. The head difference is enhanced by a large diameter well connected to a siphon (GeoSiphon™) or to open channel (GeoFlow) which empties into an outfall ditch. The technology is still passive but the induced flow allows remediation to progress much faster than the other PRB configurations. The cell containing reactive materials can be placed anywhere along the flow path of contaminant plume, be removable or permanent, contain any reactive materials and be *in situ* or *ex situ* (Ott 2000; EnviroMetal Technologies Inc. 2005).

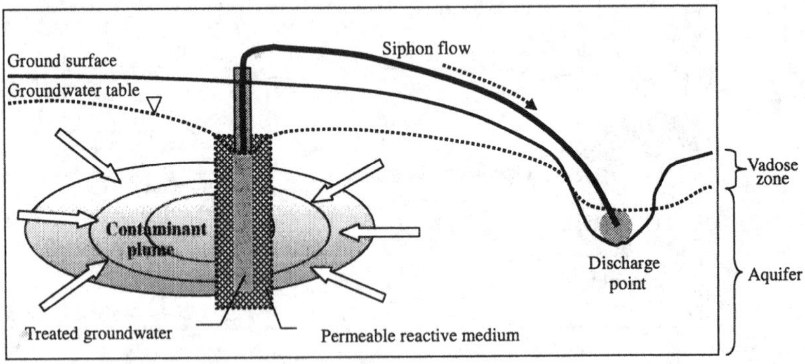

Fig. 13-5. Schematic diagram of GeoSiphon™ technology for *in situ* groundwater remediation (Phifer et al. 2002)

13.3 Emplacement Techniques for Permeable Reactive Barriers

Emplacement of PRB into subsurface mainly involves two steps, which are the emplacement of permeable treatment zone and installation of an impermeable funnel. Table 13-1 summarizes the conventional and innovative techniques for the PRB installations as well as their costs and limitations. Because of the need to key the barrier into the aquitard, the aquifer depth is the prime parameter governing the selection of the emplacement method. Moreover, required reactive barrier permeability and thickness, site topography, site access and work space, geotechnical constraints of a contaminated site, soil characteristics, disposal requirements of contaminated trench spoils, and costs are the factors deciding the type of emplacement method used (US Army Corps of Engineers 1997). Customarily, the emplacement method selected should be able to maintain the dimension of the installation to avoid waste of costly reactive materials, avoid dewatering and subsequent treatment of contaminated groundwater during installation, and ensure a rapid and simple construction sequence so as to control the installation cost (Day et al. 1999).

Table 13-1. Conventional and Innovative Techniques for the Emplacement of Permeable Reactive Barriers (adapted from Gavaskar et al. 2000, <http://www.itrcweb.org/Documents/PRB-2b.pdf>)

Emplacement techniques	Maximum depth (ft)	Vendor-quoted costs	Comments
Techniques for permeable treatment zone emplacement			
Caisson-based emplacement	45+	na	No personnel entry into excavation and relatively inexpensive
Mandrel-based emplacement	190	$\$7/ft^2$	Relatively inexpensive, fast production rate, requiring of multiple void spaces to constitute a permeable treatment zone
Continuous trenching	35-40	$\$5\text{-}12/ft^2$	High production rate and high mobilization cost
Techniques for impermeable funnel emplacement			
Soil-bentonite slurry wall			Requiring of a large working area for mixing of backfill, generation of some trench spoil and relatively inexpensive when using a backhoe
• By standard backhoe excavation	30	$\$2\text{-}8/ft^2$	
• By modified backhoe excavation	80	$\$2\text{-}8/ft^2$	
• By clamshell excavation	150	$\$6\text{-}15/ft^2$	
Cement-bentonite slurry wall			Generation of large quantities of trench spoil and more expensive than other slurry walls
• By standard backhoe excavation	30	$\$4\text{-}20/ft^2$	
• By modified backhoe excavation	80	$\$4\text{-}20/ft^2$	
• By clamshell excavation	200	$\$16\text{-}50/ft^2$	
Composite slurry wall	100+	na	Multiple-barrier wall
Steel sheet piles	60	$\$17\text{-}65/ft^2$	No spoil production
Innovative techniques for permeable treatment zone and impermeable funnel emplacement			
Jetting	200	$\$40\text{-}200/ft^2$	Capable of installing a barrier around existing buried utilities
Hydraulic fracturing	80-120	\$2300/fracture	Capable of emplacing a barrier at deep sites and generation of fractures only up to 3 inches thick
Deep soil mixing	150	$\$3\text{-}7/ft^3$	Probably not cost effective for the emplacement of permeable reactive barriers and generation of columns with diameter 3 to 5 ft

Note: na refers to not available.

13.3.1 Emplacement Techniques for Permeable Treatment Zone

The permeable treatment zone of a PRB basically is the portion of an aquifer, which is modified to contain reactive materials and specifically allow the flow of contaminant plume. Conventionally, permeable treatment zone is bounded on both upgradient and downgradient sides by thin sections of pea gravels particularly when the surrounding aquifer is highly heterogeneous. The pea gravels can increase the hydraulic conductivity surrounding the permeable treatment zone and uniformly draw groundwater flow into the reactive medium therein through homogeneous materials. Heretofore, trench excavation, caisson-based emplacement, mandrel-based emplacement and continuous trenching are the conventional techniques for the permeable treatment zone installation (Gavaskar et al. 1998).

13.3.1.1 Trench Excavation

This emplacement technique involves excavation of a trench, which houses reactive materials. Backhoe and clamshell are the most common types of equipment used for the excavation (see Fig. 13-6). For backhoe excavation, the digging apparatus is staged on a crawler-mounted vehicle and consists of a boom and a dipper stick with a mounted bucket. Generally, it is used for excavating shallow trench down to 30 ft but it can be modified with extended dipper stick to reach a depth up to 80 ft (US Army Corps of Engineers 1997). To enhance the stability of the trench wall during the installation, cofferdam approach and trench box are used. Cofferdam approach involves driving of steel sheet piles into the subsurface along the boundaries of the intended permeable treatment zone prior to the excavation. They are then reinforced with bracing when the trench is excavated. If it is necessary, sheet piles can also be used temporarily to separate reactive materials and pea gravel sections within the trench. Another way of the trench stabilization is the use of trench box. It can be fabricated above ground in advance using interlocking steel sheet piles and then inserted into an open trench. After backfilling of reactive materials, the sheet piles used in cofferdam approach or as trench box are then removed (Gavarskar et al. 2000). To install permeable treatment zone down to a depth 200 ft, a clamshell

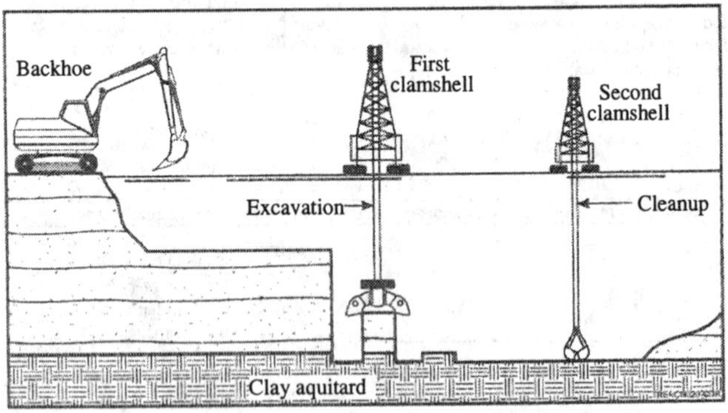

Fig. 13-6. Trench excavation using backhoe and clamshell (adapted from Gavaskar et al. 1998, <http://www.itrcweb.org/Documents/PRB-2b.pdf>)

bucket is commonly used. A cable suspended mechanical clamshell is crane-operated grabbing tool relying on gravity for accurate excavation and closure of the grab. Besides, hydraulic clamshell equips with a kelly bar to help guide and control the vertical line in addition to providing weight (US Army Corps of Engineers 1997).

13.3.1.2 Caisson-based Emplacement

Caisson-based emplacement involves first pushing or vibrating a hollow and load-bearing caisson vertically into subsurface (Fig. 13-7a). Once the caisson reaches the intended depth, the soil within the caisson is augered out and replaced with reactive materials. Upon completion of the emplacement, the caisson is pulled straight out to allow the flow of contaminated groundwater (Gavaskar et al. 2000). Since it is not economical to drive a caisson with diameter larger than 8 ft into subsurface, funnel-and-gate system with multiple caisson treatment zone should be used to provide adequate residence time for remediation of a site with wide plume, high level of contaminant and high groundwater velocity. Besides, because of the absence of any internal bracing, the caisson can be simply installed from the ground surface and completed without requiring personnel to enter into the excavation. Further, dewatering of excavation is not required (US Army Corps of Engineers 1997).

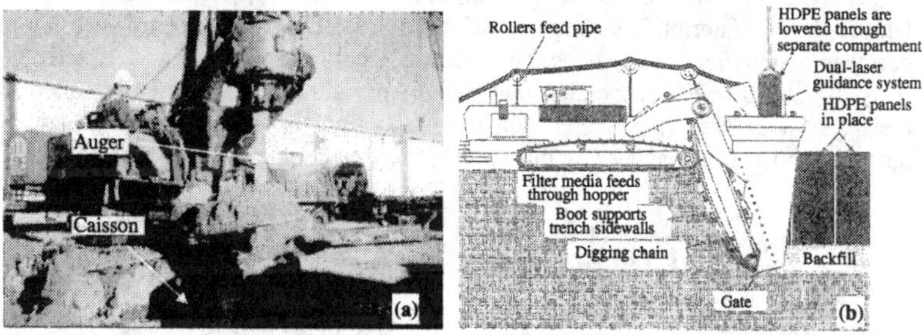

Fig. 13-7. (a) Augering out the native soil from the emplaced caisson for subsequent filling of reactive materials and (b) schematic diagram showing the operation of continuous trencher (adapted from Gavaskar et al. 2000, <http://www.itrcweb.org/Documents/PRB-2b.pdf>)

13.3.1.3 Mandrel-based Emplacement

Instead of emplacing a caisson into an aquifer, mandrel-based emplacement uses a hollow steel shaft to create a vertical void space in subsurface for the purpose of emplacing reactive materials. Prior to being hammered down into an aquifer using a vibratory hammer, a sacrificial drive shoe is placed over the bottom end of a mandrel. When void space is created, reactive materials can be simply poured into the void via a tremie tube. Since the typical size of mandrel is approximately 2 inch × 5 inch, a series of mandrel-emplaced voids rather than a single insertion is required to constitute a permeable treatment zone (Gavaskar et al. 1998).

13.3.1.4 Continuous Trenching

Continuous trenching machine allows simultaneous excavation and backfilling of reactive materials without an open trench. It can be seen from Fig. 13-7b that the trencher operates by cutting through soil using a chain-saw type apparatus attached to the boom of a crawler-mounted vehicle. Simultaneously, reactive materials are fed from an attached and overhead hopper into the trailing end of the excavated trench to stabilize the trench wall. Another compartment in the hopper is also capable of unrolling a continuous high-density polyethylene (HDPE) liner if necessary (Blowes et al. 1999a; Blowes et al. 1999b; Day et al. 1999). The continuous trencher can excavate in a water-filled trench without dewatering and having to install sheet piles for trench wall stabilization. Further, as shown in Fig. 13-7b, the boom of the trencher can be positioned almost vertically into the ground, thereby minimizing the amount of trenched spoils generated and avoiding sloping end (Gavaskar 1999). However, it may be problematic if continuous trencher is used for excavating wet and unconsolidated materials since there is a difficulty in bringing the trench spoils to the ground surface.

Generally, after the excavation, reactive materials are backfilled into the permeable treatment zone either using suspended bag (see Fig. 13-8a) or front-end loader (Gavaskar et al. 2000). Recent studies by the University of Texas and University of Waterloo have shown that reactive materials can be emplaced into the permeable treatment zone using biodegradable polymer slurry, such as guar gum (Day et al. 1999). Fe^0 suspended into biopolymer slurry can be directly added into the trench (see Fig. 13-8b) and then the biopolymer is broken down by naturally occurring microorganisms or by introducing enzyme compounds.

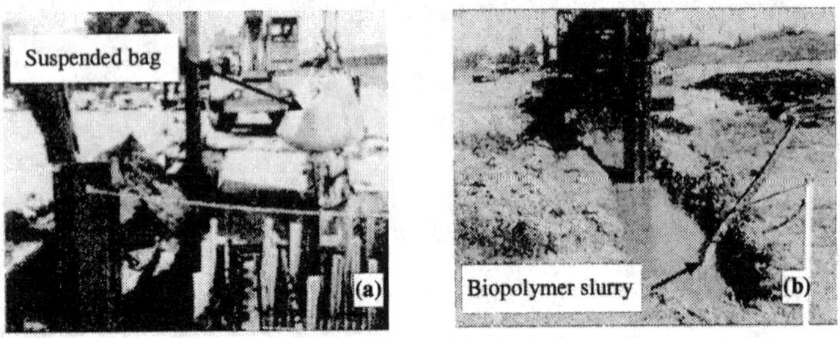

Fig. 13-8. (a) Backfilling of zero-valent iron from suspended bag (adapted from Gavaskar et al. 2000, <http://www.itrcweb.org/Documents/PRB-2b.pdf>) and (b) addition of biopolymer slurry containing zero-valent iron into an open trench (adapted from RTDF 2002)

13.3.2 Emplacement Techniques for Impermeable Funnels

13.3.2.1 Steel Sheet Piling

Steel sheet piling barrier and slurry wall are the two most common types of impermeable barrier used as an impermeable funnel. Sheet piles are typically 40 ft long but it can be lengthened by welding several pieces of sheet piles together (see Fig. 13-9a). Before driving into subsurface by a drop hammer or vibrating hammer, sheet piles are connected at their edge interlocks to form a cutoff wall (Gavaskar et al. 1998).

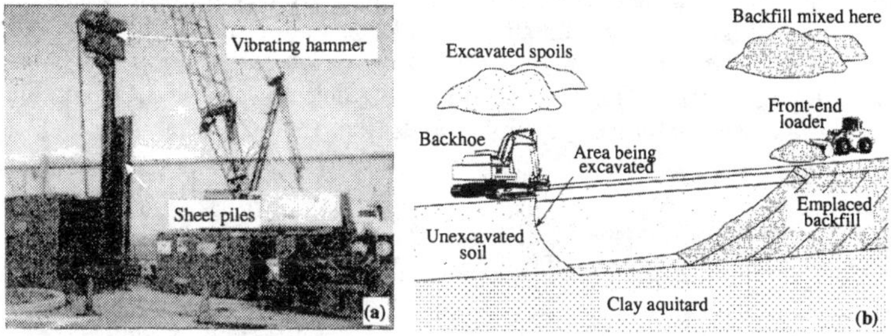

Fig. 13-9. (a) Emplacement of sheet piles using a vibrating hammer and (b) schematic diagram indicating the installation of soil-bentonite slurry trench (excavation and backfilling) (adapted from Gavaskar et al. 2000, <http://www.itrcweb.org/Documents/PRB-2b.pdf>)

13.3.2.2 Slurry Wall

Slurry wall is constructed by first excavating a slurry trench with backhoe, modified backhoe or clamshell, depending upon the depth (see Fig. 13-9b). Then slurry, which is usually a mixture of bentonite and water, is placed into the trench. As the slurry permeates into the sides of the excavation, a fully hydrated filter cake of bentonite is formed along the sides, which helps stabilize and maintain the integrity of the trench. Finally, a soil-bentonite backfill (soil-bentonite slurry wall), cement-bentonite backfill (cement-bentonite slurry wall), or a mixture of water, bentonite, cement and aggregate (plastic concrete slurry wall) is quickly placed into the trench, which are then left to harden (Gavaskar 1999). Composite barrier slurry wall is a multiple-layer barrier which offers three walls of defense, each with increasing chemical resistance and decreasing permeability. It is composed of an outer bentonite filter cake, soil-bentonite, cement-bentonite or plastic-concrete middle layer and an inner HDPE geomembrane. The installation of the outer and middle layers in the composite barrier slurry wall is similar to those of the soil-bentonite, cement-bentonite and plastic concrete slurry walls. Afterwards, the geomembrane envelope is inserted vertically into the slurry trench by either mounting it onto a detachable and removable frame, pulling it down using weights affixed to the membrane bottom or driving it down using a pile driver (Gavaskar et al. 2000).

13.3.3 Innovative Technologies for the Emplacement of Permeable Treatment Zone and Impermeable Funnels

Types of innovative emplacement technique for the installation of a permeable treatment zone and an impermeable funnel include jetting, hydraulic fracturing and deep soil mixing. Since these techniques do not involve excavation equipment, they have considerable potential to minimize health and safety issues. However, the requirement of specialized equipment leads higher operating and maintenance costs than the conventional emplacement techniques (US Army Corps of Engineers 1997).

13.3.3.1 Jetting

Jetting is a technique in which grout or reactive materials suspended in biodegradable slurry is injected into subsurface at high pressure through the nozzles of a drill stem as it is raised up through the soil. Depending upon whether or not the drill stem rotates as it is raised, the resulting barrier can be either a thin diaphragm wall (Fig. 13-10a) or a column in which two or three rows of overlapping and interlocking columns can form an effective barrier (Fig. 13-10b).

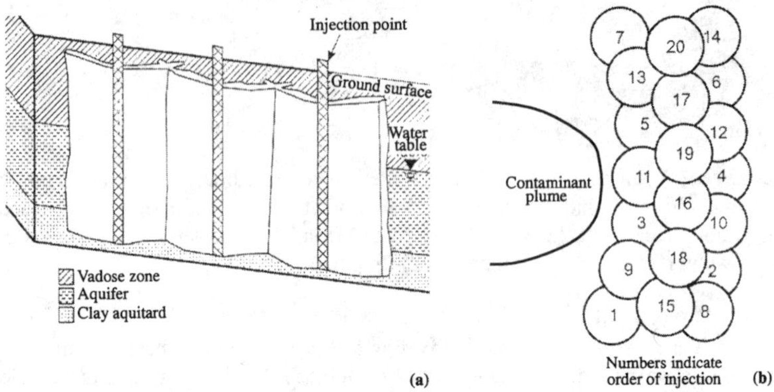

Fig. 13-10. (a) Diagram of a vertical thin diaphragm wall and (b) plan view of a grouted impermeable barrier installed using jetting technique (adapted from US Army Corps of Engineers 1997, <http://www.usace.army.mil/inet/usace-docs/design-guides/dg1110-345-117/entire.pdf>)

13.3.3.2 Hydraulic Fracturing

Hydraulic fracturing is the intentional fracturing of a subsurface formation. It begins by slowly pumping a fracturing solution customarily composed of water and guar gum into a sealed portion at the bottom of a cased borehole. Under a high pressure, fractures open and propagate out laterally from an initiation point previously notched out of the casing. A fracture-fill slurry either composed of reactive materials or grout can then be injected into the fractures to form a barrier. A series of horizontally stacked fractures can form an effective reactive barrier or impermeable funnel (Fig. 13-11a).

13.3.3.3 Deep Soil Mixing

This innovative technique involves direct injection of the bentonite or cement slurry, or reactive materials suspended in biodegradable slurry into the soil. As shown in Fig. 13-11b, the soil in subsurface is mixed up by a special rotating auger connected in series and equipped with mixing paddles. Simultaneously, the slurry is injected into the subsurface through a hollow drill stem as the auger retreats back to the surface. A permeable treatment zone or an impermeable funnel is formed by successive overlapping penetrations made with the deep soil mixer, resulting in a series of columns. Unlike jetting and hydraulic fracturing replacing soil with reactive materials, they are mixed with soil. Therefore, only 40 to 60% of reactive materials are present in a completed column (Gavaskar et al. 2000).

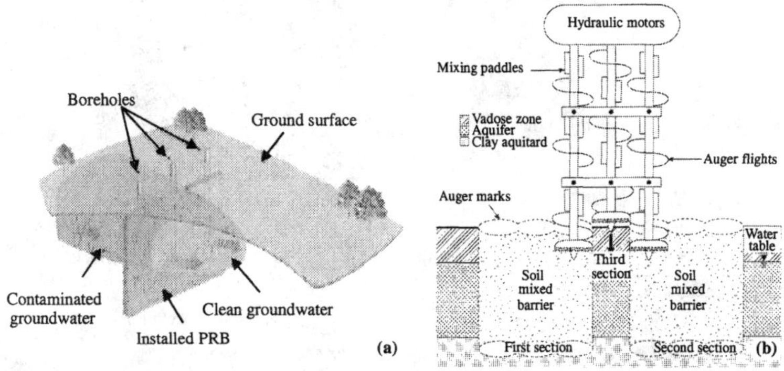

Fig. 13-11. (a) Permeable reactive barrier installed by hydraulic fracturing (Adapted from EnviroMetal Technologies Inc. 2005, <http://www.eti.ca/>) and (b) the technique of deep soil mixing (adapted from Gavaskar et al. 2000, <http://www.itrcweb.org/Documents/PRB-2b.pdf>)

13.4 Case Studies

13.4.1 Configuration and Construction of a Permeable Reactive Barrier at Vapokon Site, Denmark

The Vapokon site is located in the northern part of Fuen Island in Denmark. Its geological features are a layer of loam and soil filings extending approximately 1.5 m below the ground surface. Just below the vadose zone, there is a water-saturated sandy aquifer approximately 9 m in depth underlain by a relatively impermeable clay layer with a minimum thickness of about 15 m (Lo et al. 2003; Lai et al. 2006a). Naturally, the groundwater flows in a southeastern direction at a seepage velocity of 400 to 500 m/yr towards a creek and rain water pond nearby (Lai et al. 2006b). From 1976 to 1997, because of a spill from an underground concrete tank in which the used solvent was reloaded from drums, there has been heavy soil and groundwater contamination with CAHs and aromatic hydrocarbons (e.g., BTEX), resulting in a 80 × 200 m contaminant plume in Vapokon site (see Fig. 13-12a).

For the sake of remediating the CAH contaminated groundwater, a funnel-and-gate Fe^0 PRB composed of two 110 to 130 m long sheet piles (the funnel) and a permeable treatment zone containing granular Fe^0 was installed (Fig. 13-12b). As illustrated in Fig. 13-13, an approach of trench excavation was applied for the permeable treatment zone installation, whereas steel sheet piling barriers were selected as impermeable funnels. Before the excavation, interlocked steel sheet piles were driven into the subsurface along the boundaries of the intended permeable treatment zone to stabilize the trench wall (see Fig. 13-13a). To ensure a contact between the Fe^0 barrier and the clay layer beneath, the 110 to 130 m long sheet piling barriers and the sheet piling case surrounding the permeable treatment zone were

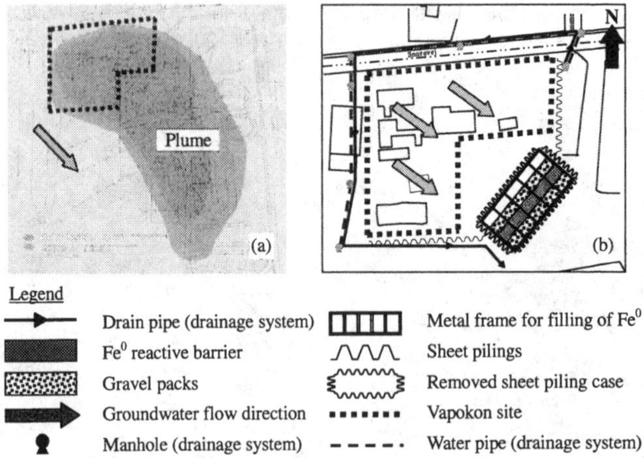

Fig. 13-12. (a) Contaminant plume and (b) plan views of the funnel-and-gate iron reactive barrier and drainage system installed in Vapokon site, Denmark (adapted from Lai et al. 2006a)

Fig. 13-13. (a) Driving of steel sheet piles into the subsurface along the boundary of the intended permeable treatment zone, (b) excavation of the trench using backhoe, (c) fabrication of metal frame, (d) backfilling of granular zero-valent iron into the trench and (e) permeable treatment zone of the funnel-and-gate permeable reactive barrier at Vapokon site, Denmark

rammed to at least 1 m into the till underneath the aquifer. After the insertion of sheet piling case, backhoe was applied for excavation (Fig. 13-13b). Metal frame was then constructed for filling of Fe^0 and bracing the sheet piling case (Fig. 13-13c). Finally, 270 tons of granular Fe^0 was backfilled into the trench using suspended bag (Fig. 13-13d) and sheet piling case was removed, resulting in a Fe^0 barrier with a 0.8 m thick, 14.5 m wide and 9 m deep (Fig. 13-13e) (Lo et al. 2003; Lai et al. 2006a).

13.4.2 Configuration and Construction of a Permeable Reactive Barrier at United States Coast Guard (USCG) Support Centre in Elizabeth City, North Carolina

Continuous Fe^0 PRB was installed in 1996 at United States Coast Guard (USCG) Support Center in Elizabeth City to treat overlapping plumes of hexavalent chromium [Cr(VI)] and TCE released from an electroplating shop between 1950s and 1980s (Blowes et al. 1999a; Blowes et al. 1999b; Puls et al. 1999a; Puls et al. 1999b; Wilkin et al. 2001). As is evident in Figs. 13-14a and 13-14b, Cr(VI) and TCE plumes extending approximately 65 m from the source with a width between 35 and 60 m were detected in USCG Support Center. The maximum observed Cr(VI) and TCE concentrations exceeded 10 and 19 mg/L, respectively (Blowes et al. 1999a).

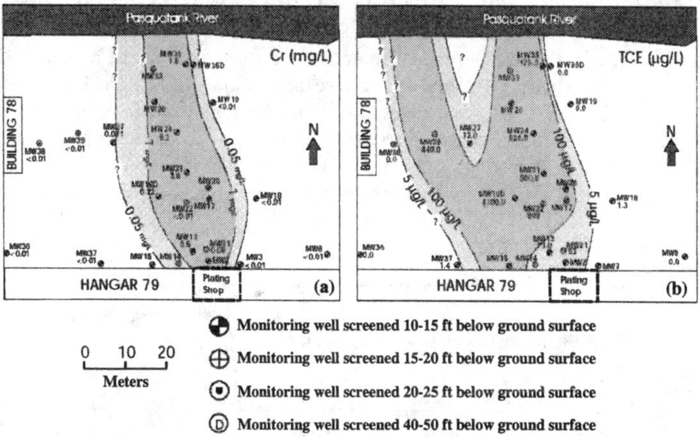

Fig. 13-14. Plan views of (a) Cr(VI) plume and (b) TCE plume in the aquifer at United States Coast Guard (USCG) Support Center in Elizabeth City measured in 1994 (Adapted from Blowes et al. 1999a, EPA/600/R-99/095a)

A continuous PRB with a 46 m long, 7.3 m deep and 0.6 m thick composing of 100% Fe^0 was installed at USCG Support Center (Fig. 13-15a) using continuous trenching machine. As illustrated in Fig. 13-15b, continuous trenching machine used to install the continuous iron reactive barrier simultaneously removed aquifer materials and backfilled Fe^0. Aquifer materials were brought to the ground surface using the digging chain, while granular Fe^0 was backfilled into the trench from the hopper.

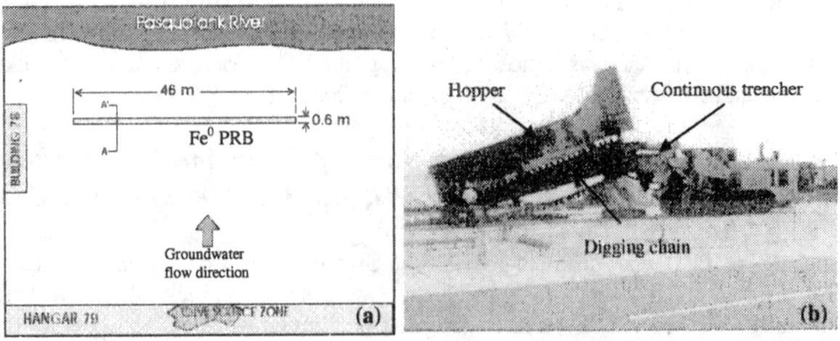

Fig. 13-15. (a) Plan view of the continuous iron reactive barrier installed at United States Coast Guard (USCG) Support Center jn Elizabeth City, North Carolina and (b) installation of the continuous iron barrier using continuous trenching machine (adapted from Blowes et al. 1999a, EPA/600/R-99/095a)

13.4.3 Configuration and Construction of a Permeable Reactive Barrier at an industrial site in Belfast, Northern Ireland

At an industrial site in Belfast, Northern Ireland, the historic spillage of chlorinated solvents led to a localized contaminant source and plume in which TCE with concentration up to 390 mg/L was detected to be the major contaminant in an aquifer. To remediate the CAH contaminated groundwater, a caisson PRB or circular *in situ* reactive vessel was installed to a depth of about 40 ft at the industrial site (see Fig. 13-16). Two 100 ft long slurry walls were installed as impermeable funnels to direct

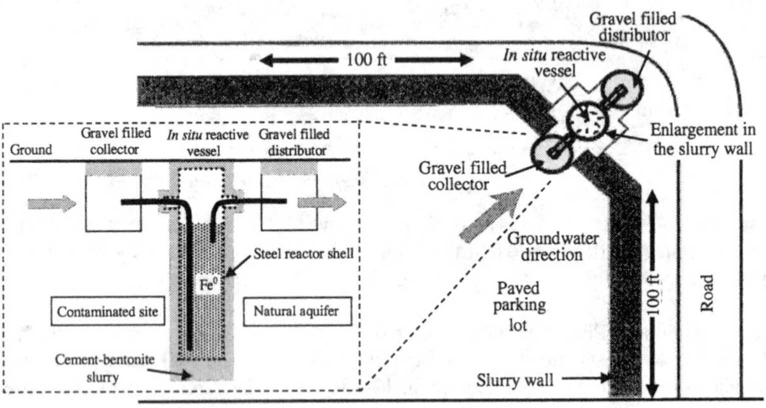

Fig. 13-16. Caisson permeable reactive barrier or *in situ* reactive vessel installed at an industrial site in Belfast, Northern Ireland (redrawn from Powell et al. 1998, EPA/600/R-98/125)

contaminated groundwater to the gravel filled collector or inlet of the steel reactive vessel containing Fe^0. The steel reactive vessel was placed in an enlargement in the cement-bentonite slurry wall so that it was located between the boundary of the contaminated site and the natural aquifer. As seen from Fig. 13-16, contaminated groundwater flowed vertically along the *in situ* reactive vessel before releasing into the gravel filled distributor so that the full depth of the Fe^0 packed medium could be saturated whatever the seasonal variation in groundwater level. Both cement-bentonite slurry wall and the enlargement for steel reactive vessel were keyed into an aquiclude layer so as to prevent underflow of contaminated groundwater (NATO 1998; Powell et al. 1998).

13.5 Summary

The key components of the PRB technology are the reactive materials and impermeable funnels emplaced in subsurface and various configurations of the emplaced reactive materials and funnels lead to different PRB configurations. Heretofore, continuous PRB and funnel-and-gate PRB are the two most common configurations used in PRB sites. However, there is also an increase in the popularity of the application of caisson PRB, trench PRB and GeoSiphon™/GeoFlow for *in situ* groundwater remediation.

Installation of iron reactive barriers into the contaminated aquifers mainly involves the emplacement of a permeable treatment zone and an impermeable funnel. Generally, trench excavation, caisson- and mandrel-based emplacements as well as continuous trenching are the types of emplacement techniques conventionally used for the permeable treatment zone installation. Besides, steel sheet piling and slurry walls, including soil-bentonite, cement-bentonite, plastic concrete and composite barrier slurry walls, are the impermeable barriers commonly used as impermeable funnels in PRB technology. The advent of innovative emplacement techniques, such as jetting, hydraulic fracturing and deep soil mixing, gives alternatives to the conventional emplacement techniques. They can potentially minimize health and safety issues happened during the PRB construction since these innovative techniques do not require personnel entry into the excavation.

13.6 References

Bedient, P. B., Rifai, H. S., and Newell, C. J. (1999). *Ground water contamination transport and remediation*, 2nd Ed., Prentice Hall PTR, Upper Saddle River, NJ 07458, United States.

Blowes, D. W., Gillham, R. W., Ptacek, C. J., Puls, R. W., Bennett, T. A., O'Hannesin, S. F., Hanton-Fong, C. J., and Bain, J. G. (1999a). *An in situ permeable reactive barrier for the treatment of hexavalent chromium and trichloroethylene in ground water Volume 1 design and installation*, EPA/600/R-99/095a, Office of Research and Development, United States Environmental Protection Agency, Washington, D. C. 20460, United States.

Blowes, D. W., Puls, R. W., Gillham, R. W., Ptacek, C. J., Bennett, T. A., Bain, J. G., Hanton-Fong, C. J., and Paul, C. J. (1999b). *An in situ permeable reactive barrier for the treatment of hexavalent chromium and trichloroethylene in ground water: Volume 2 performance monitoring*, EPA/600/R-99/095b, Office of Research and Development, United States Environmental Protection Agency, Washington, D. C. 20460, United States.

Day, S. R., O'Hannesin, S. F., and Marsden, L. (1999). "Geotechnical techniques for the construction of reactive barriers." *J. Hazard. Mater.*, B67(3), 285-297.

Elder, C. R. (2000). "Evaluation and design of permeable reactive barrier amidst heterogeneity." Ph.D. Dissertation, Department of Civil and Environmental Engineering, University of Wisconsin – Madison, United States.

EnviroMetal Technologies Inc. (1998). *Bench-scale treatability report of the EnviroMetal process at the Vapokon site, Fyn, Denmark*, ETI 31438.10, Guelph, Ontario, Canada N1K1S6.

EnviroMetal Technologies Inc. (2005). *EnviroMetal Technologies Inc.*, < http://www.eti.ca/> (Mar. 3, 2005).

ESTCP (2003). *ESTCP cost and performance report –Evaluating the longevity and hydraulic performance of permeable reactive barriers at Department of Defense Sites*, CU-9907, Environmental Security Technology Certification Program, United States Department of Defense, United States,

Gavaskar, A. R., Gupta, N., Sass, B. M., Janosy, R. J., and O'Sullivan, D. (1998). *Permeable barriers for groundwater remediation - Design, construction, and monitoring*, Battelle Press, Columbus.

Gavaskar, A. R. (1999). "Design and construction techniques for permeable reactive barriers." *J. Hazard. Mater.*, 68, 41-71.

Gavaskar, A., Gupta, N., Sass, B., Janosy, R., and Hicks, J. (2000). *Design guidance for application of permeable reactive barriers for groundwater remediation*, Strategic Environmental Research and Development Program, Battelle Columbus, Ohio, <http://www.itrcweb.org/Documents/PRB-2a.pdf>, <http://www.itrcweb.org/Documents/PRB-2b.pdf>, <http://www.itrcweb.org/Documents/PRB-2c.pdf>.

Gillham, R. W., and O'Hannesin, S. F. (1992). "Metal-calalysed abiotic degradation of halogenated organic compounds." In *1992 IAH Conference "Modern Trends in Hydrogeology"*, May 10-13, Hamilton, Ontario, Canada.

Gu, B., Liang, L., Dickey, M. J., Yin, X., and Dai, S. (1998). "Reductive precipitation of uranium(VI) by zero-valent iron." *Environ. Sci. & Technol.*, 32(21), 3366-3373.

Lai, K. C.-K., and Lo, I. M.-C. (2002). "Bench-scale study of the effects of seepage velocity on the dechlorination of TCE and PCE by zero-valent iron." In *Proceedings of the 6th International Symposium on Environmental Geotechnology*, July 2-5, Seoul, Korea.

Lai, K. C.-K., Lo, I. M.-C., Birkelund, V., and Kjeldsen, P. (2006a). "Field Monitoring of a Permeable Reactive Barrier for Removal of Chlorinated Organics", *J. of Environ. Eng. ASCE*, 132(2), 199-210.

Lai, K. C.-K., Lo, I. M.-C., and Kjeldsen, P. (2006b). "Natural gradient tracer test for the permeable reactive barrier in Denmark 1. Field study of tracer movement." *J. of Hazardous, Toxic and Radioactive Waste Management, ASCE*, in press.

Lo, I. M.-C., Lai, K. C.-K., and Kjeldsen, P. (2003). "Performance of permeable reactive barrier on remedying aliphatic chlorinated organics contaminated groundwater." In *Proceedings of the International Conference on Contaminated Soils and Groundwater in Arid Countries*, Muscat, Oman.

Lo, I. M.-C., Lam, C. S.-C., and Lai, K. C.-K. (2005). "Competitive effects of TCE on Cr(VI) removal by zero-valent iron." *J. Environ. Eng. ASCE*, 131(11), 1598-1606.

NATO (1998). *NATO/CCMS pilot study – Evaluation of demonstrated and emerging technologies for the treatment of contaminated land and groundwater (phase III)*, 1998 special session, treatment walls and permeable reactive barriers, Vol. 229, North Atlantic Treaty Organization.

O'Hannesin, S. F., and Gillham, R. W. (1992). "Permeable reaction wall for in situ degradation of halogenated organic compounds." In *Proceedings of the 45th Canadian Geotechnical Conference*, Toronto, Ontario, Canada.

O'Hannesin, S. F., and Gillham, R. W. (1998). "Long-term performance of an In Situ "Iron Wall" for remediation of VOCs." *Ground Water*, 36(1), 164-170.

Ott, N. (2000). *Permeable reactive barriers for inorganics*, National Network of Environmental Management Studies Fellow, Office of Solid Waste and Emergency Response, United States Environmental Protection Agency, Washington, D. C. 20460, United States.

Phifer, M. A., Nichols, R. L., Sappington, F. C., Steimke, J. L., and Jones, W. E. (2002). *GeoSiphon™ ground water remediation system hydraulics*, WSRC-MS-2002-00260, Westinghouse Savannah River Company, DE-AC09-96SR18500, U.S. Department of Energy.

Powell, R. M., and Puls, R. W. (1997). *Permeable reactive subsurface barriers for the interception and remediation of chlorinated hydrocarbon and chromium (VI) plumes in ground water*, EPA/600/F-97/008, U. S. EPA Remedial Technology Fact Sheet, United States Environmental Protection Agency, Washington, D. C. 20460, United States.

Powell, R. M., Blowes, D. W., Gillham, R. W., Schultz, D., Sivavec, T., Puls, R. W., Vogan, J. L., Powell, P. D., and Landis, R. (1998). *Permeable reactive barrier technologies for contaminant remediation*, EPA/600/R-98/125, Office of Research and Development, United States Environmental Protection Agency, Washington, D. C. 20460, United States.

Puls, R. W., Blowes, D. W., and Gillham, R. W. (1999a). "Long-term performance monitoring for a permeable reactive barrier at the U. S. Coast Guard Support Center, Elizabeth City North Carolina." *J. Hazard. Mater.*, 68(1-2), 109-124.

Puls, R. W., Paul, C. J., and Powell, R. M. (1999b). "The application of in situ permeable reactive (zero-valent iron) barrier technology for the remediation of chromate-contaminated groundwater: A field test." *Applied Geochemistry*, 14(8), 989-1000.

RTDF (2002). *Summary of the remediation technologies development forum permeable reactive barriers action team meeting*, Remediation Technologies Development Forum, < http://www.rtdf.org/public/permbarr/minutes/110702 /summary.html>.

Senzaki, T., and Kumagai, Y. (1989). "Removal of organochloro compounds by reduction treatment – Treatment of trichloroethylene with iron powder." *Kogyo Yosui*, 369, 19-24 (Japanese).

Sweeney, K. H. (1972). *Reductive degradation of halogenated pesticides*, United States Patent No. 3,640,821.

US Army Corps of Engineers (1997). *Design guidance for application of permeable barriers to remediate dissolved chlorinated solvents*, DG 1110-345-117, Armstrong Laboratory, Environics Directorate, US Air Force, <http://www.usace. army.mil/inet/usace-docs/design-guides/dg1110-345-117/entire.pdf>.

USDOE (2002). *Innovative technology summary report – Passive reactive barrier*, DOE/EM-0623, Office of Science and Technology, Office of Environmental Management, United States Department of Energy.

USEPA (2002). *Field applications of in situ remediation technologies: Permeable reactive barriers*, EPA/68/W-00/084, Technology Innovation Office, Office of Solid Waste and Emergency Response, United States Environmental Protection Agency, Washington, D. C. 20460, United States.

Westerhoff, P. (2003). "Reduction of nitrate, bromate, and chlorate by zero-valent iron (Fe^0)." *J. of Environ. Eng. ASCE*, 129(1), 10-16.

Wilkin, R. T., Puls, R. W., and Sewell, G. W. (2001). *Long-term performance of permeable reactive barriers using zero-valent iron: An evaluation at two sites*, Office of Research and Development, Subsurface Protection and Remediation Division, United States Environmental Protection Agency, Ada, OK, United States.

CHAPTER 14

Design Methodology for the Application of a Permeable Reactive Barrier for Groundwater Remediation

Irene M. C. Lo[1], Keith C. K. Lai[2], and Rao Surampalli[3]

Abstract: Application of a permeable reactive barrier (PRB) in a contaminated site for groundwater remediation requires proper design of its reactivity, dimension and configuration as well as the selection of an appropriate construction method based upon the geotechnical constraints of the site. The overall design methodology for the application of a PRB involves preliminary assessment, site characterization, reactive media selection, treatability testing, hydrogeologic and geochemical modelings, configuration and construction method selection, monitoring plan preparation, and finally PRB cost evaluation. After showing the suitability of the contaminated site for PRB application through the preliminary assessment, the location, dimension and configuration of the reactive barrier as well as the reactive material used in the PRB are evaluated in the treatability and modeling studies. Finally, by estimating the capital, operating and maintenance costs for the PRB application, the economic advantage of the PRB technology over other conventional remediation approaches can be evaluated.

CE Database subject headings: Barriers; Degradation; Design; Ground-water pollution; Iron; Methodology; TCE.

14.1 Introduction

In comparison with a pump-and-treat system, permeable reactive barrier (PRB) is relatively permanent structure (Lo et al. 2003). Whereas for a pump-and-treat system, the locations of pumping wells, pumping rates, and aboveground treatment methods can be readily changed or modified, PRB is difficult to relocate and change once it is emplaced. Therefore, proper design of a reactive barrier prior to the installation in subsurface is essential in order to achieve the expected efficiency of a PRB on groundwater remediation (Lo and Lai 2002). Basically, the overall design methodology for the application of a PRB in a contaminated site involves several steps (Fig. 14-1). They are preliminary assessment, site characterization, reactive media selection, treatability testing, hydrogeologic and geochemical modelings, selection of a suitable PRB configuration and construction method, monitoring plan

[1]Associate Professor, Dept. of Civil Engineering, The Hong Kong University of Science and Technology, Clear Water Bay, Kowloon, Hong Kong (corresponding author). E-mail: cemclo@ust.hk
[2]Research Associate, Dept. of Civil Engineering, The Hong Kong University of Science and Technology, Clear Water Bay, Kowloon, Hong Kong.
[3]Engineer Director, United States Environmental Protection Agency, Kanasas City, Kansas 66117, United States

preparation, and economic evaluation (US Army Corps of Engineers 1997).

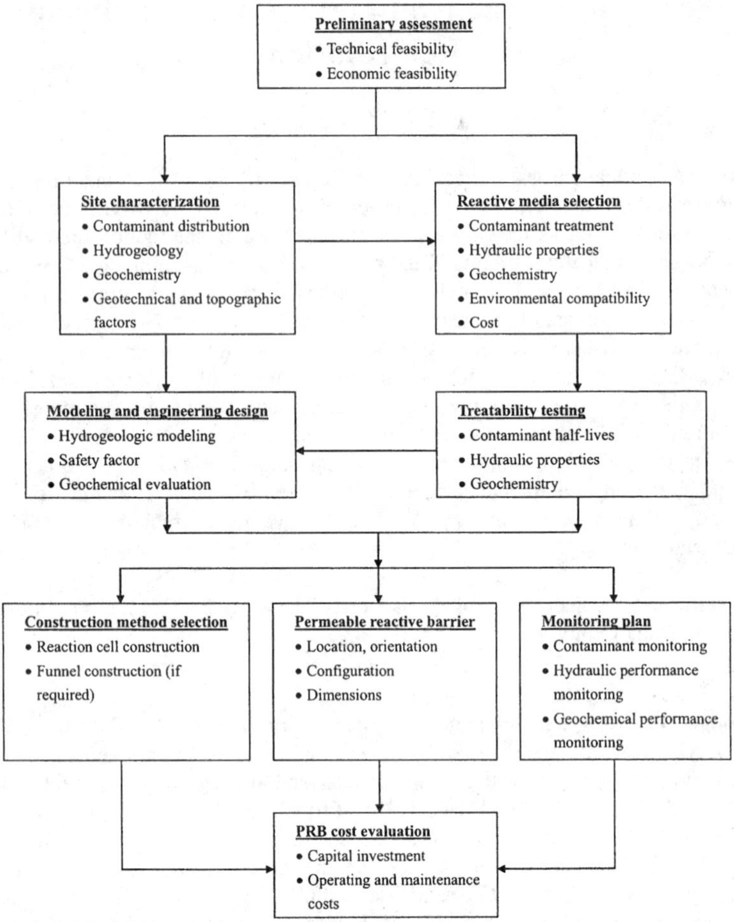

Fig. 14-1. Design methodology for a PRB application (redrawn from Gavaskar et al. 2000, <http://www.itrcweb.org/Documents/PRB-2a.pdf>)

Generally, before going into the detailed design steps, preliminary assessment is first conducted to evaluate the applicability of the PRB technology to a contaminated site. If the contaminated site is evaluated to be suitable for the application of the PRB technology for groundwater remediation, additional site characterization will then be implemented if the available site characterization data is inadequate. The site characterization data is generally used for the modelings, reactive media and emplacement method selections. After selecting the reactive material and determining the required PRB thickness from the treatability testing, the location, configuration and dimension of the PRB can be evaluated according to the results of hydrogeologic and geochemical modelings as well as site characterization data. Based upon the selected configuration, designed PRB dimension and the site characterization, appropriate construction method is chosen. Thereafter, performance monitoring plan is designed for the determination whether the operating performance of the constructed PRB is consistent with the design objectives. Finally, the capital, operating and maintenance costs for a PRB application are estimated and compared with the costs of other conventional remediation approaches (Gavaskar et al. 1998). Since most PRB sites are hitherto installed for the remediation of groundwater polluted by chlorinated aliphatic hydrocarbons (CAHs) such as trichloroethylene (TCE), this chapter primarily focuses on the design methodology for the application of a PRB to remediate CAHs in groundwater (USEPA 2002).

14.2 Preliminary Assessment

The purpose of implementing preliminary assessment is to decide whether or not a contaminated site is technically and economically suitable for a PRB technology before performing other design steps. Factors influencing the technical and economic feasibility of a PRB technology include contaminant type, plume size and distribution, depth of an aquifer, geotechnical constraints in the contaminated site, aquitard competence for reactive barrier installation as well as groundwater velocity (Gavaskar et al. 2000). The contaminants found in the site must be amenable to degradation by suitable reactive media. Table 14-1 summarizes the contaminants to be treatable or not treatable by reactive materials in PRBs (Powell et al. 1998). Besides, it is costly for a PRB application if contaminant plume is too large and deep. Further, if the aquifer is very deep and the reactive barrier must be keyed into the impermeable layer beneath, the construction cost can be high.

Table 14-1. Contaminants Treatable and Not Treatable by Reactive Materials in Permeable Reactive Barriers (modified from Powell et al. 1998, EPA/600/R-98/125)

Organic compounds		**Inorganic compounds**	
Treatable contaminants			
Methanes	Tetrachloromethane and Trichloromethane	Trace metals	Chromium, nickel lead, uranium, technetium, iron manganese, selenium copper, cobalt, cadmium and zinc
Ethanes	Hexachloroethane, 1,1,1-trichloroethane, 1,1,2-trichloroethane and 1,1-dichloroethane	Anionic contaminants	Nitrate, [b]bromate, [b]chlorate, sulphate, phosphate and arsenic
Ethylenes	Tetrachloroethylene, trichloroethylene, cis-dichloroethylene, trans-dichloroethylene, 1,1-dichloroethylene and vinyl chloride		

Table 14-1 (*Continued*)

Organic compounds		Inorganic compounds
Treatable contaminants		
Propanes	1,2,3-Trichloropropane and 1,2-dichloropropane	
Others	Hexachlorobutadiene, 1,2-dibromoethane, Freon 113, N-nitrosodimethylamine and [a]polychlorinated biphenyls	
Non treatable contaminants		
Dichloromethane, 1,2-dichloroethane, chloroethane and chloromethane		Chloride and perchlorate

Note: Most of the information is quoted from Powell et al. (1998).
[a]The information is obtained from Doyle et al. (1998) and Kim et al. (2004).
[b]The information is obtained from Westerhoff (2003).

Table 14-2. Groundwater Velocity in PRB Sites

Field sites	Groundwater velocities (m/yr)
[a]Lowry Air Force Base, CO	111
[a]Cape Canaveral Air Station, FL	11.1 to 55.6
[a]Savannah River Site TNX Area, Aiken, SC	334
[a]Watervliet Arsenal, Albany, NY	16.7
[a]Public School, Langton, Ontario, Canada	101
[a]Fry Canyon, UT	167
[b]Nickel Rim Mine Site, Sudbury, Ontario, Canada	10 to 20
[b]Marzone Superfund Site	3.0 to 6.1
[b]Kansas City Plant	126
[c]Superfund Site in Iowa	3340 to 6680
[c]Monticello, Utah	223 to 1450
[d]Dupont's Kinston Plant, NC	11.1
[e]Denver Federal Centre, Denver, Colorado	14.6 to 36.5
[f]Canadian Forces Base, Borden, Ontario	29.9
[g]NAS Moffett Field Mountain View, California	54.8 to 110
[h]Former Industrial Facility in Upstate New York	73 to 256
[i]Northwestern Tip of Alameda Island	21.9
[j]US Coast Guard Support Center, Elizabeth City, NC	47.5 to 65.7
[k]Vapokon Site, Fuen Island, Denmark	99.5

[a]USEPA (1999)
[b]RTDF (2000)
[c]RTDF (2001)
[d]RTDF (2002)
[e]McMahon et al. (1999)
[f]O'Hannesin and Gillham (1998)
[g]Reeter et al. (1998)
[h]Vogan et al. (1999)
[i]Morkin et al. (2000)
[j]Blowes et al. (1999a)
[k]Lai et al. (2006c)

The presence of consolidated sediments, or large gravels or rocks in the aquifer as well as aboveground structures, such as buildings, may increase the difficulty and cost of the barrier construction. Another important point is that the possibility of migration of contaminants into the lower aquifer zone because of the breach of the aquitard during the PRB installation should also be considered. In addition, the groundwater velocity in the site should not be too high since the thickness of the reactive barrier required to obtain desired design residence time may be high (US Army Corps of Engineers 1997). As illustrated in Table 14-2, a PRB has been installed at a Superfund site in Iowa with groundwater velocity as high as 6680 m/yr or 18 m/d (RTDF 2001; Lai 2004).

14.3 Site Characterization

Once the preliminary assessment shows that the site is suitable for a PRB application, the next step is to check whether the available site characterization data are sufficient to locate and design a PRB. Otherwise, additional site characterization may be required. Site information related to aquifer characteristics, organic and inorganic compositions, and physical characteristics of contaminated groundwater as well as geotechnical and topographic data of the site are the important information required (Gavaskar et al. 2000).

Aquifer characteristics including groundwater depth, depth to aquitard, aquitard thickness and continuity, groundwater velocity, site stratigraphy or heterogeneity, hydraulic conductivities of different stratigraphic layers, porosity and dimension of contaminant plume are needed to know. This information is required for subsequent implementing hydrogeologic and geochemical modelings to reliably locate and design a reactive barrier in the contaminated site. In addition, understanding the types and concentration of groundwater contaminants assists the selection of an appropriate reactive material, the implementation of treatability test and the thickness design of a PRB. Knowledge of the groundwater physical characteristics (e.g., pH and redox potential) as well as the concentration of hardness, alkalinity, iron and sulfate etc in groundwater helps evaluate the potential of mineral precipitation within a PRB, which may affect the barrier's long-term reactivity and hydraulic performance. Another important point is that the information of underground (e.g., utility lines or rocks) and aboveground structures (e.g., buildings) in the site helps select an appropriate PRB configuration and emplacement technique (Gavaskar et al. 1998).

14.4 Reactive Media Selection

Another step of designing a PRB is to select candidate reactive materials, which are potentially used in a reactive barrier, for the further study in treatability testing. The criteria affecting the selection include the reactivity, hydraulic performance, long-term stability, as well as availability and price of reactive materials. Moreover, the environmental compatibility of the byproducts generated during the degradation is needed to consider (US Army Corps of Engineers 1997). The candidate reactive materials being selected should be able to degrade the target contaminants with an acceptable residence time. Basically, the shorter the degradation half-life of contaminants within a given reactive material or the higher the degradation rate constant, the better the reactive material. Table 14-3 summarizes the degradation half-life of various groundwater contaminants using different reactive materials.

Although small particle size of reactive materials exerts high reactivity, high permeability of reactive media contrarily requires large particle size. Therefore, the trade-off between the reactivity and hydraulic conductivity should be taken into account during the selection of particle size of reactive materials. With regard to the

Table 14-3. Degradation Half-life of Various Groundwater Contaminants Using Different Reactive Materials (All the Half-lives are Normalized to 1 m^2 Surface Area Per Litre Solution, i.e., $t_{1/2-N}$)

Contaminants	Reactive materials	Half-lives, $t_{1/2-N}$ ($hr\ m^2\ L^{-1}$)
Hexavalent chromium	Fe^0	[a]0.8 to 2.7 [b]4.4 to 5.2
Uranium (VI)	Master-builder Fe^0	[c]5.80
	Peerless Fe^0	[c]0.80 to 1.54
	Cercona cast Fe^0	[c]0.24
Nitrate	Fisher Fe^0	[a]408 to 518
Nitrite	Fisher Fe^0	[a]13.2 to 57.8
Tetrachloroethylene	Fe^0	[d]330
Trichloroethylene		[d]1777
cis-Dichloroethylene		[d]16902
trans-Dichloroethylene		[d]5775
1,1-Dichloroethylene		[d]10828
Vinyl chloride		[d]13860
Hexachloroethane		[d]22.4
1,1,2,2-Tetrachloroethane		[d]53.3
1,1,1,2-Tetrachloroethane		[d]49.5
1,1,1-Trichloroethane		[d]63.0
Chloroform		[d]753
Tetrachloroethylene	Green rust	[e]4278
Trichloroethylene		[e]8115

Contaminants	Reactive materials	half-lives, $t_{1/2-N}$ ($hr\ m^2\ L^{-1}$)
cis-Dichloroethylene	Green rust	[e]13378
Vinyl chloride		[e]8925
Tetrachloroethylene	Iron sulfide	[f]608
Trichloroethylene		[f]233
4-Monochlorinated biphenyl	Palladized iron	[g]116
3-Monochlorinated biphenyl		[g]154
2-Monochlorinated biphenyl		[g]173
3,4-Monochlorinated biphenyl		[g]355
2,4-Monochlorinated biphenyl		[g]447
2,3-Monochlorinated biphenyl		[g]478
4-Monochlorinated biphenyl	Palladized zinc	[g]14.6
3-Monochlorinated biphenyl		[g]25.2
2-Monochlorinated biphenyl		[g]28.9
Carbon tetrachloride	Pure Fe^0	[h]20, [i]3, [j]23
Bromoform		[h]41
Hexachloroethane		[h]13
Nitrobenzene		[k]8
1,2,3-Trichloropropane	Commercial Fe^0	[l]24000
1,2-Dichloropropane		[l]4500
1,3-Dichloropropane		[l]2200

[a]Alowitz and Scherer (2002) [d]Johnson et al. (1996) [g]Kim et al. (2004) [j]Lipczynska-Kochany et al. (1994)
[b]Cantrell et al. (1995) [e]Lee and Batchelor (2002) [h]Gillham and O'Hannesin (1994) [k]Agrawal and Tratnyek (1994)
[c]Gu et al. (1998) [f]Butler and Hayes (1999) [i]Matheson and Tratnyek (1994) [l]Focht (1994)

long-term stability, the candidate reactive materials should be able to address the inorganic components of the site groundwater and thereby is capable of retaining their reactivities and hydraulic conductivities over time. Moreover, the reactive materials should be easily available in large quantity at reasonable price. Another important point is whether the byproducts generated during the degradation have deleterious impact to natural environment. For instance, degradation of TCE by granular zero-valent iron (Fe^0) generates small amount of toxic byproducts, such as dichloroethylene (DCE) isomers and vinyl chloride (VC). Since both DCE isomers and VC are also treatable by Fe^0, it is an eligible reactive material used in a PRB for remediation of TCE contaminated groundwater (Lai et al. 2006a).

Currently, Fe^0 is the most common reactive material used in a PRB. Granular Fe^0 was first applied by Gillham and O'Hannesin (1992) for remediating CAH contaminated groundwater. Orth and Gillham (1996) found that pseudo first-order kinetics with respect to contaminant concentration were applicable for describing the degradation process of CAHs by Fe^0 [i.e., Eq. (14.1)]. The observed degradation rate constant (i.e., k_{obs}) was also found to be relatively insensitive to initial contaminant concentration. Gotpagar et al. (1997) reported that the observed degradation rate constant increased linearly with the ratio of Fe^0 surface area to solution volume (i.e., λ_{SA}), which is depicted in Eq. (14.2).

$$\frac{d[C]}{dt} = -k_{obs}[C] \tag{14.1}$$

$$k_{obs} = k_{SA}\lambda_{SA} \tag{14.2}$$

where [C] is contaminant concentration (mg L^{-1}); k_{obs} is the observed degradation rate constant (h^{-1}); k_{SA} is the normalized degradation rate constant (L m^{-2} h^{-1}); t is time (h); and λ_{SA} is the ratio of Fe^0 surface area to solution volume (m^2 L^{-1}).

In addition to Fe^0, other zero-valent metals have also been investigated for their potentials to reduce chlorinated hydrocarbons (Gavaskar et al. 1998). Su and Puls (1999) reported that zero-valent tin (Sn^0) was as effective as Fe^0 on the degradation of TCE and greater temperature effect was observed for Sn^0 than for Fe^0 on TCE degradation. Effective degradation of tetrachloroethylene (PCE), TCE and cis-dichloroethylene (cis-DCE) by zero-valent zinc (Zn^0) was also reported by Arnold and Roberts (1998) in which reductive-β-elimination [Eq. (14.3)] rather than hydrogenolysis [Eq. (14.4)] was the dominant degradation pathway.

$$RHCl{=}RHCl + 2e^- \Leftrightarrow RH{\equiv}RH + 2Cl^- \tag{14.3}$$

$$RHCl{=}RHCl + H^+ + 2e^- \Leftrightarrow RH_2{=}RHCl + Cl^- \tag{14.4}$$

As shown in Eqs. (14.5) to (14.7), oxidation of Fe^0 with water under both aerobic and anaerobic conditions leads an increase in pH inside a Fe^0 PRB (Lai et al. 2002; Lo et al. 2005), which results in precipitation of minerals and thereby consequently deteriorate the long-term performance of the Fe^0 PRB (Yabusaki et al. 2001; Andrea et al. 2005). However, addition of pyrite (FeS_2) to granular Fe^0 helps moderate the pH since the oxidation of pyrite can produce hydrogen ion [Eq. (14.8)], which can neutralize the hydroxide ions released [Eqs. (14.5) and (14.6)] and offset the hydrogen ions consumed during the oxidation of ferrous iron into ferric iron [Eq. (14.7)] (Burris et al. 1995; Holser et al. 1995).

$$Fe^0_{(s)} + 2H_2O_{(l)} \Leftrightarrow H_{2(g)} + Fe^{2+} + 2OH^- \tag{14.5}$$
$$2Fe^0_{(s)} + O_{2(aq)} + 2H_2O_{(l)} \Leftrightarrow 2Fe^{2+} + 4OH^- \tag{14.6}$$
$$4Fe^{2+} + 4H^+ + O_{2(aq)} \Leftrightarrow 4Fe^{3+} + 2H_2O_{(l)} \tag{14.7}$$
$$FeS_{2(s)} + 7/2\ O_{2(aq)} + H_2O_{(l)} \Leftrightarrow Fe^{2+} + 2H^+ + 2SO_4^{2-} \tag{14.8}$$

A number of bimetallic reactive materials, which involve plating of various metals onto the granular Fe^0, have been shown to be capable of reducing CAHs at significantly higher rate than Fe^0 itself. For example, laboratory studies of palladized Fe^0 (Pd-Fe^0) demonstrated that the rate of TCE degradation could be increased by two orders of magnitude over that of Fe^0 alone since palladium can catalyze the abiotic reduction of TCE by Fe^0 (Korte et al. 1995; Muftikian et al. 1995; Orth and McKenzie 1995). In addition, it was reported that Pd-Fe^0 allowed effective reduction of recalcitrant compounds, such as dichloromethane and polychlorinated biphenyls (PCBs) (US Army Corps of Engineers 1997; Kim et al. 2004). However, long-term investigation of Pd-Fe^0 indicated severe deterioration of the reactivity over time (Gavaskar et al. 2000).

Cercona^{TM} iron foam made by *Cercona Inc.* is an innovative reactive material for PRB technology (Bostick et al. 1996). Fabrication of *Cercona*^{TM} iron foam is mainly based upon gelation of soluble silicates with soluble aluminates. These two solutions are combined with aggregates or powdered materials and additive (i.e., Fe^0) under specific ratio, solution concentration and temperature. Unlike conventional reactive materials (e.g., granular Fe^0) possessing a trade-off between surface area (reactivity) and porosity (hydraulic conductivity), *Cercona*^{TM} iron foam can potentially achieve high surface area and high porosity at the same time (Gavaskar et al. 1998).

14.5 Treatability Testing

Following the site characterization and identification of candidate reactive materials for further study, bench-scale treatability test is subsequently conducted to choose the most suitable reactive material and determine some of the important parameters for a PRB application. Treatability test usually consists of two main parts, which are batch test and column test (Gavaskar et al. 1998). The former is usually used as an initial screening tool for evaluating the candidate reactive materials or for assessing the degradability of the contaminants. Generally, it is relatively simple and inexpensive compared to column test. However, to get more accurate design information, such as degradation half-life ($t_{1/2}$), and residence time or required thickness of a PRB as well as to select the most suitable reactive material, column test is a favoured method since it is more representative of dynamic field conditions (Gavaskar et al. 1998).

Batch test involves putting the candidate reactive material and contaminant -spiked deionized water in septum-capped vials with no headspace. By sacrificing a vial at each sampling time, water sample is temporally collected from the vials so as to investigate the rate of degradation of contaminant. Column test is performed by feeding contaminated groundwater into a column packed with the candidate reactive material (Fig. 14-2). Collapsible Teflon bag is chosen as an influent reservoir in the

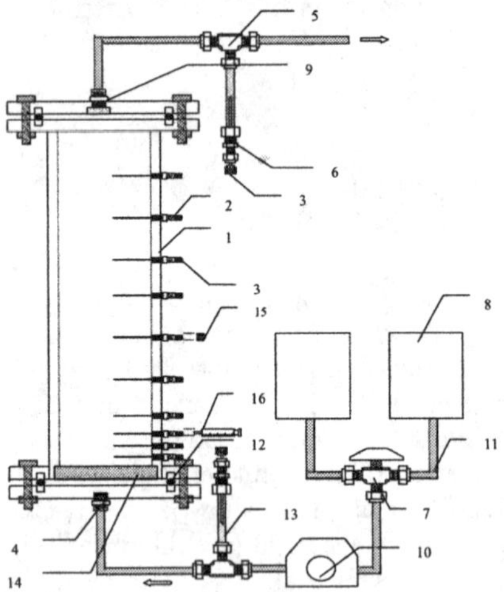

1. Plexiglass^{TM} column
2. SS Swagelok tube fitting (SS-100-1-2)
3. SS Luer-Lok^{TM} syringe needle (18G)
4. SS Swagelok tube fitting (SS-200-1-4)
5. Union tee Swagelok tube fitting (SS-200-3)
6. Swagelok SS reducing union
7. SS 3-way valve (SS-41XS2)
8. 5-L Teflon bag
9. Pyrex filter disc (1 cm in diameter)
10. Peristaltic pump with L/S 13 viton tubing
11. Teflon tubing (3.2 mm in I.D.)
12. O-ring
13. Teflon tubing (1.6 mm in I.D.)
14. Pyrex filter disc (5 cm in diameter)
15. Teflon Luer-Lok^{TM} plug
16. 5-mL glass syringe

*SS-stainless steel, I.D.-internal diameter

Fig. 14-2. Schematic diagram of the experimental setup for the column test (Adapted from Lai 2004)

column test to prevent any loss of CAHs to the headspace. Contaminated groundwater collected from the contaminated site is preferably used as feeding solution in the test since the native inorganic constituents in the site groundwater play an important role in affecting the reactivity and longevity of the reactive barrier. Besides, column test should be performed at the groundwater velocity and groundwater temperature similar to those in the site in order to obtain more reliable and representative results. This is because the rate of CAH degradation by Fe^0 is mainly dependent upon the rate of mass transfer of CAHs between bulk solution and Fe^0 surface via the stagnant fluid layer surrounding the Fe^0 (i.e., boundary layer) as well as the rate of electron transfer from Fe^0 to CAHs (Fogler 1992), which are a function of groundwater velocity and groundwater temperature, respectively. Groundwater velocity can affect the thickness of the boundary layer and thereby can potentially influence the rate of the mass transfer, whereas temperature is a major factor affecting the rate constant for the electron transfer from Fe^0 to CAHs (Fogler 1992; Su and Puls 1999). Optionally, a section of sand or pea gravel may be placed above and below the reactive medium in the column to ensure good flow distribution.

To investigate the degradation of CAHs along the column, groundwater samples were collected from the influent, effluent and intermediate sampling ports for organic analysis so as to obtain the concentration profile of CAHs along the column. Groundwater sampling along the column should be conducted every 5 to 10 pore volumes until there is no significant change of the concentration profile over time (i.e., obtaining steady state concentration profiles). It is clear to see that in Fig. 14-3, steady state concentration profile of PCE along a Fe^0 packed column is achieved after running for approximately 50 pore volumes. Physical and inorganic analyses of the collected groundwater samples are also implemented to assess the geochemical variation of groundwater characteristics along the column.

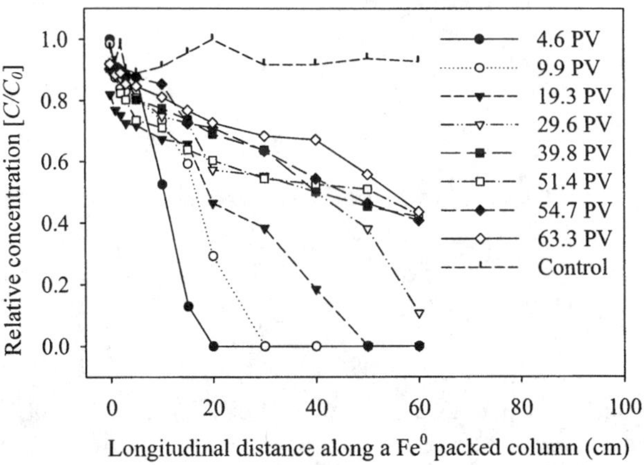

Fig. 14-3. Temporal variation of PCE concentration profile along a Fe^0 packed column. PV is the abbreviation of pore volume.

CAH concentration profiles are initially plotted as a function of longitudinal distance along the reactive column (see Fig. 14-3). However, by knowing the flowrate and porosity therein, the distance along the column can be readily converted to residence time. Then a graph of CAH concentration versus residence time can be generated. k_{obs} of each type of CAHs can subsequently be calculated by fitting the corresponding steady state concentration profile into a first-order equation [i.e., Eq. (14.1)] using mathematical program, such as Scientist® (MicroMath 1995). However, to determine the k_{obs} of the degradation byproducts, such as cis-DCE, 1,1-dichloroethylene (1,1-DCE) and VC which are the byproducts generated from the PCE and TCE degradation by Fe^0 (Fig. 14-4), first-order equations considering concurrent degradation and generation of the byproducts are used. Eqs. (14.9) to (14.13) show the first-order equations involved in the sequential degradation of PCE to TCE, cis-DCE, 1,1-DCE and VC by Fe^0 (EnviroMetal Technologies Inc. 1998). Once the k_{obs} is known, the corresponding $t_{1/2}$ for each type of CAHs can then be estimated using Eq. (14.14). Among various candidate reactive materials, the reactive medium requiring the shortest residence time or possessing the highest rate constant for the degradation of CAHs is finally selected as a reactive material for the PRB.

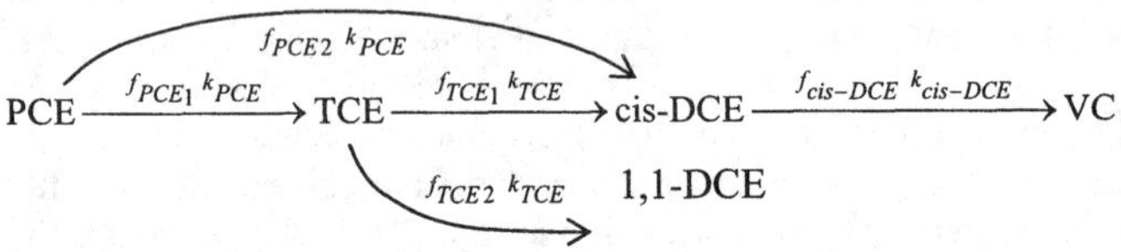

Fig. 14-4. Sequential degradation of PCE by Fe^0 (redrawn from EnviroMetal Technologies Inc. 1998)

$$\frac{d[PCE]}{dt} = -k_{PCE}[PCE] \tag{14.9}$$

$$\frac{d[TCE]}{dt} = f_{PCE1}k_{PCE}[PCE] - k_{TCE}[TCE] \tag{14.10}$$

$$\frac{d[cis-DCE]}{dt} = f_{PCE2}k_{PCE}[PCE] + f_{TCE1}k_{TCE}[TCE] - k_{cis-DCE}[cis-DCE] \tag{14.11}$$

$$\frac{d[VC]}{dt} = f_{cis-DCE}k_{cis-DCE}[cis-DCE] - k_{VC}[VC] \tag{14.12}$$

$$\frac{d[1,1-DCE]}{dt} = f_{TCE2}k_{TCE}[TCE] - k_{1,1-DCE}[1,1-DCE] \tag{14.13}$$

$$t_{1/2} = \frac{\ln 2}{k_{obs}} \tag{14.14}$$

where f_{PCE1} is molar fraction of PCE converting into TCE; f_{PCE2} is the molar fraction of PCE converting into cis-DCE; f_{TCE1} is the molar fraction of TCE converting into cis-DCE; f_{TCE2} is the molar fraction of TCE converting into 1,1-DCE; $f_{cis-DCE}$ is the molar fraction of cis-DCE converting into VC; $k_{cis-DCE}$, $k_{1,1-DCE}$, k_{PCE}, k_{TCE}, and k_{VC} are the k_{obs} of cis-DCE, 1,1-DCE, PCE, TCE and VC, respectively (h^{-1}); [cis-DCE], [1,1-DCE], [PCE], [TCE] and [VC] are the concentration of cis-DCE, 1,1-DCE, PCE, TCE and VC along the Fe^0 medium, respectively (mg L^{-1}); $t_{1/2}$ is the degradation half-life (h).

After selecting the reactive material for the PRB and determining the corresponding k_{obs} for the degradation of CAHs, the k_{obs} and initial concentration of CAHs are substituted back into Eqs. (14.9) to (14.13) to recalculate the concentration profiles of all the CAHs along the reactive medium, as illustrated in Fig. 14-5. Therefore, the required residence time for degrading different CAHs down to their maximum contaminant levels (MCLs) in the downgradient of the reactive barrier can be simply obtained from these recalculated concentration profiles, which is equal to the shortest residence time allowing all the CAHs to be degraded down to their MCLs. As you can see from Fig. 14-5, the required residence time for all the CAH concentrations drops below 10 μg/L (i.e., MCLs) is about 64 hrs (EnviroMetal Technologies Inc. 1998) so that the required thickness of the PRB can be readily determined by multiplying the required residence time (i.e., 64 hrs) by the groundwater velocity.

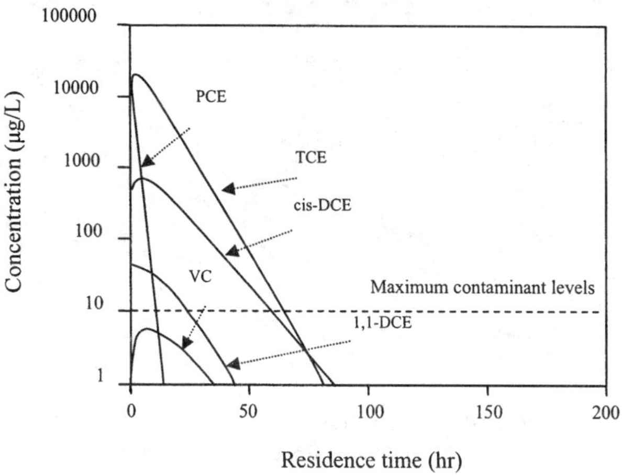

Fig. 14-5. Recalculated concentration profiles of CAHs computed during the design of the Fe^0-based reactive barrier at Vapokon site (redrawn from EnviroMetal Technologies Inc. 1998)

The methods mentioned above for the determination of the required thickness or residence time of the reactive barrier certainly rely on the k_{obs} obtained from the column test, which should be operated at the groundwater velocity and groundwater temperature similar to those in the site. Otherwise, correction factors for the groundwater velocity and groundwater temperature are required to correct the k_{obs} before being used to determine the required PRB thickness. Fig. 14-6 shows the correlation of the degradation rate constant of TCE and PCE by Fe^0 with the groundwater velocity (Lai and Lo, unpublished manuscript, 2006), whereas Fig. 14-7 correlates the degradation rate constant with the groundwater temperature.

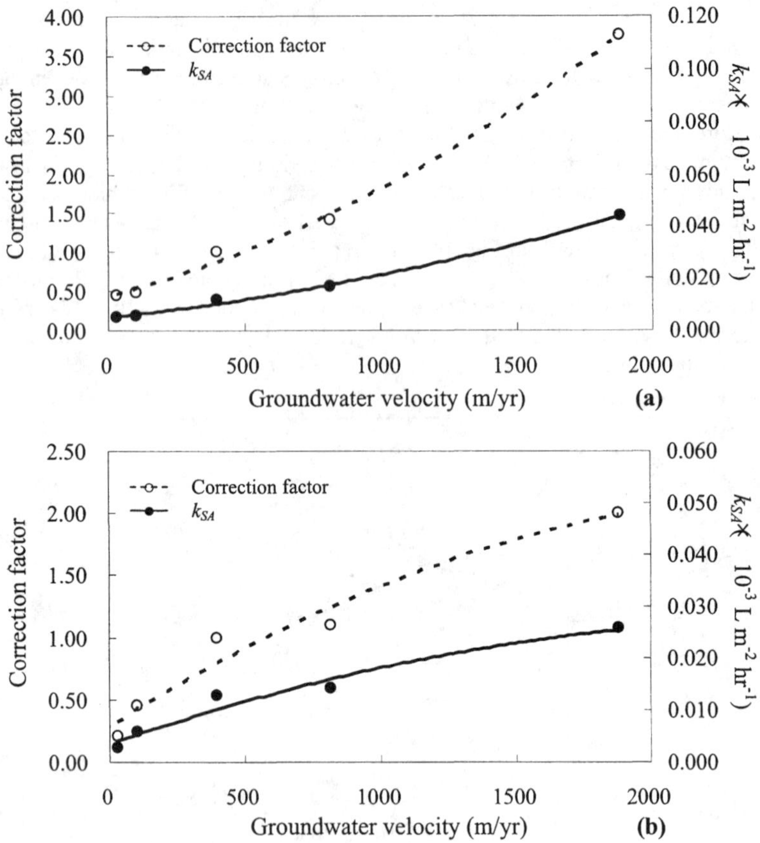

Fig. 14-6. Correlation of the degradation rate constant of (a) TCE and (b) PCE by Fe^0 with groundwater velocity at 10 ^{0}C

14.6 Hydrogeologic and Geochemical Modelings

The purposes of performing hydrogeologic and geochemical modelings are to determine an approximate location and configuration of the PRB with respect to the groundwater flow and plume movement as well as to estimate the expected groundwater flow velocity through the reactive medium. Determination of the width of the reactive medium or the width of the funnel for a funnel-and-gate PRB, estimation of the hydraulic capture zone of the PRB as well as evaluation of the hydraulic effect of potential loss in porosity over the long-term and the potential for underflow or overflow of contaminated groundwater across the PRB can also be achieved through the simulation of hydraulic and geochemical behaviours of the PRB (US Army Corps of Engineers 1997).

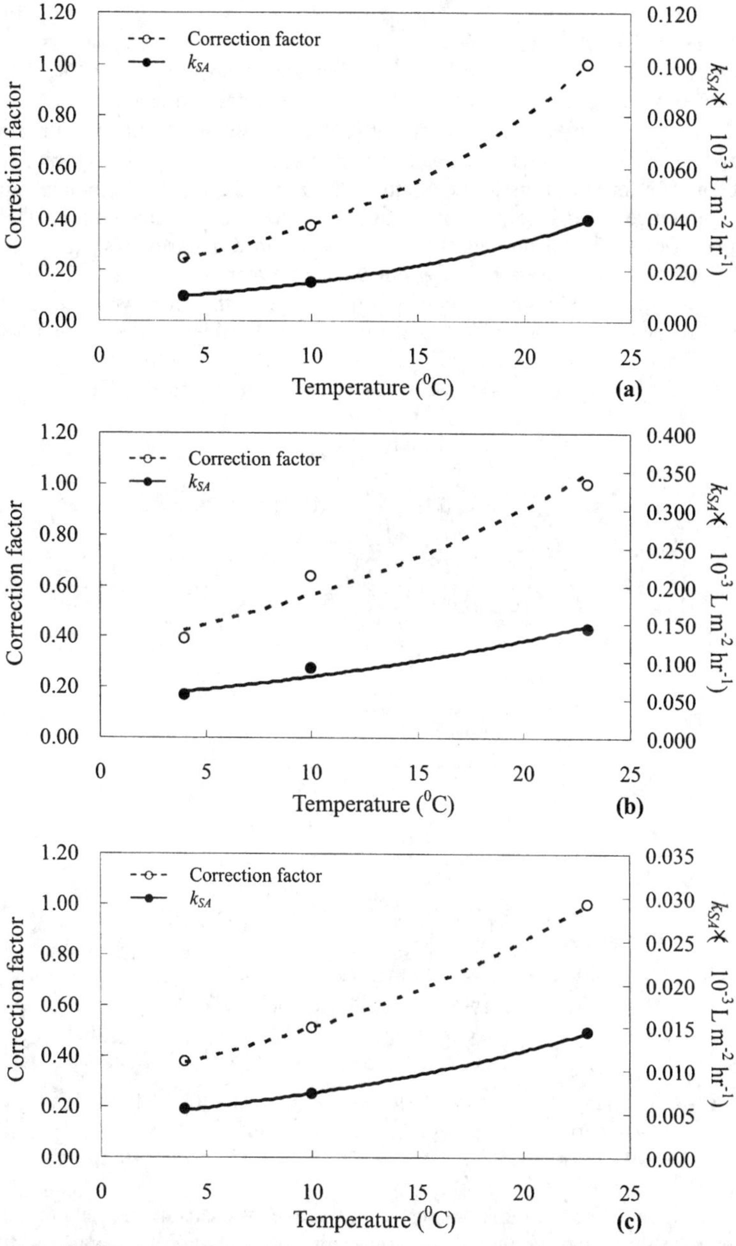

Fig. 14-7. Correlation of the degradation rate constant of (a) chloroform (TCM), (b) 1,1,1-trichloroethane (1,1,1-TCA) and (c) TCE by Fe^0 with groundwater temperature at 400 m/yr of groundwater velocity

Hydrogeologic modeling can evaluate several design configurations, performance and longevity scenarios of the reactive barrier as well as the site parameters. Heretofore, widely available and validated models, such as FRAC3D and MODFLOW, are generally sufficient to achieve PRB design objective (Blowes et al. 1999a; US Army Corps of Engineers 1997). When designing a full-scale PRB, width of hydraulic capture zone and residence time are two independent parameters of concern. The former refers to the width of the zone of groundwater which can pass through the reactive medium and it can be maximized by maximizing the flow volume of groundwater through the reactive medium. Contrarily, residence time can be lengthened by minimizing the groundwater flow through the reactive medium. Hydrogeologic modeling helps figure out the design of the PRB, which is capable of balancing the capture zone width against the required residence time. Figs. 14-8a and 14-8b illustrate the simulated groundwater flow and capture zones for funnel-and-gate and continuous PRBs, respectively (Blowes et al. 1999a).

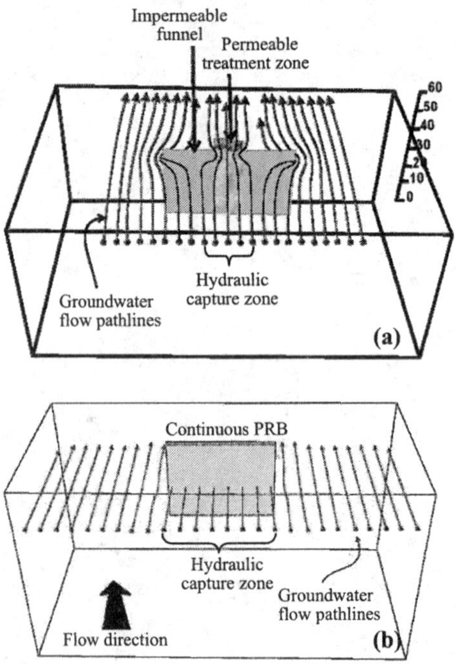

Fig. 14-8. Simulated groundwater flow and capture zone for (a) funnel-and-gate and (b) continuous permeable reactive barriers (adapted from Blowes et al. 1999a, EPA/600/R-99/095a)

Geochemical modeling involves a qualitative assessment of the potential for precipitate formation within the reactive medium based upon the data from the site characterization and treatability test. Commercially available models, such as PHREEQ and MINTEQA2, are generally sufficient for fulfilling this purpose

(Blowes et al. 1999a; US Army Corps of Engineers 1997). Basically, most available geochemical models are equilibrium model in which reactive kinetics are not incorporated. However, the mass of precipitating and dissolving compounds along a known flowpath can be back-calculated using the inverse modeling codes in the models. After selecting the reactive material, which is effective on contaminant degradation and unfavourable for mineral precipitation as well as figuring out the PRB configuration and dimension giving the maximum capture zone width and residence time for the degradation, emplacement method of the PRB is selected. By taking account of the aquifer depth, reactive barrier thickness and geotechnical constraints of the contaminated site, emplacement techniques including trench excavation, continuous trenching, caisson- or mandrel-based emplacement, jetting, hydraulic fracturing, or deep soil mixing is chosen for the reactive medium emplacement (Gavaskar et al. 1998).

14.7 Monitoring Plan

Unlike other groundwater remediation approaches, PRB is a relatively permanent structure. Once the PRB is installed, it is hard and costly to relocate and modify. Therefore, to ensure that the installed PRB can treat the contaminated groundwater to meet the regulatory standards and adequately capture the contaminant plume as designed, periodic monitoring of the operating performance of the PRB is required. Generally, there are two main types of monitoring, namely contaminant monitoring and performance monitoring. The former aims to check the current operating status of the PRB, whereas the latter is implemented to evaluate whether the hydraulic and geochemical conditions created by the reactive medium can allow good performance of PRB currently and in the future (Gavaskar et al. 2000).

Contaminant monitoring involves periodic sampling of groundwater from the upgradient and downgradient locations of the PRB, and the reactive medium for the measurement of the concentration of contaminants (e.g., PCE and TCE) and potential degradation by-products (cis-DCE and VC) (Lai et al. 2006a). Customarily, long-screen wells or piezometers are installed for groundwater sampling if the contaminant distribution in an aquifer is relatively homogeneous. Otherwise, multilevel samplers (MLSs) are used for collecting groundwater samples from a relatively heterogeneous aquifer. Each MLS consists of a number of sampling tubes, which normally screen at discrete depth intervals along the aquifer or PRB (Lai et al. 2006b). In addition, to minimize the disturbance to the groundwater flow in the aquifer or PRB, the groundwater sampling rate should be less than 1 L/min or even less than 100 mL/min, depending on the transmissivity of the medium (Powell and Puls 1997).

Performance monitoring of PRB generally requires the measurement of the water level, and analysis of field parameters (i.e., dissolved oxygen, pH, redox potential and specific conductivity) and inorganic constituents of the groundwater in the PRB and its vicinity. Performance monitoring includes both hydraulic and geochemical performance monitoring. The purpose of implementing hydraulic performance monitoring is to evaluate whether the installed PRB can completely capture the contaminant plume as designed and to estimate the actual residence time available for groundwater contaminants inside the PRB. Because of an unanticipated seasonal variation in the groundwater flow direction and velocity (Gasvaskar et al. 2000), as well as continuous decrease in the porosity of the PRB (Lai et al. 2006a) due to mineral precipitation (Roh et al., 2000; Lo et al. 2006), microbial growth (Gu et al. 2002) and accumulation of hydrogen gas (Mackenzie et al. 1999; Zhang and Gillham 2005) within the reactive medium, the actual hydraulic capture zone of the PRB and the residence time inside the installed PRB may adversely differ from what they have been designed. To implement the hydraulic performance monitoring, the approach of water level measurement (Gavaskar et al. 2000), calculation using Darcy's law (Bedient et al. 1999), *in situ* groundwater velocity sensor (Ballard 1996) or tracer test (Focht et al. 1997; Vogan et al. 1999; Lai et al. 2006b; Lai et al. 2006c) is applied to delineate the groundwater flow pattern and determine the groundwater velocity inside and around the PRB.

Geochemical performance monitoring of the PRB can evaluate how long the reactive medium inside the PRB can provide the desired performance for groundwater remediation. It can also indicate how well the installed PRB matches the geochemical condition predicted in the design stage. Hitherto, there are three main methods for monitoring the geochemistry of the PRB: (1) groundwater monitoring for field parameters and inorganic species; (2) geochemical modeling; and (3) reactive medium core extraction and analysis (Gavaskar et al. 2000). Groundwater monitoring is of utmost importance to understand the geochemical condition inside the PRB and is also a prerequisite for groundwater modeling. It involves on-site field parameter measurement and inorganic chemical analysis of groundwater samples collected along the PRB. Field parameters, such as dissolved oxygen, redox potential, pH and specific conductivity in groundwater, are measured on-site using probes, which are either configured for downhole submersion or equipped to a flow cell for aboveground measurement (Lai et al. 2006a). Inorganic species including calcium and magnesium hardness, alkalinity, sulfate, and total dissolved iron in the groundwater samples are analyzed to evaluate the potential for mineral precipitation inside the PRB (O'Hannesin and Gillham 1998; Wilkin and Puls 2003). Based upon the geochemical condition of the PRB obtained from the groundwater monitoring, geochemical modeling can simulate the reactions between the native groundwater and the reactive medium, and thereby can qualitatively and quantitatively evaluate the mineral precipitates formed inside the PRB.

Reactive medium core sampling is not common at most PRB sites. However, it can provide essential geochemical information to evaluate the longevity of the reactive medium. Surface characterization of the reactive medium core samples using Raman spectroscopy (Bonin et al. 2000), scanning electron microscopy (SEM) (Roh et al. 2000; Lo et al. 2005), Mössbauer spectroscopy (Refait et al. 1998) and x-ray diffraction (XRD) (Lo et al. 2006) can show the morphology of the reactive medium and directly identify the type of mineral precipitates, such as iron oxides, hydroxides (Liang et al. 2000; Mackenzie et al. 1999; Ritter et al. 2002) and carbonates (Naftz et al. 2002), formed on the reactive medium. Therefore, this information is useful for inferring the influence of mineral precipitation on the reactivity and hydraulic performance of the reactive medium.

Table 14-4. Estimating capital investment for an application of a full-scale Fe^0 PRB at Dover AFB (adapted from Gavaskar et al 2000, <http://www.itrcweb.org /Documents/PRB-2c.pdf>)

Items	Descriptions	Bases	Costs
Phase I: Preconstruction activities			
Preliminary site assessment	Historical site data evaluation	Other reports procurement and evaluation as well as site meeting	$15,000
Site characterization	Characterization plan, fieldwork and laboratory analysis	Analysis of water samples for contaminants, selection of samples for geotechnical analysis, slug tests and ground-penetrating radar survey	$200,000
Column tests	Two column tests	Column tests, laboratory analysis of water samples and report writing	$50,000
Design, procurement of subcontractors, and regulatory review	Data evaluation, modeling, engineering design, design plan, procurement of subcontractors and interactions with regulators	Characterization/column test data evaluation, hydrogeologic modeling, geochemical evaluation, engineering design, report writing, procurement process and regulatory interactions	$100,000
		Subtotal costs	**$365,000**
Phase II: PRB construction activities			
Site preparation	Utilities clearances, arrangements for equipment/media storage and debris disposal	Coordination with regulators and site facilities staff	$10,000
Reactive media procurement	*Connelly* iron and shipping	Iron: 108 tons @ $360 per ton Shipping: $9,000	$48,000
PRB construction	Mobilization/demobilization, installation of four 8 ft-diameter caisson gates to a depth of 40 ft and 120 ft long sheet pile funnel as well as asphalt parking lot restoration	Mobilization/demobilization: $60,000 Gates: $266,000 Monitoring wells: $25,000 Funnel: $102,000 Surface restoration: $34,000	$487,000
Monitoring system construction	Thirty-four PVC aquifer wells installed for monitoring the pilot-scale PRB	Aquifer wells: $37,000	$37,000
		Subtotal costs	**$582,000**
		Total capital costs	**$947,000**

14.8 Permeable Reactive Barrier Economics

One of the important driving forces behind the widespread interest in PRB technology is its economic benefit. It is a passive technology requiring no annual energy or labor input other than for site monitoring. Therefore, it has pronounced cost advantages over other conventional groundwater remediation technologies, such as pump-and-treat system and air sparging.

The capital cost of the application of a PRB includes the costs of reactive medium, emplacement, disposal of spoils and restoration of ground surface as well as technology licensing cost. The cost of reactive medium depends upon the type of reactive material selected and the amount of reactive material required which are functions of the type and concentration of contaminants, regulatory treatment criteria, groundwater velocity as well as distribution of groundwater flow and contaminant. Besides, plume and aquifer depths, plume width and geotechnical constraints of the site, such as the presence of rocks or highly consolidated sediments, are the factors affecting the total emplacement cost. The cost of spoil disposal depends upon the technique of emplacement used and location of the PRB. If the PRB has to be installed within the contaminant plume, the spoils generated have to be disposed as hazardous waste and it is costly. Restoration cost is the cost for restoring the site surface which includes repaving the ground surface for built-up site. Concerning the technology licensing cost, *EnviroMetal Technologies Inc.* holds the patent for Fe^0 technology and charges a licensing fee of up to 15% of the capital costs for the application of the technology (US Army Corps of Engineers 1997). Table 14-4 illustrates the estimating capital cost for an application of a full-scale Fe^0 PRB at Dover Air Force Base (AFB) (Gavaskar et al. 2000).

Table 14-5. Estimating operating and maintenance costs for an application of a full-scale Fe^0 PRB at Dover AFB (adapted from Gavaskar et al 2000, <http://www.itrcweb.org/Documents/PRB-2c.pdf>)

Items	Descriptions	Bases	Costs
Annual monitoring activities			
Groundwater sampling	Quarterly, labour, materials and travel	Forty wells	$80,000
Contaminant analysis	Quarterly, forty wells	44 per quarter @ $120 per sample	$20,000
Inorganic analysis	Annual, 20 wells	22@ $200 per sample	$4,000
Water-level survey	Quarterly and labour	40 wells per quarter	$4,000
Data analysis, report writing and regulatory review	Quarterly and labour	4 times per year	$40,000
		Annual operating cost	**$148,000**
Maintenance activities (assumed once every ten years)			
Site preparation	Permitting and clearances	labour	$10,000
Reactive media procurement	*Connelly* iron and shipping	Iron: 108 tons@ $360 per ton Shipping: $9,000	$48,000
Removal/replacement of gates	Mobilization/demobilization, installation of four 8 ft-diameter caisson gates to a depth of 39 ft and asphalt parking lot restoration	Mobilization/demobilization: $38,000 Gates: $266,000 Monitoring wells: $25,000 Surface restoration: $34,000	$363,000
		Periodic maintenance cost (assumed once every ten years)	**$421,000**

The operating and maintenance costs of a PRB technology are the recurring or periodic costs incurred during the operating life of the system. They include the costs of compliance monitoring, additional performance monitoring and periodic maintenance. Compliance monitoring cost depends upon the regulatory requirements, number of monitoring wells and frequency of samplings. Additional performance monitoring is usually conducted to achieve other performance evaluation objectives. Periodic maintenance cost includes the cost of flushing the reactive medium with reagent and replacement of reactive medium due to the precipitation of minerals. Table 14-5 shows the estimating operating and maintenance costs for an application of a full-scale Fe^0 PRB at Dover AFB (Gavaskar et al. 2000).

14.9 Summary

Proper design of the PRB prior to the emplacement in subsurface is of utmost importance to achieve the expected efficiency of the reactive barrier on groundwater remediation since it is a relatively permanent structure and is difficult to change and relocate after the installation. The design methodology for the application of PRB first involves preliminary assessment, which evaluates whether or not the contaminated site is suitable for a PRB application. If it is so, additional site characterization may subsequently be implemented to provide information for the further hydrogeologic and geochemical modelings as well as selection of the reactive barrier configuration and emplacement method. After screening the candidate reactive materials possibly for PRB, treatability test is conducted to primarily select the most suitable reactive material and determine the required thickness of reactive barrier. With the application of the site characterization data, the location, dimension and configuration of the PRB giving large enough capture zone width and long enough residence time can then be simulated using the hydrogeologic and geochemical modelings. Afterwards, the most cost-effective technique for the emplacement of the reactive barrier can be selected based upon the PRB dimension and configuration as well as the geotechnical constraints in the site. Monitoring strategy is also prepared for PRB to ensure that once the PRB is installed, its operating performance is consistent with the design objectives. Finally, the cost advantage of PRB technology over other conventional remediation approaches for groundwater remediation is assessed by estimating the capital, operating and maintenance costs for the application of PRB in the contaminated site.

14.10 References

Agrawal, A., and Tratnyek, P. G. (1994). "Abiotic remediation of nitro-aromatic groundwater contaminants by zero-valent iron." In *Proceedings of the 207th American Chemical Society National Meeting,* San Diego, CA, 34(1), 492-494.

Alowitz, M. J., and Scherer, M. M. (2002). "Kinetics of nitrate, nitrite, and Cr(VI) reduction by iron metal." *Environ. Sci. & Technol.*, 36(3), 299-306.

Andrea, P. D., Lai K. C.-K., Kjeldsen, P., and Lo, I. M.-C. (2005). "Effect of groundwater inorganics on the reductive dechlorination of TCE by zero-valent iron." *Water, Air, and Soil Pollut'n.*, 162, 401-420.

Arnold, W. A., and Roberts, A. L. (1998). "Pathways of chlorinated ethylene and chlorinated acetylene reaction with Zn(0)." *Environ. Sci. & Technol.,* 32(19), 3017-3025.

Ballard, S. (1996). "The in situ permeable flow sensor: A ground-water flow velocity meter." *Ground Water*, 34(2), 231-240.

Blowes, D. W., Gillham, R. W., Ptacek, C. J., Puls, R. W., Bennett, T. A., O'Hannesin, S. F., Hanton-Fong, C. J., and Bain, J. G. (1999a). *An in situ permeable reactive barrier for the treatment of hexavalent chromium and trichloroethylene in ground water: Volume 1 design and installation*, EPA/600/R-99/095a, Office of Research and Development, United States Environmental Protection Agency, Washington, D. C. 20460, United States.

Blowes, D. W., Puls, R. W., Gillham, R. W., Ptacek, C. J., Bennett, T. A., Bain, J. G., Hanton-Fong, C. J., and Paul, C. J. (1999b). *An in situ permeable reactive barrier for the treatment of hexavalent chromium and trichloroethylene in ground water: Volume 2 performance monitoring*, EPA/600/R-99/095b, Office of Research and Development, United States Environmental Protection Agency, Washington, D. C. 20460, United States.

Bostick, W. D., Jarabek, R. J., Slover, W. A., Fiedor, J. N., Farrell, J., and Heferich, R. (1996). *Zero-valent iron and metal oxides for the removal of soluble regulated metals in contaminated groundwater at a DOE site, K/TSO-35-P,* Lockheed Martin Energy Systems, Inc., Oak Ridge, TN.

Bonin, P. M. L., Odziemkowski, M. S., Reardon, E. J., and Gillham, R. W. (2000). "In situ identification of carbonate-containing green rust on iron electrodes in solutions simulating groundwater." *J. Solution Chem.*, 29(10), 1061-1074.

Burris, D. R., Campbell, T. J., and Manoranjan, V. S. (1995). "Sorption of trichlroethylene and tetrachloroethylene in a batch reactive metallic iron-water systems." *Environ. Sci. & Technol.*, 29(11), 2850-2855.

Butler, E. C., and Hayes, K. F. (1999). "Kinetics of the transformation of trichloroethylene and tetrachloroethylene by iron sulfide." *Environ. Sci. & Technol.*, 33(12), 2021-2027.

Cantrell, K. J., Kaplan, D. I., and Wietsma, T. W. (1995). "Zero-valent iron for the in situ remediation of selected metals in groundwater." *J. Hazard. Mater.*, 42(2), 201-212.

Doyle, J. G., Miles, T. A., Parker, E., and Cheng, I. F. (1998). "Quantification of total polychlorinated biphenyl by dechlorination to biphenyl by Pd/Fe and Pd/Mg bimetallic particles." *Microchem. J.*, 60(3), 290-295.

EnviroMetal Technologies Inc. (1998). *Bench-scale treatability report of the EnviroMetal process at the Vapokon site, Fyn, Denmark*, ETI 31438.10, Guelph, Ontario, Canada N1K1S6.

Fiorenza, S., Oubre, C. L., and Ward, C. H. (2000). *Sequenced reactive barriers for groundwater remediation,* Lewis Publishers, Boca Raton, FL.

Freyberg, D. L. (1986). "A natural gradient experiment on solute transport in a sand aquifer 2. Spatial moments and the advection and dispersion of nonreactive tracers." *Wat. Resour. Res.*, 22(13), 2031-2046.

Fogler, H. S. (1992). *Elements of chemical reaction engineering*, 2nd Ed., Prentice-Hall Inc., Englewood Cliffs, New Jersey.

Fotch, R. M. (1994). *Bench-scale treatability testing to evaluate the applicability of metallic iron for above ground remediation of 1,2,3-trichloropropane contaminated groundwater*, Master Thesis, Department of Earth Sciences, University of Waterloo, Ontario.

Focht, R. M., Vogan, J. L., and O'Hannesin, S. F. (1997). "Hydraulic studies of in-situ permeable reactive barriers." In *Proceedings of the International Containment Technology Conference and Exhibition*, St. Petersburg, Florida.

Garabedian, S. P., LeBlanc, D. R., Gelhar, L. W., and Celia, M. A. (1991). "Large-scale natural gradient tracer test in sand and gravel, Cape Cod, Massachusetts 2. Analysis of spatial moments for a nonreactive tracer." *Wat. Resour. Res.*, 27(5), 911-924.

Gavaskar, A. R., Gupta, N., Sass, B. M., Janosy, R. J., and O'Sullivan, D. (1998). *Permeable barriers for groundwater remediation - Design, construction, and monitoring*, Battelle Press, Columbus.

Gavaskar, A., Gupta, N., Sass, B., Janosy, R., and Hicks, J. (2000). *Design guidance for application of permeable reactive barriers for groundwater remediation*, Strategic Environmental Research and Development Program, Battelle Columbus, Ohio, <http://www.itrcweb.org/Documents/PRB-2a.pdf>, <http://www.itrcweb.org/Documents/PRB-2b.pdf>, <http://www.itrcweb.org/Documents/PRB-2c.pdf>.

Gillham, R. W., and O'Hannesin, S. F. (1992). "Metal-catalyzed abiotic degradation of halogenated organic compounds." In *Proceedings of the IAH Conference Modern Trends in Hydrogeology*, May 10-13, Hamilton, Ontario, Canada.

Gillham, R. W., and O'Hannesin, S. F. (1994). "Enhanced degradation of halogenated aliphatics by zero-valent iron." *Ground Water*, 32, 958-967.

Gotpagar, J., Grulke, E., Tsang, T., and Bhattacharyya, D. (1997). "Reductive dehalogenation of trichloroethylene using zero-valent iron." *Environmental Progress*, 16(2), 137-143.

Gu, B., Liang, L., Dickey, M. J., Yin, X., and Dai, S. (1998). "Reductive precipitation of uranium(VI) by zero-valent iron." *Environ. Sci. & Technol.*, 32(21), 3366-3373.

Gu, B., Watson, D. B., Wu, L. Y., Phillips, D. H., White, D. C., and Zhou, J. Z. (2002). "Microbiological characteristics in a zero-valent iron reactive barrier, *Environ. Monit. Assess.*, 77, 293-309.

Holser, R. A., McCutcheon, S. C., and Wolfe, N. L. (1995). "Mass transfer effects on the dehalogenation of trichloroethylene by iron/pyrite mistures." In *Proceedings of the 209th ACS National Meeting, Anaheim, Cal.*, Anaheim, CA.

Johnson, T. L., Scherer, M. M., and Tratnyek, P. G. (1996). "Kinetics of halogenated organic compound degradation by iron metal." *Environ. Sci. & Technol.*, 30(8), 2634 – 2640.

Kearl, P. M., Korte, N. E., Stites, M., and Baker, J. (1994). "Field comparison of micropurging vs traditional ground water sampling." *Ground Water Monitoring and Remediation*, 14(4), 183-190.

Kim, Y. H., Shin, W. S., and Ko, S. O. (2004). "Reductive dechlorination of chlorinated biphenyls by palladized zero-valent metals." *Journal of Environmental Science & Health Part A-Toxic/Hazardous Substances & Environmental Engineering*, 39(5), 1177-1188.

Korte, N. E., Liang, L., and Clausen, J. (1995). "The use of palladized iron as a means of treating chlorinated contaminants." In *Emerging technologies in Hazardous Waste Management VII, Extended Abstracts for the Special Symposium*, Atlanta, GA.

Lai, K. C.-K., and Lo, I. M.-C. (2002). "Bench-scale study of the effects of seepage velocity on the dechlorination of TCE and PCE by zero-valent iron." In *Proceedings of the 6th International Symposium on Environmental Geotechnology*, July 2-5, Seoul, Korea.

Lai, K. C.-K. (2004). *Laboratory and full-scale studies of a permeable reactive barrier on the dechlorination of chlorinated aliphatic hydrocarbons*, Ph.D. Thesis, Civil Engineering Department, The Hong Kong University of Science and Technology.

Lai, K. C.-K., Lo, I. M.-C., Birkelund, V., and Kjeldsen, P. (2006a). "Field monitoring of a permeable reactive barrier for removal of chlorinated organics." *J. of Environ. Eng. ASCE*, 132(2), 199-210.

Lai, K. C.-K., Lo, I. M. C., and Kjeldsen, P. (2006b). "Natural gradient tracer test for the permeable reactive barrier in Denmark 1. Field study of tracer movement." *J. of Hazardous, Toxic and Radioactive Waste Management, ASCE*, in press.

Lai, K. C.-K., Lo, I. M.-C., and Kjeldsen, P. (2006c). "Natural gradient tracer test for the permeable reactive barrier in Denmark 2. Spatial Moments Analysis and Dispersion of Conservative Tracer" *J. of Hazardous, Toxic and Radioactive Waste Management, ASCE*, in press.

LeBlanc, D. R., Garabedian, S. P., Hess, K. M., Gelhar, L. W., Quadri, R. D., Stollenwerk, K. G., and Wood, W. W. (1991). "Large-scale natural gradient tracer test in sand and gravel, Cape Cod, Massachusetts 1. Experimental design and observed tracer movement." *Wat. Resour. Res.*, 27(5), 895-910.

Lee, W. J., and Batchelor, B. (2002). "Abiotic reductive dechlorination of chlorinated ethylenes by iron-bearing soil minerals. 2. Green rust." *Environ. Sci. & Technol.*, 36(24), 5348-5354.

Liang, L. Y., Korte, N., Gu, B., Puls, R., and Reeter, C. (2000). "Geochemical and microbial reactions affecting the long-term performance of in situ iron barriers." *Adv. Environ. Res.*, 4(4), 273-286.

Lipczynska-Kpchany, E., Harms, S., Milburn, R., Sprah, G., and Nadarajah, N. (1994). "Degradation of carbon tetrachloride in the presence of iron and sulphur containing compounds." *Chemosphere*, 29, 1477-1489.

Lo, I. M.-C., and Lai, K. C.-K. (2002). "Preliminary design of permeable reactive wall based on column studies." In *Proceedings of the 6th international Congress on Environmental Geotechnics*, ISEG, August 11-15, Rio de Janeiro, Brazil.

Lo, I. M.-C., Lai, K. C.-K., and Kjeldsen, P. (2003). "Performance of permeable reactive barrier on remedying aliphatic chlorinated organic contaminated groundwater." In *Proceedings of the International Conference on Soil and Groundwater Contamination & Cleanup in Arid Countries*, ICSGCC, January 20-23, Muscat, Sultanate of Oman.

Lo, I. M.-C., Lam, C. S.-C., and Lai, K. C.-K. (2005). "Competitive effects of TCE on Cr(VI) removal by zero-valent iron." *J. of Environ. Eng. ASCE*, 131(11), 1598-1606.

Lo, I. M.-C., Lam, C. S. C., and Lai, K. C.-K. (2006). "Hardness and carbonate effects on the reactivity of zero-valent iron for Cr(VI) removal." *Wat. Res.*, 40, 595-605.

Mackay, D. M., Freyberg, D. L., Roberts, P. V., and Cherry, J. A. (1986). A natural gradient experiment on solute transport in a sand aquifer 1. Approach and overview of plume movement, *Wat. Resour. Res.*, **22**(13), 2017-2029.

Mackenzie, P. D., Horney, D. P., and Sivavec, T. M. (1999). "Mineral precipitation and porosity losses in granular iron columns." *J. Hazard. Mater.*, 68, 1-17.

Matheson, L. J., and Tratnyek, P. G. (1994). "Reductive dehalogenation of chlorinated methanes by iron metal." *Environ. Sci. & Technol.*, 28(12), 2045-2053.

McMahon, P. B., Dennehy, K. F., and Sandstrom, M. W. (1999). "Hydraulic and geochemical performance of a permeable reactive barrier containing zero-valent iron, Denver Federal Center." *Ground Water*, 37(3), 396-404.

MicroMath Scientist® (1995). *Experimental data fitting/Microsoft Windows version 2.0*, Salt Lake City, Utah, 84121.

Morkin, M., Devlin, J. F., Barker, J. F., and Buler, B. J. (2000). "In situ sequential treatment of a mixed contaminant plume." *J. Contam. Hydrol.*, 45(3-4), 283-302.

Muftikian, R., Fernando, Q., and Korte, N. (1995). "A method for the rapid dechlorination of low molecular weight chlorinated hydrocarbons in Water." *Wat. Res.*, 29(10), 2434-2439.

Naftz, D. L., Morrison, S. J., Davis, J. A., and Fuller, C. C. (2002). *Handbook of groundwater remediation using permeable reactive barriers: Applications to radionuclides, trace metals, and nutrients*, Amsterdam, Academic Press.

O'Hannesin, S. F., and Gillham, G. W. (1998). "Long-term performance of an in situ "iron wall" for remediation of VOCs." *Ground Water*, 36(1), 164-170.

Orth, W. S., and Gillham, R. W. (1996). "Dechlorination of trichloroethene in aqueous solution using Fe^0." *Environ. Sci. & Technol.*, 30(1), 66-71.

Orth, R. G., and McKenzie, D. E. (1995). "Reductive dechlorination of chlorinated alkanes and alkenes by iron metal and metal mixtures." In *Extended abstract from the Special Symposium Emerging Technologies in Hazardous Waste Management VII*, American Chemical Society, Atlanta, GA.

Powell, R. M., Blowes, D. W., Gillham, R. W., Schultz, D., Sivavec, T., Puls, R. W., Vogan, J. L., Powell, P. D., and Landis, R. (1998). *Permeable reactive barrier technologies for contaminant remediation*, EPA/600/R-98/125, Office of Research and Development, United States Environmental Protection Agency, Washington, D. C. 20460, United States.

Reeter, C., Gavaskar, A., Gupta, N., and Sass, B. (1998). "Permeable reactive wall remediation of chlorinated hydrocarbons in groundwater, NAS Moffett Field, Mountain View, California." In *Proceedings of the 2nd International Water Resources Engineering Conference*, Memphis, TN, United States.

Refait, Ph., Abdelmoula, M., and Génin, J.-M. R. (1998). "Mechanisms of formation and structure of green rust one in aqueous corrosion of iron in the presence of chloride ions." *Corros. Sci.*, 40(9), 1547-1560.

Ritter, K., Odziemkowski, M. S., and Gillham, R. W. (2002). "An in situ study of the role of surface films on granular iron in the permeable iron wall technology." *J. Contam. Hydrol.*, 55(1-2), 87-111.

Roh, Y., Lee, S., Y. and Elless, M. P. (2000). "Characterization of corrosion products in the permeable reactive barriers." *Environ. Geol.*, 40(1-2), 184-194.

RTDF (2000). *Summary of the remediation technologies development forum permeable reactive barriers action team meeting*, Remediation Technologies Development Forum, Feb 16-17, Melbourne, Florida, United States.

RTDF (2001). *Summary of the remediation technologies development forum permeable reactive barriers action team meeting*, Remediation Technologies Development Forum, Jun 12, Orlando, Florida, United States.

RTDF (2002). *Summary of the remediation technologies development forum permeable reactive barriers action team meeting*, Remediation Technologies Development Forum, Nov 6-7, Washington, D.C., United States.

Su, C., and Puls, R. W. (1999). "Kinetics of trichloroethene reduction by zero-valent iron and tin: Pretreatment effect, apparent activation energy, and intermediate products." *Environ. Sci. & Technol.*, 33(1), 163-168.

US Army Corps of Engineers (1997). *Design guidance for application of permeable barriers to remediate dissolved chlorinated solvents*, DG 1110-345-117, Armstrong Laboratory, Environics Directorate, US Air Force, <http://www.usace.army.mil/inet/usace-docs/design-guides/dg1110-345-117/entire.pdf>.

USEPA (1999). *Field Applications of in situ remediation technologies: Permeable Reactive Barrier*, EPA-542-R-99-002, Solid Waste and Emergency Response (5102G), United States Environmental Protection Agency, Washington, D. C. 20460, United States.

USEPA (2002). *Field applications of in situ remediation technologies: Permeable reactive barriers*, EPA/68/W-00/084, Technology Innovation Office, Office of Solid Waste and Emergency Response, United States Environmental Protection Agency, Washington, D. C. 20460, United States.

Vogan, J. L., Focht, R. M., Clark, D. K., and Graham, S. L. (1999). "Performance evaluation of a permeable reactive barrier for remediation of dissolved chlorinated solvents in groundwater." *J. Hazard. Mater.*, 68(1-2), 97-108.

Westerhoff, P. (2003). "Reduction of nitrate, bromate, and chlorate by zero-valent iron (Fe^0)." *J. of Environ. Eng. ASCE*, 129(1), 10-16.

Wilkin, R. T., and Puls, R. W. (2003). *Capstone report on the application, monitoring, and performance of permeable reactive barriers for ground-water remediation: Volume 1 – performance evaluations at two sites*, EPA/600/R-03/045a, Office of Research and Development, United States Environmental Protection Agency, Cincinnati, OH 45268, USA.

Yabusaki, S., Cantrell, K., Sass, B., and Steefel, C. (2001). "Multicomponent reactive transport in an in situ zero-valent iron cell." *Environ. Sci. & Technol.*, 35(7), 1493-1503.

Zhang, Y., and Gillham, R. W. (2005). "Effects of gas generation and precipitates on performance of Fe^0 PRBs." *Ground Water*, 43(1), 113-121.

CHAPTER 15

Hydraulic Issues Related to Granular Iron Permeable Reactive Barriers

John L. Vogan[1], and Stephanie F. O'Hannesin[2]

Abstract: Granular iron permeable reactive barriers (PRBs) are accepted as a viable alternative for the *in situ* remediation of chlorinated solvents in groundwater. Hydraulic issues involving unrecognized variability in plume characteristics and to a lesser extent, artifacts of PRB construction have been the cause of poor PRB performance. To date, no PRB has shown evidence of hydraulic plugging due to inorganic precipitate formation thus requiring rejuvenation. Additional plume characterization along the proposed alignment, the inorganic geochemistry of the plume, and choice of construction method should all be taken into consideration when assessing the possible hydraulic performance of the PRB. The relationship between groundwater flow velocity and iron PRB thickness (and PRB cost) is linear. Depending on the hydraulic characteristics of the reactive material, the flow velocity in the PRB may be reduced relative to that in the aquifer. Therefore, use of the aquifer groundwater flow velocity in PRB design may lead to an overdesign of the PRB thickness. Depending on the confidence level associated with aquifer parameters, such relationships may be used to optimize the PRB design.

CE Database subject headings: Contamination; Design; Ground water; Hydraulics; Iron.

15.1 Introduction

Granular iron permeable reactive barriers (PRBs) are accepted as a viable alternative for the *in situ* remediation of chlorinated solvents in groundwater, with over 125 applications around the globe and a 10-year track record of successful field performance. PRBs are remediating contaminated groundwater in United States, Europe, Japan, Canada and Australia. Indeed, Wilkin and Puls (2003) has stated that "The use of PRBs for the restoration of contaminated groundwater has evolved from innovative to accepted standard practice, for the containment and treatment of a variety of contaminants in groundwater". Innovative design configurations and construction methods have evolved to reflect site-specific parameters such as site geology, hydrogeology and contaminant distribution, thus further reducing PRB costs.

A recent review of 55 full-scale PRB applications (Fig. 15-1) indicated that hydraulic issues were the main cause of observed lack of performance of this

---

[1]President, EnviroMetal Technologies Inc., 745 Bridge Street West, Suite 7, Waterloo, Ontario, Canada N2V 2G6.

[2]Principal Hydrogeologist, EnviroMetal Technologies Inc., 745 Bridge Street West, Suite 7, Waterloo, Ontario, Canada N2V 2G6 (corresponding author). E-mail: sohannesin@eti.ca

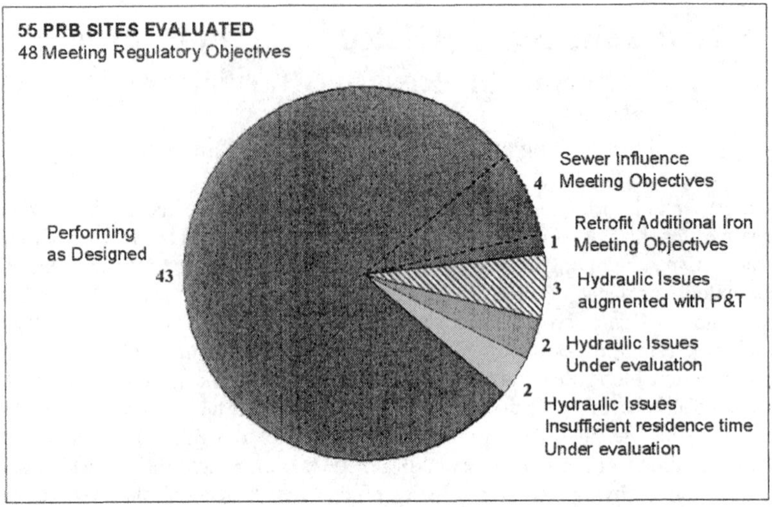

Fig. 15-1. Evaluation of PRB performance at 55 installations

technology, although 48 (87%) of those sites were still meeting regulatory objectives. It is the purpose of this paper to review hydraulic issues related to PRBs in detail, as a means of identifying the most appropriate approach for optimizing the hydraulic performance of future applications.

15.2 Hydraulic Characteristics of Granular Iron and Impact on PRB Design

PRB technology has benefited from the use of a homogeneous granular iron material that exhibits relatively uniform hydraulic characteristics. Typical grain size curves for granular iron used in excavation based (coarser grained, 2.0 to 0.25 mm) applications and common injection-based (finer grained, 1.2 to 0.17 mm) construction methods are shown in Fig. 15-2. The porosity of both of these materials is quite high, ranging from 40 to 54% as measured in laboratory columns (Table 15-1). The hydraulic conductivity of the coarser material ranges from about 5 x 10^{-2} cm/sec to 10^{-1} cm/sec, while the finer grained "injectable" granular iron is only slightly less permeable, about 10^{-2} cm/sec (Gavaskar et al. 2000). These material properties, in particular the porosity, may have a significant impact on PRB design.

One of the key parameters in granular iron PRB design is iron flow-through thickness, which can be estimated from the anticipated groundwater velocity in the PRB, multiplied by the residence time or contact time required for treatment of contaminants. Groundwater velocity in the PRB therefore has a significant impact on the amount of iron required and cost of installing a PRB remedy. In certain instances, the groundwater velocity in the PRB will not only be controlled by the groundwater

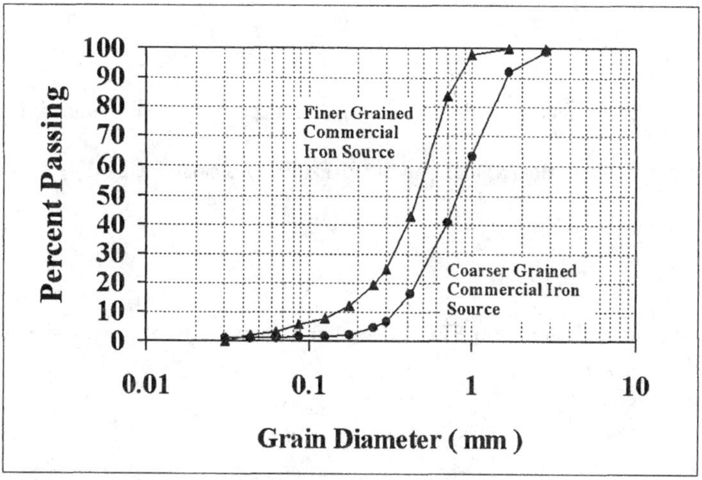

Fig. 15-2. Typical grain size curves for finer and coarser grained commercial granular iron sources

Table 15-1. Porosity Measurements of Common 100% Iron Sources Obtained from Laboratory Treatability Testing

Iron (100%) Source	Average Porosity	Number of Measurements
Coarser Grained (2.0 to 0.25 mm)		
Connelly	0.56	61
Peerless	0.55	18
Master Builders	0.44	19
Finer Grained (1.2 to 0.17 mm)		
Connelly	0.50	10
Peerless	0.48	2
Master Builders	0.52	2

velocity in the aquifer, but also by the difference in porosity of the iron media relative to that of the aquifer. This can be shown using Darcy's law:

Using the conventions of Freeze and Cherry (1980), Darcy's law is stated as:

$$v = -Ki \tag{15.1}$$

where v = specific discharge, K = hydraulic conductivity, and i = hydraulic gradient.

The groundwater flow velocity in the aquifer ($\overline{v}_{aq}$) is calculated by:

$$\overline{v}_{aq} = \frac{v_{aq}}{n_{aq}} \tag{15.2}$$

where specific discharge in the aquifer (v_{aq}) and porosity in the aquifer (n_{aq}) are known.

The groundwater flow velocity in the PRB ($\overline{v}_{PRB}$) is calculated by:

$$\overline{v}_{PRB} = \frac{v_{PRB}}{n_{PRB}} \tag{15.3}$$

where specific discharge in the PRB (v_{PRB}) and the porosity of the reactive material in the PRB (n_{PRB}),

and, since

$$v_{aq} = v_{PRB} \tag{15.4}$$

so substituting:

$$\overline{v}_{PRB} = \frac{n_{aq}\overline{v}_{aq}}{n_{PRB}} \tag{15.5}$$

Therefore, the groundwater flow velocity in the PRB equals the groundwater flow velocity in the aquifer times the ratio of the aquifer porosity to the PRB porosity.

Given the range in granular iron (100% content) porosity in Table 15-1, in contrast to the porosity of most aquifer materials (within the range of about 25% to 35%), from Eq. (15-5), the groundwater flow velocity in the PRB could be within the range of about 50% to 75% of the aquifer velocity.

The use of this porosity ratio in PRB design depends on the overall design objective and economics of the application. In the most conservative approach, the aquifer velocity is used to determine iron thickness. This approach treats any differences in porosity as a form of safety factor in the design. However, this maximizes the amount of iron needed in the application. A less conservative approach uses the difference in porosity between the aquifer and granular iron (100% content) to reduce the velocity in the iron treatment zone by a corresponding factor, thereby reducing the required iron flow-through thickness, and volume of iron required.

Parameters to consider when evaluating this aspect of PRB design include:

- lower porosities typically occur in iron/sand mixtures used with some construction methods, relative to 100% iron content;
- as discussed later in this paper, carbonate precipitation over time may cause a decline in porosity relative to the original porosity, depending on the carbonate flux through the system;

- if the PRB has a higher hydraulic conductivity, it will actually draw some water in from the surrounding aquifer as well as that water directly flowing perpendicular to the influent face, causing an increase in flow rate;
- the accuracy of the aquifer groundwater velocity estimate; and
- whether a probabilistic design approach incorporating these uncertainties is most appropriate for the application being considered.

It is estimated that the design(s) of over 90% of installed PRBs have incorporated a conservative approach when assessing the impact of hydraulic properties of the iron media. That is, aquifer parameters only were used to estimate the velocity through the PRB. However, there are sites working effectively where the porosity difference between the iron media and the aquifer was incorporated into the design (Heneman et al. 1999). The decision as to the groundwater velocity to be incorporated into the PRB design is site specific, and a number of technical factors, as well as iron costs and regulatory objectives, need to be considered when determining the most appropriate PRB velocity for design purposes.

15.3 Influence of Inadequate Characterization of Plume Hydrogeology on Hydraulic Performance

Numerous design documents (e.g., ITRC 2005; Wilkin and Puls 2003) have stressed the need to adequately characterize the plume as part of the PRB design effort. Of the 55 sites examined 48 sites (87%) are meeting regulatory objectives (Fig. 15-1). Included in this total are 43 PRB sites are performing as designed, 1 site where system expansion (additional iron) has been required, mainly to address the lack of complete plume capture achieved using the original design and 4 additional sites where utilities were influencing flow, while overall still meeting site objectives. Of the 7 remaining sites, 4 are older applications which employed a funnel and gate configuration and have hydraulic issues or inadequate residence time which are still under evaluation, and the remaining 3 sites where pump and treat alternatives have been implemented to ensure capture of the portion of the plume bypassing the PRB.

In addition to lack of adequate plume delineation, other common problems in this regard include a lack of recognition in seasonal groundwater flow direction, which may cause periodic bypass around the ends of a PRB, and a lack of recognition of the variability in hydraulic conductivity of the aquifer. Post-construction subsurface investigations at two of the sites shown in Fig. 15-1, indicated the presence of higher hydraulic conductivity zones (i.e., high groundwater velocity zones) that were not identified during the design phase of the PRB. These zones led to insufficient contaminant residence time in the reactive material. The effect of variability in aquifer hydraulic conductivity has been modelled using theoretical, stochastic approaches by several researchers (e.g., Elder et al. 2001). While these studies tend to present "worst case" type of scenarios in terms of negative effects, spatial heterogeneities along the line of PRB installation should always be considered during design. This is particularly important for larger systems (e.g., Heneman et al. 2001), and may influence the basic choice of a deterministic versus probabilistic design approach in certain heterogeneous environments (ITRC 2005).

15.4 Influence of Construction Methods on Hydraulic Performance

The first PRBs constructed were shallow, and often deployed sheet piling to form a reactive, rectangular gate flanked on either side by sheet piling or slurry walls to funnel groundwater through the treatment section. At several sites, the use of sheet piling to create the treatment zone caused smearing of fine-grained materials present in the aquifer sediments across the face of the treatment section impeding flow. The recognition of this shortcoming led to the development of novel construction methods that do not involve sheet piling, especially for deeper applications.

Most current PRB applications employ a continuous wall configuration rather than a funnel-and-gate configuration (Fig. 15-3). In funnel-and-gate applications, low permeability slurry walls or sheet piling is used to direct groundwater towards a central treatment zone. As noted above, a disproportionate number of hydraulic problems have arisen with funnel-and-gate applications, including:

- inadequate funnel lengths;
- flow over reactive material placed in gates;
- establishment of a low hydraulic conductivity smear zone at the gate interface; and
- flow beneath or over the funnel sections.

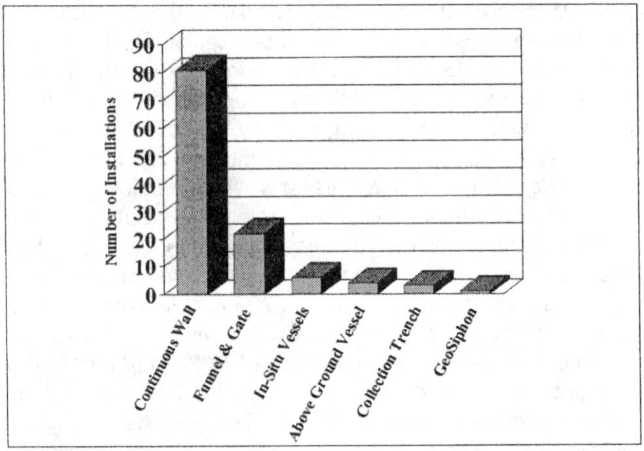

Fig. 15-3. Number of installations involving various design configurations

This has led to a continuous, permeable reactive wall becoming the "default option" when PRB configuration are considered.

Buried utilities have also influenced groundwater flow patterns through shallow PRBs. Ideally, groundwater flows perpendicular to the PRB, but the highly conductive bedding materials for sewer pipes and other utilities has caused deviation of this perpendicular flow at several sites (Tappert 2001; Goodman et al. 2003). Fig. 15-4 shows how high hydraulic conductivity bedding causes groundwater to flow along the PRB rather than perpendicular to the PRB (Goodman et al. 2003).

Fortunately, degradation rates in the PRB at this site were sufficient to allow regulatory goals to be met, in spite of this unforeseen variation in the flow field.

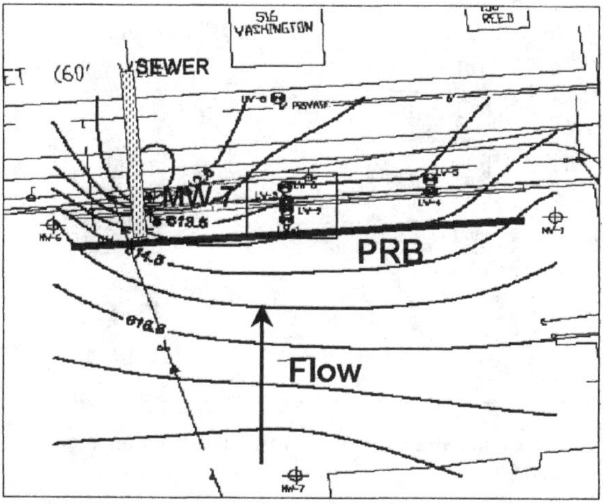

Fig. 15-4. Observations of groundwater flow along PRB rather than perpendicular due to sewer line influence (Goodman et al. 2003)

Continuous trenching machines allow simultaneous excavation and backfilling without an open trench. Excavation is performed by a cutting chain immediately in front of a trench-box (boot) that extends the width and depth of the finished treatment zone. As the trencher moves forward, granular iron or an iron-sand mixture is added to the boot creating a continuous treatment zone. Trenchers are available to install treatment zones from 0.5 m to 0.9 m in width to depths of 11 m. Due to the nature of the excavation, which minimizes smearing, there have been minimal hydraulic issues associated with this construction method. Some mixing of iron media and aquifer sediments should be anticipated at the aquifer/PRB interface, which may influence the hydraulic properties of this zone.

The most significant advancement in PRB construction has been the use of biodegradable slurry formulations called biopolymer (BP) to support trench walls during construction, or as a carrier fluid to allow iron to be injected into the subsurface (Fig. 15-5). In construction, the term BP usually refers to a viscous water solution that can be broken down by microbes along with enzyme addition. There are several compounds that serve as a basis for BP (e.g. guar gum, xanthan gum, polyacrylamide, etc.). The most common BP used in PRB construction has been guar gum, a powder milled from guar beans, that when mixed with water hydrates to form a viscous solution consisting of long chain carbohydrates. As part of these construction approaches, an enzyme added to the BP solution facilitates breakdown of the BP so that it becomes less viscous, allowing groundwater to flow freely through the PRB.

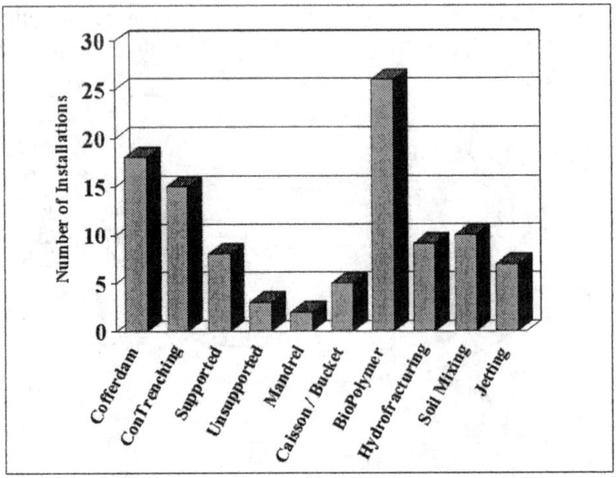

Fig. 15-5. Number of installations of various construction methods used for PRB application

Since 1999, BP construction methods (BP trenching, vertical hydraulic fracturing, jetting and soil mixing) have accounted for 45% of the granular iron PRBs constructed. These construction methods have proven to be cost competitive with other methods for shallow plumes and have also allowed PRBs to be constructed to depths greater than 35 m. Several references are available which discuss these methods (Day et al. 1999; Hocking and Wells 2002; Richards 2002).

BP trenching methods (Fig. 15-6) have not caused any negative hydraulic effects when properly deployed for PRB applications. However, a few early applications exhibited lower hydraulic conductivities than expected due to incomplete breakdown

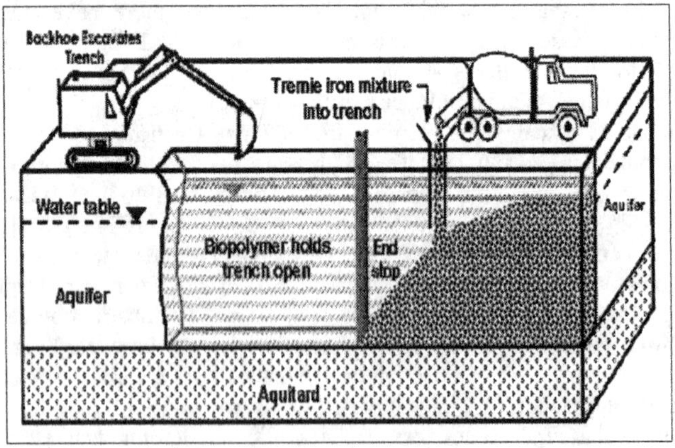

Fig. 15-6. Schematic of biopolymer trenching construction method showing granular iron placement (Courtesy of GeoSyntec Consultants)

of the BP. As discussed by Moylan and Buechler (2003) for a site in Missouri where a BP trenching technique was used. Problems maintaining the viscosity of the BP and in placement of the sand and granular iron backfill resulted in a significant reduction of permeability of the PRB, due to mixing of fine grain aquifer sediments with the granular iron in the BP trench. Most field data has been similar to that reported by Focht et al. (2001) and Sivavec et al. (2003), where hydraulic conductivity measurements in a BP trench PRB were similar to previous laboratory measurements on granular iron using falling head permeameter tests. Hocking and Wells (2002) report hydraulic test results from an injected iron PRB constructed using vertical hydraulic fracturing (Fig. 15-7) which show no influence of the PRB on aquifer hydraulic responses following installation.

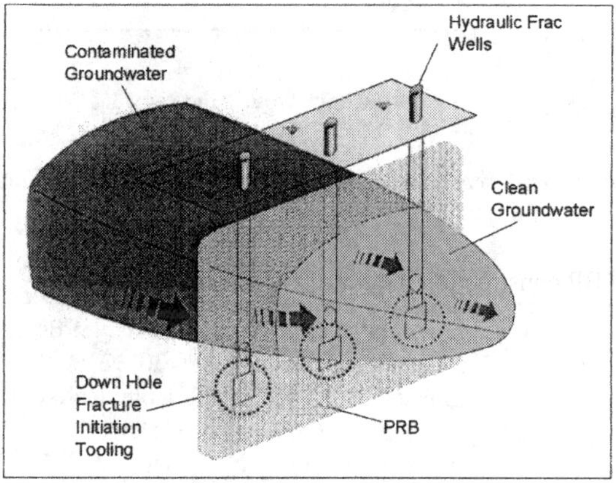

Fig. 15-7. Schematic of vertical hydraulic fracturing construction method involving iron injection (Courtesy of GeoSierra, LLC)

15.5 Influence of Long-Term Geochemical Changes on Hydraulic Performance

The precipitation of carbonate, hydroxide and sulphide minerals in PRBs as a result of iron corrosion has been well documented (e.g., Wilkin et al. 2002), and the resulting impact of mineral precipitation on the hydraulic properties of PRB has been investigated at the laboratory and field scale. For example, Zhang and Gillham (2005) showed in a long-term column study, that calcium carbonate precipitation occurs as a moving front through the iron as evidenced by the temporal changes in profiles of bicarbonate (Fig. 15-8). The maximum loss in porosity was about 7% of the initial porosity, followed by no further accumulation. Lin et al. (2005) used geochemical and transport modeling to simulate the long-term change in hydraulic properties in iron PRBs. Assuming a typical calcium and alkalinity concentration range and

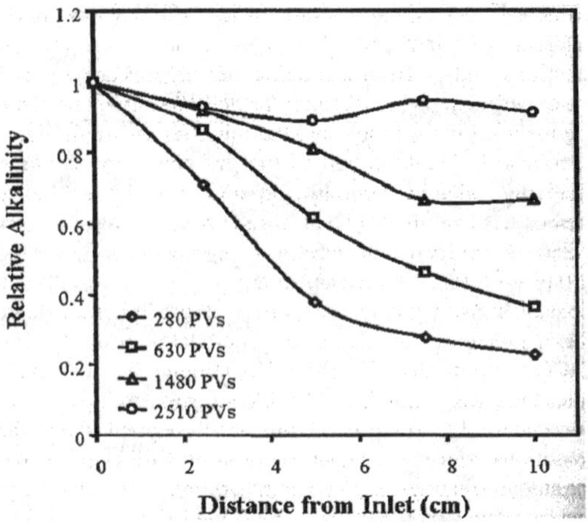

Fig. 15-8. Changes in relative alkalinity over time versus column distance from inlet (Zhang and Gillham 2005)

groundwater velocity up to 0.2 m/day, the model showed that precipitate formation resulted in only subtle changes in PRB's porosity and hydraulic conductivity within the first 10 years of operation, and predicted that the most significant changes would not occur until the PRB had operated for at least 30 years. Significantly, most laboratory column testing studies have shown considerable decline in VOC reactivity within zones of carbonate precipitation (Vikesland et al. 2003; Zhang and Gillham 2005). Other laboratory column tests (e.g. Kamolpornwijit et al. 2003; Gillham et al. 2001) have shown that under certain conditions, precipitates could lead to non-uniform flow within the iron zone. Tracer tests at a funnel and gate installation in Denmark (Lai et al. 2006a and 2006b) showed non-uniform flow within the iron treatment gate which was attributed to mineral precipitation. However, the reactivity of this system appeared to be unaffected by those precipitates over a three year period (Lo et al. 2004).

Field experience to date indicates, that at most sites, calcium carbonate represents the largest volume of precipitates. The first recorded core analyses, from pilot-scale systems operating for 18 months and 2 years, containing 100% granular iron revealed porosity losses in the upgradient few centimetres of iron in the range of 10% of the initial porosity (Fig. 15-9), with losses declining sharply over the first 30 cm to below 2% (Vogan et al. 1998; Vogan et al. 1999). These porosity losses were calculated based on carbonate analyses of retrieved cores, and were consistent with losses predicted on the basis of changes in the inorganic water chemistry. Assuming an initial porosity of 0.5, the porosity in the first few inches of the iron zones had declined to about 0.45. Concurrent field data (VOC and groundwater velocity measurements) indicated that the precipitates had not adversely affected system hydraulics and iron reactivity. Similar or lower levels of precipitate have been observed at other sites operating for similar periods of time (Blowes et al. 1999).

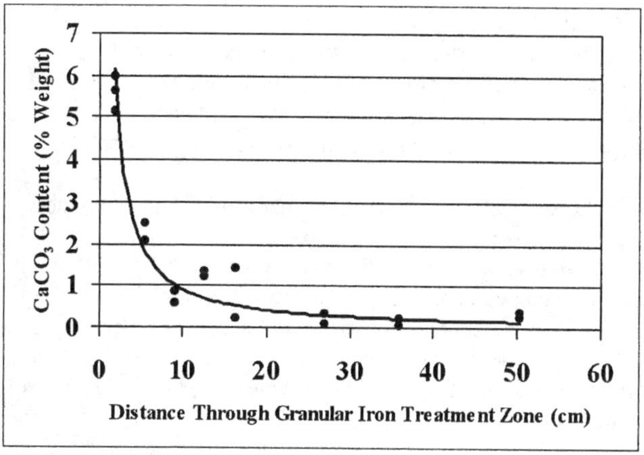

Fig. 15-9. Calcium carbonate content (% weight) versus distance through a granular iron treatment zone (Vogan et al. 1999)

Extrapolation from these early coring results indicated that the porosity loss in a thin zone at the upgradient interface of the PRB over a ±10 year period would necessitate some form of rehabilitation to restore the reactivity and/or hydraulics of the system. It was not known how much total porosity loss would occur as the systems "aged". However, recent cores taken from one of the same sites as described in Vogan et al. (1998) showed no more precipitate build-up than originally observed (Gavaskar et al. 2002). This would appear to support the work of Gillham et al. (2001) that shows "levelling off" of the amount of precipitate forming at any one point in the iron. Cores taken from the original University of Waterloo test site at CFB Borden, Ontario, Canada, 10 years after installation (Reynolds 2002) also showed no large degree of precipitate build-up since it was cored, 6 years previously (O'Hannesin and Gillham 1998).

Several PRBs have now been performing for 10 years or more (e.g., O'Hannesin 2003; Warner et al. 2004), with no apparent decline in reactive performance. Of over 125 installations, no site that we are aware of has needed rehabilitation because of loss of iron reactivity. Two PRBs evaluated by the U.S. DoD indicate that these PRBs are currently performing as designed and were predicted to perform acceptably for at least 30 years (NFESC 2002).

Given the preponderance of field data, PRBs at most sites should be able to last at least 15 years with no need for rehabilitation to address the adverse hydraulic affects of mineral precipitation. Economic analyses (e.g., Powell et al. 2002) have shown that if rehabilitation activities only need to occur at intervals of about 10 to 15 years, then PRBs will be extremely cost-competitive versus pump-and-treat systems.

15.6 Summary

The hydraulic properties of the reactive material used in iron PRBs are conducive to the longevity and hydraulic performance of the system. However, to ensure optimal PRB hydraulic performance, consideration must be given to the following:

(i) the degree to which the high porosity of iron material should be incorporated into PRB design;
(ii) whether the groundwater plume has been adequately characterized in three dimensions to avoid contaminant by-pass;
(iii) the impact, if any, of existing utilities on flow through the PRB; and
(iv) whether the planned construction approach could lead to reduced flow through the PRB. If implemented properly, the use of BP trenching and injection methods will not negatively affect the hydraulics of PRB operation;
(v) the expected flux of inorganic constituents (e.g., carbonate) through the PRB over time.

With respect to (v) above, while this may be an issue in some high total dissolved solids groundwaters, it appears PRBs will last 10 to 15 years in most geochemical environments before their hydraulic properties are compromised due to mineral precipitation in the reactive media.

15.7 References

Blowes, D. W., Puls, R. W., Gillham, R. W., Ptacek, C. J., Bennett, T. A., Bain, J. G., Hanton-Fong, C. J., and Paul, C. J. (1999). *An in situ permeable reactive barrier for the treatment of hexavalent chromium and trichloroethylene in ground water: Volume 2, performance monitoring*, EPA/600/R-99/095b, United States Environmental Protection Agency, United States.

Day, S. R., O'Hannesin, S. F., and Marsden, L. (1999). "Geotechnical techniques for the construction of reactive barriers." *J. Hazard. Mater.*, B67, 285-297.

Elder, C. R., Benson, C. H., and Eykholt, G. R. (2001). "Economics and performance based design of monitoring systems for PRBs." In *Proceedings of the 2001 International Containment and Remediation Technology Conference* (CD-ROM), June 11-13, Orlando, Florida.

Focht, R. M., Vogan, J. L., and Krug, T. A. (2001). "BioPolymer construction techniques for installation of permeable reactive barriers containing granular iron for groundwater remediation." In *American Chemical Society National Meeting*, San Diego, California. April 1-5, 41(1), 1120-1125.

Freeze, R. A., and Cherry, J. A. (1979). *Groundwater*, Prentice-Hall, New Jersey.

Gavaskar, A., Gupta, N., Sass, B., Janosy, R., and Hicks, J. (2000). *Design guidance for application of permeable barriers for remediate dissolved chlorinated solvents*, Battelle Press, Ohio.

Gavaskar, A., Sass, B., Gupta, N., Drescher, E., Yoon, W-S., Sminchak, J., Hicks, J., and Condit, W. (2002). *Evaluating the longevity and hydraulic performance of permeable reactive barriers at the Department of Defense sites*, Battelle Press, Ohio.

Gillham, R. W., Ritter, K., Zhang, Y., and Odziemkowski, M. S. (2001). "Factors in the long-term performance of granular iron PRBs." In *Proceedings of the 2001 International Containment and Remediation Technology Conference* (CD-ROM), June 11-13, Orlando, Florida.

Goodman, W. M., Ignaszak, K. D., Storonsky, M. P., and Duchene, M. (2003). "Influence of sewers on groundwater flow in the vicinity of a PRB in an urban setting, Geneva, NY." *RTDF permeable reactive barriers (PRBs) action team meeting*, October 15-16, Niagara Falls, New York, 43-54.

Heneman, F. C., May, M. R., Powers, B. G., and Perez, E. J. (2001). "The End of a Trilogy: The F.E. Warren AFB Spill Site iron-filings permeable reactive barrier is working." In *Proceedings of the National Defense Industrial Association 27th Environmental Symposium and Exhibition*, Austin, Texas.

Heneman, F. C., Moylan, J. E., and May, M. R. (1999). "Application of in-situ iron-filings treatment wall technology at Spill Site 7, F.E. Warren Air Force Base, Cheyenne (a case study)." In *Proceedings of the National Defense Industrial Association (NDIA) 25th Environmental Symposium & Exhibition*, April 1, NDIA Compendium, Arlington, Virginia, Denver, Colorado.

Hocking, G., and Wells, S. (2002). "Groundwater performance monitoring of an iron permeable reactive barrier." In *Proceedings of the 3rd International Conference on Remediation of Chlorinated and Recalcitrant Compounds* (CD-ROM), May 20-23, Monterey, California.

Interstate Technology & Regulatory Council (ITRC) (2005). *Permeable reactive barriers: Lessons learned/new directions*, PRB-4, Permeable Reactive Barriers Team, District of Columbia, Washington. <www.itrcweb.org>.

Kamolpornwijit, W., Liang, L., West, O., Moline, G., and Sullivan, A. (2003). "Preferential flow path development and its influence on long-term PRB performance: column study." *J. Contam. Hydrol.*, 66(3-4), 161-178.

Lai, K. C.-K., Lo, I. M. C., and Kjeldsen, P. (2006a). "Natural gradient tracer test for the permeable reactive barrier in Denmark 1. Field study of tracer movement." *J. of Hazardous, Toxic and Radioactive Waste Management, ASCE*, in press.

Lai, K. C.-K., Lo, I. M.-C., and Kjeldsen, P. (2006b). "Natural gradient tracer test for the permeable reactive barrier in Denmark 2. Spatial Moments Analysis and Dispersion of Conservative Tracer" *J. of Hazardous, Toxic and Radioactive Waste Management, ASCE*, in press.

Lin, L., Benson, C. H., and Mergener, E. A. (2005). "Impact of mineral fouling on hydraulic behaviour of permeable reactive barriers." *Ground Water*, in press.

Lo, I. M.-C., Lai, K. C.-K., and Kjeldsen, P. (2004). "Field monitoring of Fe^0 PRB for removal of chlorinated hydrocarbons." In *Proceedings of the 4th International Conference on Remediation of Chlorinated and Recalcitrant Compounds* (CD-ROM), May 24-27, Monterey, California.

Moylan, J., and Buechler, T. (2003). "Evaluation of the permeable reactive wall at the northeast corner operable unit, Lake City Army Ammunition Plant (LCAAP), MO." *RTDF permeable reactive barriers (PRBs) action team meeting*, Oct 15-16, Niagara Falls, New York, 23-41.

Naval Facilities Engineering Command (2002). *Advances in permeable reactive barrier technologies*, TDS-2089-ENV, August, NFESC Tech Data Sheet.

O'Hannesin, S. F. (2003). "10 years of North American experience in granular iron PRB technology for VOC groundwater remediation." In *Proceedings of the 8th International FZK/TNO Conference on Contaminated Soil*, Consoil, May 12-16, Gent, Belgium, 1568-1571.

O'Hannesin, S. F., and Gillham, R. W. (1998). "Long-term performance of an in-situ "iron wall" for remediation of VOCs." *Ground Water*, 36(1), 164-170.

Powell, R. M., Powell, P. D., and Puls, D. W. (2002). *Economic analysis of the implementation of permeable reactive barriers for remediation of contaminated groundwater*, EPA/600/R-02/034, United States Environmental Protection Agency, United States.

Reynolds, T. J. (2002). *An evaluation of the ten year performance of a permeable reactive barrier including an assessment of the in situ microcosm technique*, Master Thesis, Department of Earth Sciences, University of Waterloo, Waterloo, Ontario, Canada.

Richards, P. (2002). "Construction of a permeable reactive barrier in a residential neighborhood." *Remediation Journal*, 12(4), 65-79.

Sivavec, T., Krug, T. A., Berry-Spark, K., and Focht, R. M. (2003). "Performance monitoring of a permeable reactive barrier at the Somersworth, NH Landfill Superfund site." In *Chlorinated solvent and DNAPL remediation: Innovative strategies for subsurface cleanup*, American Chemical Society Symposium Series 837, 259-277.

Tappert, S. E. (2001). "A permeable reactive barrier installation in New Jersey: Notes from the underground." In *Proceedings of the 1st International Conference on Oxidation and Reduction Technologies for In-Situ Treatment of Soil and Groundwater*, June 25-29, Niagara Falls, Ontario, Canada.

Vikesland, P. J., Klausen, J., Zimmermann, H., Roberts, A., and Ball, W. P. (2003). "Longevity of granular iron in groundwater treatment processes: Changes in solute transport properties over time." *J. Contam. Hydrol.*,, 64(1-2), 3-33.

Vogan, J. L., Butler, B. G., Odziemkowski, M. K., Friday, G., and Gillham, R. W. (1998). "Laboratory evaluation of cores from permeable reactive barriers." In *Proceedings of the 1st International Conference on Remediation of Chlorinated and Recalcitrant Compounds*, May 18-21, Monterey, California, Battelle Press, Columbus, Ohio, Vol. C1-6, 163-168.

Vogan, J. L., Focht, R. M., Clark, D. K., and Graham, S. L. (1999). "Performance of a permeable reactive barrier for remediation of dissolved chlorinated solvents in groundwater." *J. Hazard. Mater.*, 68, 97-108.

Warner, S. D., Longino, B. L., Zhang, M., Bennett, P., and Szerdy, F. S. (2004). "Ten-year performance assessment of the first commercial permeable reactive barrier composed of granular iron." In *Proceedings of the 1st International Symposium on Permeable Reactive Barriers*, March 14-16, Belfast, Ireland.

Wilkin, R. T., and Puls, R. W. (2003). *Capstone report on the application, monitoring, and performance of permeable reactive barriers for ground-water remediation: Volume 1- Performance evaluations at two sites*, EPA/600/R-03/045a, United States Environmental Protection Agency, United States.

Wilkin, R. T., Puls, R. W., and Sewell, G. W. (2002). *Long-term performance of permeable reactive barriers using zero-valent iron: an evaluation at two sites*, EPA/600/S-02/001, United States Environmental Protection Agency, United States.

Zhang, Y., and Gillham, R. W. (2005). "Effects of gas generation and precipitates on performance of Fe^0 PRBs." *Ground Water*, 43(1), 113-121.

CHAPTER 16

Tracer Experiments in Zero-Valent Iron Permeable Reactive Barriers

Markus Ebert[1], Ralf Köber[2], Anika Parbs[3], and Andreas Dahmke[4]

Abstract: The use of reactive tracers combined with conservative tracers offers potential for reactivity information integrated along a certain flow path. In column experiments, several substances were tested to examine their transport behaviours inside a porous medium packed with zero-valent iron (Fe^0). The results showed that tritium was only conservatively transported and that salt tracers commonly used in field testing can be significantly retarded. Therefore, the combination of weakly sorbing anions and cations is suggested for use in tracer testing. Natural gradient tracer tests conducted at permeable reactive barrier (PRB) sites have uncovered more complex flow patterns than those derived from other monitoring methods and have been proved to be the only method for a detailed examination of the flow velocity in PRBs. The results of forced gradient tracer testing presented here and completed on the basis of conservative and reactive tracers at two PRB sites indicate the potential of this method for uncovering reactivity issues. A small degree of reactivity is responsible for partial performance limitations in one section of the Rheine site PRB, probably due to a desegregation of iron and gravel. Preferential flow and flow bypassing are responsible for the performance limitations of the Tübingen site PRB uncovered by forced gradient and natural gradient tests.

CE Database subject headings: Barriers; Field tests; Ground-water flow; Hydraulic conductivity; Iron; Laboratory tests; Monitoring; Porosity; Tracers.

16.1 Introduction

Groundwater flow through a permeable reactive barrier (PRB) is a basic requirement for the remediation technology. For this reason, the hydraulic performance of PRBs must be continually evaluated to monitor any potential changes in groundwater hydraulics which may affect the performance of a PRB. The hydraulics of a PRB can change over time for several reasons, including the precipitation of minerals within the PRB and a decrease in the hydraulic conductivity of the reactive media. Typical minerals found in zero-valent iron (Fe^0) reactive zones are green rust minerals (Klausen et al. 2003; Wilkin and Puls 2003), siderite (Agrawal and Tratnyek 1996;

[1] Assistant Professor, Institute of Geosciences, Christian-Albrechts-Universität zu Kiel, Ludewig-Meyn Straße 10, 24118 Kiel, Germany (corresponding author). E-mail: me@gpi.uni-kiel.de
[2] Scientist, Institute of Geosciences, Christian-Albrechts-Universität zu Kiel, Ludewig-Meyn Straße 10, 24118 Kiel, Germany.
[3] Ph.D. Student, Institute of Geosciences, Christian-Albrechts-Universität zu Kiel, Ludewig-Meyn Straße 10, 24118 Kiel, Germany.
[4] Professor and Chair of Applied Geosciences, Institute of Geosciences, Christian-Albrechts-Universität zu Kiel Ludewig-Meyn Straße 10, 24118 Kiel, Germany.

Phillips et al. 2000), calcite and aragonite (Roh et al. 2000; Wilkin et al. 2003), iron sulphides (Gu et al. 1999; Phillips et al. 2003; Wilkin et al. 2003), and mixed Fe(II/III) (hydr)oxides, such as magnetite (Farrell et al. 2000; Klausen and Tröber 1995; Odziemkowski et al. 1998), lepidocrocite (Gu et al. 1999), or akaganite (Phillips et al. 2003). The formation of these mineral phases reduces the void space of the reactive media, and their effect on porosity and hydraulic conductivity can be only estimated by the molar volume of pure minerals and correlations such as the Kozeny-Carmen equation (e.g. Mackenzie et al. 1999; Morrison 2003; Wilkin and Puls 2003; Li et al. 2005). In addition, the formation of hydrogen gas bubbles through the anaerobic corrosion of Fe^0 (Matheson and Tratnyek 1994; Reardon 1995; Ebert et al. 1999; Gillham et al. 2001; Liang et al. 2001; Sorel et al. 2001) may decrease porosity more than mineral formation does and is usually more difficult to detect. Tracer experiments using conservative tracers are the best method for determining effective porosity in porous media, even in laboratory column experiments or under field conditions (Stephens et al. 1998).

Construction methods also may cause unexpected poor hydraulic performance of a PRB, including smearing of low-permeability materials across the interface of the PRB during construction, mixing of aquifer material with the reactive media, or the incomplete degradation of a bio-polymer (Vogan 2003). Moreover, other external factors can affect the hydraulic performance of a PRB, including the addition or shutdown of groundwater extraction in the area of the PRB, changes in regional groundwater flow (excessive rainfall, drought, etc.), and changes in land use (e.g. property development which reduces or changes infiltration and groundwater flow) (ITRC 2005). In general, water-level measurements are the most cost-effective tool for characterizing the bulk flow of groundwater at a PRB site. Limitations of this method arise primarily from small groundwater gradients and small distances between monitoring wells. If the water table can be characterized by water-level measurements, the capture zone of a PRB may be determined, but hydraulic conductivity and effective porosity must be known to calculate flow velocity within the PRB and in its vicinity as well. The monitoring of the reactivity of a PRB requires flow velocity because residence time must be calculated. At many PRB sites several methods were evaluated to determine hydraulic conductivity or flow velocity, including slug or pump tests (e.g. Blowes et al. 1999; Gavaskar et al. 2002; Juhasz 2001; Moylan 2003), *in situ* and down-hole flow measurements (e.g. Gavaskar et al. 2000b; Gavaskar et al. 2002; Liang et al. 2001), tracer experiments (e.g. Lai et al. 2005b; Li et al. 2005; Muza et al. 2003; Parbs et al. 2003; Puls et al. 1996; Reeter et al. 1999; Vogan et al. 1999; Zhang et al. 2001), and groundwater modelling (e.g. Gavaskar et al. 2002; Juhasz 2001; Laase et al. 2000; Li et al. 2005; Mayer et al. 2001; Naftz et al. 2002). Overall, tracer tests have proved to be a very useful tool for determining flow velocity, flow direction and the extent of groundwater captured by PRBs. A recent report on PRB technology stated that "Tracer tests provide the best flow information but are difficult and expensive to conduct" (ITRC 2005).

Advantages of tracer tests are primarily the detailed examination of flow patterns including vertical or horizontal heterogeneities, the assessment of residence times within reactive media, a comprehensive characterization of the capture zone, and the recognition of potential flow bypasses. Disadvantages are primarily increased time expenditure, the large number of samples which must be analysed, and the possible missing of tracer peaks due to an inadequate monitoring network, too small sampling density or unexpected flow patterns. In case of salt tracers (such as bromide) continuous monitoring is possible using down-hole, ion-selective electrodes. Continuous monitoring with probes such as these increases the probability of capturing tracer peaks and reduces laboratory costs. Ion-selective probes are expensive, but their cost may be justified by reduced laboratory requirements and increased chances of success. Forced-gradient tracer tests reduce the time required for tracer breakthrough and also increase the probability of capturing tracer peaks. A forced-gradient tracer test modifies natural flow patterns and is not useful for the characterization of capture zones, but can be useful for an investigation of the hydrogeochemical properties of the reactive media (Ebert et al. 2002b; Parbs et al. 2003). The strategy for tracer testing depends on the monitoring objective. If the only objective is to determine the flow path of groundwater from a specific location upgradient of the PRB, simple tracer tests may be conducted with injection in one upgradient well and monitoring for tracer arrival in the reactive cell. However, if detailed delineation of the capture zone upgradient of the PRB is required, multiple tracer tests using various tracers and an extensive monitoring network are then needed.

Bromide is the most frequently used tracer in field studies (e.g. Liang et al. 2001; Muza et al. 2003; Puls et al. 1996; Reeter et al. 1999; Vogan et al. 1999) or laboratory investigations (e.g. Kamolpornwijit et al. 2003; Liang et al. 2001; Parbs et al. 2003). Other tracers are iodide and noble gases (Liang et al. 2001), lithium (Lai et al. 2005b) and combinations of salt tracers (Parbs et al. 2003). Increasing contaminant concentrations downgradient of a PRB are not necessarily related to hydraulic considerations, because performance limitations of a Fe^0 PRB may also arise from a loss in reactivity due to the aging of iron surfaces. Decreasing iron reactivity is well known from laboratory studies (e.g. Agrawal et al. 2002; Ebert 2004; Gillham et al. 2001; Klausen et al. 2003; Odziemkowski and Gillham 1997; Schlicker et al. 2000; Vikesland et al. 2003). Under field conditions it may be difficult to discern the underlying reasons causing an increase in downgradient contaminant concentrations. Combined tracer tests with conservative and reactive substances have proved to be a potential tool for an investigation of this subject (Ebert et al. 2003a; Parbs et al. 2005; Parbs et al. 2003). Reactive tracer tests are known from the characterization of NAPL distribution (e.g. Davis et al. 2002; Imhoff et al. 2003; Istok et al. 2002; Rao et al. 2000) or the determination of the reactive surface area of Fe(III) minerals in the subsurface (Dethlefsen et al. 2004; Veehmayer 2000). The interpretation of breakthrough curves (BTCs) of species with known interactions with the solid phase (or NAPL) as compared to conservative tracer transport yields information integrated along the flow path. This concept is also applicable for the estimation of reactive sites in Fe^0 columns, so that BTCs of oxidants, such as nitrate, chromate, or oxidizing organic substances, can be interpreted in terms of reactive sites (Ebert et al. 2003a; Parbs et al. 2003; Schlicker et al. 2000). However, at least one conservative tracer is required in each test for the determination of changing porosities and residence times or as a reference for reactive tracers. Since the transport behaviour of several substances in Fe^0 columns was investigated in laboratory experiments, the following sections first examine results from tracer experiments in laboratory columns and then present examples of tracer tests at various PRB sites.

16.2 Tracer Experiments in Laboratory Columns

Deuterium (^2H_2O) or tritium water (^3H_2O) were proved to be the most conservative or so-called ideal tracers in Fe^0 columns (Ebert et al. 2002b; Eykholt et al. 1995; Mackenzie et al. 1995; Vikesland et al. 2003). In column experiments with solutions containing various amounts of natural groundwater constituents, the reapplication of a tritium tracer showed a loss of porosity between 6% and 26% during 1,100 days due to mineral precipitation and gas bubble formation (Vikesland et al. 2003). Furthermore, the experiments indicated an increase in dispersion length from <0.2 cm to a maximum of 1 cm and a migration of the tracer into immobile pore water regions. Other column experiments with deuterium tracers showed a decrease in porosity between 5% and 10% during 210 exchanged pore volumes with 40 mg/L of $CaCO_3$ solution (Mackenzie et al. 1995). Here, the loss in porosity was evenly distributed along the entire column and was more pronounced than estimated from pore water concentration profiles and stoichiometric calculations. Experiments with fluorescent dyes (e.g. fluorescein, rhodamine), as classical groundwater tracers, have shown strong retardation or degradation (Eykholt et al. 1995; Liang et al. 2001), and these tracers are unsuitable for hydraulic studies in iron-reactive media. Particle tracers (microspheres) also appear to be unsuitable for this purpose because they are transported only in macropores, or the particles are attached to the matrix (Zhang et al. 2001). Tetrafluoroborate has shown conservative transport behaviour in column experiments at residence times below 50 minutes (Eykholt et al. 1995), but decomposes significantly to borate and fluoride in the case of longer residence times in Fe^0 columns (Wüst 2000). The most frequently used tracer, bromide, has been found to work well in an iron-reactive medium when a retardation factor of 1.2 is incorporated, which, in turn, is also suggested for field applications (Gavaskar et al. 2000a, Gavaskar et al. 1998). Other studies found a nearly conservative transport of bromide (Focht et al. 1997; Powell et al. 1998) or variable retardation (Ebert et al. 2003a; Ebert et al. 2002b; Parbs et al. 2003). The potential variability of bromide retardation is interesting because of the wide usage of bromide in hydraulic studies of iron-reactive media.

In parallel column experiments using the same iron granules (*Gotthart Maier* iron) and groundwater from the Bernau site (Hein et al. 2002), a tracer experiment with 3 mM LiBr (added to the groundwater) was performed after 154 exchanged pore volumes (Ebert and Dahmke 2001). The primary difference between the experiments was the trichloroethylene (TCE) concentration with nominal 100 mg/L and 300 mg/L in Column 1 and Column 2. With TCE half life times of 4.9 h and 9.8 h (Ebert 2004), approximately 90 mg/L and 190 mg/L TCE were degraded before the tracer injection began in Column 1 and Column 2. The interpretation of BTCs using an analytical solution of the one-dimensional transport equation (Fig. 16-1) results in an effective porosity (n_e) of 0.4 for both columns if lithium is used. The retardation factor (R) of bromide in comparison to lithium was 1.05 and 1.43 in Column 1 and Column 2, and an application of bromide BTCs would only mimic a difference in effective porosity

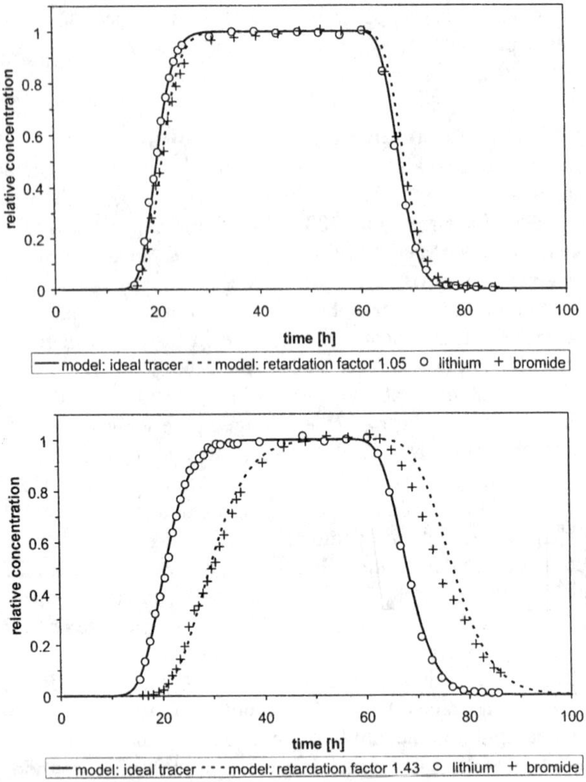

Fig. 16-1. BTC of lithium and bromide (3 mM) in parallel column experiments with Fe^0 and groundwater from a contaminated site. Bromide retardation compared to lithium was 1.05 in the experiment with 100 mg/L TCE (top) and 1.43 in the column with 300 mg/L (bottom) (modified from Ebert et al. 2002b; Ebert and Dahmke 2001).

by a factor of 1.36. Furthermore, too large or too small porosities would be calculated assuming conservative bromide transport (n_e 0.41 and 0.56) or a fixed retardation factor of 1.2 (n_e 0.35 and 0.47).

Another series of column experiments indicated a comparable variability in bromide retardation (for a more detailed description cf. Ebert et al. 2003a). Three parallel column tracer experiments were conducted using tritium at the beginning and then using a mixture of LiBr (3mM) and $NaBF_4$ (3 mM) in a background solution of 3 mM NaCl and 20 mg/L TCE. The BTCs of tritium and lithium showed a conservative transport of the small cation and indicated the same effective porosities (0.51 and 0.52), whereas bromide was retarded by a factor of 1.48 (Fig. 16-2). Chloride concentration increased and then decreased as bromide broke through the column. This chloride pre-peak suggests an anion exchange process as responsible for bromide retardation. A further tracer test using LiBr (3 mM), $NaBF_4$ (3 mM) and

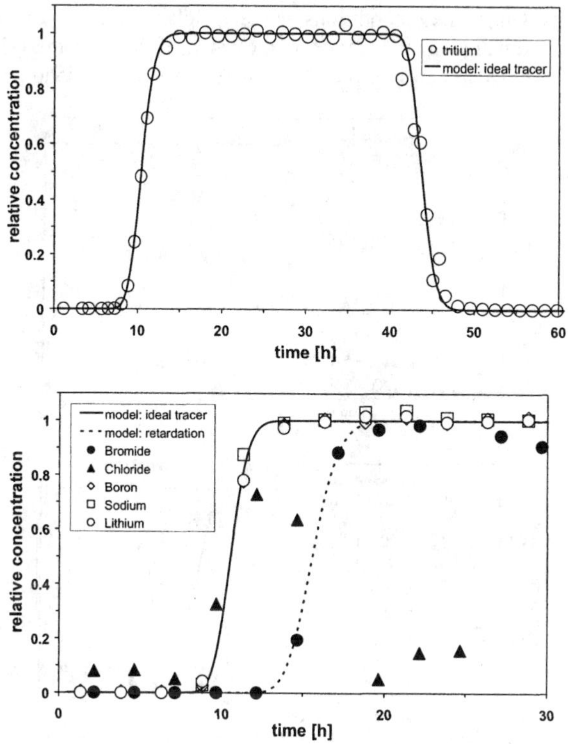

Fig. 16-2. Tritium, lithium, bromide, boron, and background-corrected sodium and chloride BTC in one of the three parallel column experiments. Analytical models use porosities of 0.51 and 0.52 and dispersivities of 0.22 for tritium and 0.1 cm and the other ions. The retardation factor for bromide is 1.48 (modified from Ebert et al. 2002b).

$NaNO_3$ (1.6 mM), and K_2CrO_4 (3 mM) was performed in the three columns after 250 exchanged pore volumes with various solutions (20 mg/L TCE in 3 mM NaCl, 3 mM chromate solution, and groundwater). The results of this tracer test show a diverse transport behaviour of the dissolved species depending on column history (Fig. 16-3). In the reference system (NaCl) bromide was retarded in comparison to lithium by a factor of 1.13, and, in the groundwater system, bromide and lithium broke through the column at the same time. Furthermore, a retardation of lithium was observed in the chromate column, while bromide, together with nitrate, broke through the column first. A one-dimensional transport model fits a significant loss in porosity of approximately 20% only for the chromate system. In this column, chromate was reduced and Cr(III) was precipitated, probably in forms of hydroxides, and Fe(III) hydroxides were precipitated (e.g. Blowes et al. 1997; Kjeldsen et al. 2001; Powell et al. 1995; Schlicker et al. 2000). During tracer experiments, pH in the chromate column was still increased, possibly due to small amounts of chromate reduction or buffering effects. Under these conditions, namely higher pH and hydroxide surfaces, typically cation exchange takes place. A comparison of the bromide BTCs from the three parallel columns and an interpretation using constant retardation factors would

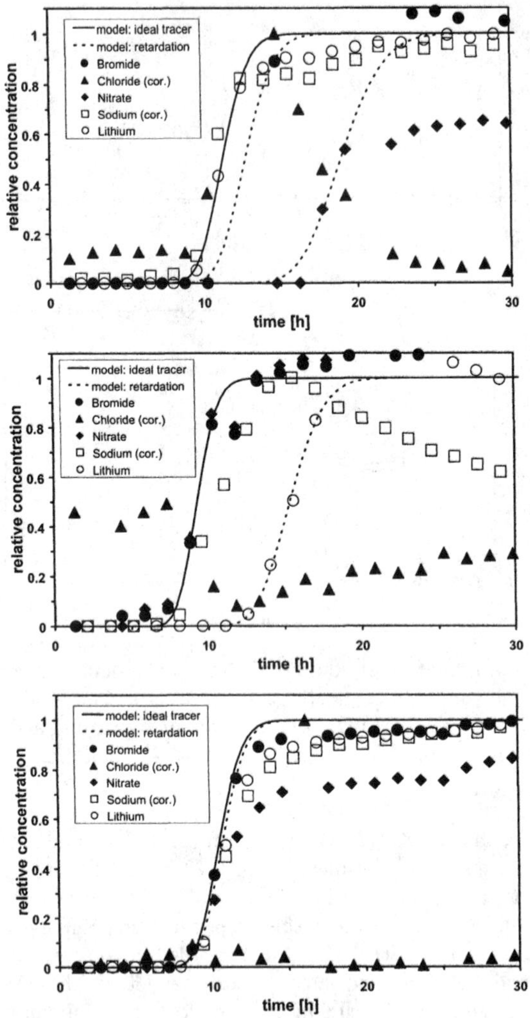

Fig. 16-3. BTCs of various tracers in the reference column (top), the chromate column (centre) and the groundwater column (bottom) after 250 exchanged pore volumes. Analytical solution of the one-dimensional transport equation is used for fitting the BTCs (modified from Ebert et al. 2002b). Ideal tracer BTCs indicate effective porosities of 0.51, 0.41, and 0.5 and an equal dispersivity of 0.2 cm in the reference, chromate, and groundwater columns.

delude a porosity reduction of 24% in the columns with NaCl background and of 32% in the columns with groundwater background. Even a decrease in effective porosity can not be ruled out for these columns due to the lack of an ideal tracer (e.g. tritium) in the second test series, this is not supported by the model fit of the BTCs of lithium or sodium, which indicate effective porosities of 0.51 and 0.5 (Fig. 16-3). Another result of these experiments was a qualitative correlation between the reactivity of iron and the retardation and abiotic reduction of nitrate (Ebert et al. 2003a). Similar correlations were also shown in other column experiments with contaminated groundwater and various types of Fe^0 granules (Parbs et al. 2003) or artificial solutions containing nitrate and TCE (Schlicker et al. 2000).

Thus far, it is not clear what causes variations in the transport behaviour of bromide in Fe^0-reactive media. These variations may be dependent on corrosion rates, surface properties of the established mineral assemblage, or simply on pH effects. In any case, bromide on its own does not seem to be an adequate tracer for the determination of hydrodynamic conditions in an iron PRB. Furthermore, in column experiments using only bromide as a tracer (e.g. Kamolpornwijit et al. 2003), an interpretation of BTCs in terms of changing porosity may lead to erroneous results. Lithium or sodium appear to be better suited as conservative tracers in operating barriers. However, it must be taken into account that high pH in effluent groundwater in field applications may lead to significant cation sorption effects with the aquifer matrix. For field applications the combination of weakly sorbing large anions and small cations was therefore suggested as an adequate tracer for monitoring hydraulics in the vicinity of a Fe^0 PRB (Ebert 2004; Parbs et al. 2005; Parbs et al. 2003).

16.3 Tracer Experiments at PRB Sites

At the NAS Moffett Field funnel-and-gate PRB (Gavaskar et al. 2002) (3 m in width, 3 m flow-through thickness with 0.6 m pea gravel zone up- and downgradient of a 1.8 m iron-reactive cell), tracer tests using sodium bromide were conducted to evaluate flow patterns within the reactive cell and within the upgradient capture zone as well (Gavaskar et al. 1998; Gupta et al. 1998). Bromide was injected into the upgradient pea gravel zone. The results showed that the tracers spread laterally within the pre-treatment zone before moving into the reactive zone, because the conductivity of the pea gravel in the pre-treatment zone was greater than that of the reactive media. At this site, the tracer tests demonstrated that flow moved in the expected downgradient direction. It also showed that flow through the reactive cell was highly heterogeneous, and residence times between 10 d and 30 d were calculated, which were well below those estimated from site characterization and modelling or from water-level measurements. However, despite very extensive monitoring, it was not possible to achieve an acceptable mass balance for the tracer, and the tracer was not detected in monitoring wells in the downgradient aquifer. Therefore, the presence of other pathways for flow were not able to be ruled out. Another tracer injection in a well in the upgradient aquifer was only of partial successes. The tracer appeared in the upgradient pea gravel zone but not in the iron-reactive cell.

At the Monticollo funnel-and-gate PRB (Liang et al. 2001; Morrison et al. 2002; Morrison et al. 2001; Muza et al. 2003) (gate width 31 m, thickness 3.5 m, height 4.1

m, pea gravel/iron-mixture zone upgradient and pea gravel zone downgradient) a multi-component tracer test was performed (Liang et al. 2001). Noble gases (helium, neon, argon) combined with bromide or iodide were injected. A simultaneous injection of various tracer mixtures directly upgradient of the barrier centre (neon, iodide and argon) and the lateral regions (bromide and helium) was chosen in order to assess variability in transport characteristics along the barrier and to examine the degree of mixing within and around the barrier. However, the gas tracers were not able to be detected in the monitoring wells, probably due to degassing phenomena during groundwater sampling. The results of the iodide and bromide BTCs suggest significant variability in transport velocities, transport directions, and residence times across the barrier. Furthermore, transport behaviour indicates significant heterogeneity within the barrier itself. Residence times varied between 22 h and 90 h and were highest in the central region of the barrier. The BTCs of iodide also indicate a vertical heterogeneity in the central barrier with faster flow velocities apparently in the shallower zone. Overall, transport rates and directions were found to be quite heterogeneous at this site and do not agree with estimations from hydraulic gradients or hydraulic modelling (20 h residence time) (Liang et al. 2001). At the Oak Ridge Y-12 drain-and-gate PRB (e.g. Liang et al. 2001; Phillips et al. 2000; Phillips et al. 2003; Watson et al. 1999; Watson et al. 2002) (gate width 8 m, thickness 0.6 m, height 9 m) tracer tests using bromide were performed 10 months, two years, and four years after construction (FRTR 2002). All tests indicated significant decreased flow in the upper and medium part in the upgradient interface region of the PRB. Coring and mineral phase characterizations suggest that extensive clogging is responsible for the flow patterns observed at that particular site (Gu et al. 2002; Phillips et al. 2000; Phillips et al. 2003). Forced iron corrosion and cementation with Fe(III) hydroxides is probably due to comparatively high groundwater flow velocity (2-3 m/d) and high nitrate background (~3 mM). 2.5 years after construction, 30% - 80 % of the Fe^0 was replaced by Fe(III) (hydr)oxides in the clogged zone (Phillips et al. 2000).

A tracer test completed at the Vapokon site (funnel-and-gate PRB with gate width 15 m, thickness 0.8 m, height 9 m, 1 m gravel packs up- and downgradient) using lithium as conservative tracer indicated a preferential flow in the reactive media, but verified the absence of bypassing flow (Lai et al. 2005b; Lo et al. 2003; Lai et al. 2005a). The natural gradient tracer test was performed over a time period of 10 months. The collection of more than 13,000 groundwater samples in multi-level screening monitoring wells made it possible to evaluate lithium transport in a sophisticated manner. Tracer injections took place in three wells directly upgradient of the gravel pack at a distance of approximately 3.5 m to the reactive cell. The results show a flow direction exactly along the centre line of the PRB, just downgradient of the injection wells. The progress of the lithium plume indicated pronounced lateral changes in the flow direction in the upgradient gravel pack before the tracer entered the iron-reactive cell. Afterwards, the tracer travelled preferentially across the PRB but showed a tendency toward a downwards direction. Overall, the tracers were injected in the shallower portion of the upgradient aquifer and emerged laterally shifted from the deeper portion of the aquifer on the downgradient side. This indicates heterogeneities in horizontal as well as vertical directions. Furthermore, first moment analysis of the lithium BTCs resulted in horizontal flow velocities of 99.5 m/y within the PRB, but only approximately 6.86 m/y upgradient in the immediate vicinity of the PRB. The formation of mineral phases probably induces preferential flow and is responsible for the decreased flow velocity in the upgradient zone suggested by the groundwater chemistry (Lai et al. 2005b). Further natural gradient tracer tests were conducted at other PRB sites with more or less success (e.g. Beck et al. 2001; Focht et al. 1997; Puls et al. 1996; Vogan et al. 1999; Zhang et al. 2001). The results, if achieved, usually indicated more complex flow patterns than those derived from water-level measurements and estimated (evenly distributed) hydraulic conductivities. Forced gradient tracer tests using conservative and reactive tracer were performed at the Rheine site PRB and the Tübingen Site PRB, with the objective of determining the reasons for the partial performance limitations observed at these sites.

16.3.1 Rheine Site PRB

The Rheine site PRB was constructed in 1998 (Ebert et al. 1999; Möller 1998) using two types of iron particles, i.e. an iron sponge (ISPAT, Hamburg, Germany, trade name ReSponge®) and a mixture of 30 vol.% grey cast iron (Gotthart Maier AG, Rheinfelden, Germany) and gravel (Fig. 16-4). The monitoring results indicate an

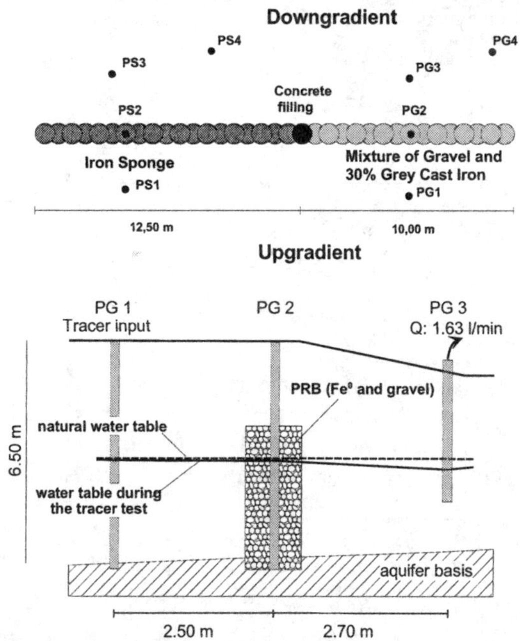

Fig. 16-4. Plan view (top) and side view (bottom) of the Rheine site PRB. Pumping in well PG 3 caused water table drawdown during the forced gradient tracer test, Δh was 1.4 cm between PG 1 and PG 2 and 25 cm between PG 2 and PG 3.

effective function of the iron sponge section for more than seven years (Fig. 16-5). Initially the downgradient contaminant concentrations were higher after installation than expected from laboratory experiments performed with iron sponge before the PRB was constructed. Tetrachloroethylene (PCE) concentrations along the iron sponge monitoring transect became coincident with the forecast from the column experiments after a period of several months (Ebert et al. 2003b). This lag in response time can be explained by desorption processes downgradient of the PRB. Currently, downgradient contaminant concentrations are below 10 µg/L. Furthermore, the overall degradation efficiency of the iron sponge was >99.5% and, thus far, decreasing reactivity has not been indicated. In contrast to the efficiently working iron sponge section, the section containing a mixture of gravel and grey cast iron showed a significantly decreasing reactivity six months after installation (Fig. 16-5). Initial degradation efficiency was >98% and decreased to ~80% one year after installation. Performance was stable, between 70% and 80%, during the past six years. Remediation efficiency is lower than expected from known material properties (Ebert 2004), even though no column experiments were performed to characterize the reactivity of this particular mixture. Normal performance monitoring was insufficient to identify probable reasons for the concentration patterns observed

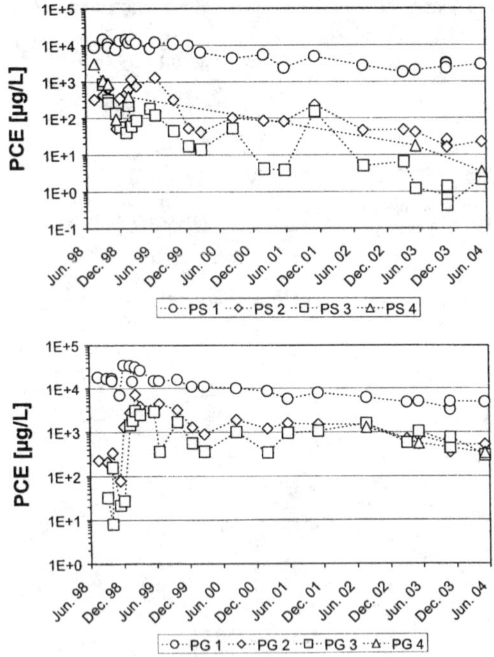

Fig. 16-5. Time series of PCE at the Rheine site PRB for the iron sponge (top) and the iron-gravel-mixture (bottom) transect. (Other chlorinated hydrocarbons are of minor importance and degradation efficiency is comparable to PCE).

(Ebert et al. 2003b). At this particular site the question arose as to whether or not downstream contaminant concentrations are caused by flow bypasses, fast passivation or by construction problems, namely, a segregation of iron particles and gravel and the formation of layers with only small amounts of iron.

A forced gradient tracer test using nitrate as a reactive tracer was conducted along the monitoring transect of interest (PG 1, PG 2, PG 3) (Parbs et al. 2005; Parbs et al. 2003) (Fig. 16-4). One kg of technical grade $NaNO_3$ (11.8 mol) and LiBr (11.5 mol) in 0.28 m^3 groundwater was injected in the upgradient monitoring well at a rate of 1.5 L/min. A recirculation pump in the injection well ensured a uniform tracer distribution over the aquifer thickness for five hours. The mean pump rate in well PG 3 was 1.6 L/min (± 0.6) over a time span of 28 d (starting 24 h before tracer injection), and sample collection took place in PG 2 at three depth levels and in the pumped water. The forced gradient technique was used because the test is then less time-consuming, and the flow system is able to be controlled more carefully. Flow velocity is faster and the tracer broke through the transect earlier than it would have under natural gradient conditions due to the higher gradient. Drawdown at the downstream monitoring well (=pumping well) increases the chance of capturing the tracer peak because of flow directions control. Furthermore, the forced gradient tracer test is, at the same time, a pumping test. Fig. 16-6 shows the BTCs of the tracers in the PRB monitoring well (in 5 m depth) as well as in the pumping well. The results indicate a similar breakthrough time for anions and cations within the PRB, but point out a retardation of the cations compared to the anions in the downgradient aquifer. Here, pH is higher than in the upgradient aquifer and suggests cation exchange conditions (Ebert et al. 2003b). An interpretation of the bromide and lithium BTCs using simple boundary conditions indicates a residence time of approximately one day within the PRB up to the monitoring point. The molar input of bromide and nitrate was nearly equal and, therefore, the difference between bromide and nitrate breakthrough shows the loss of nitrate. Nitrate reduction by Fe^0 was primarily abiotic, as indicated by increasing ammonium concentration (e.g. Cheng et al. 1997; Rahman and Agrawal 1997; Schlicker et al. 2000; Till et al. 1998; Westerhoff and James 2003). The difference between bromide and nitrate was ~25% within the PRB monitoring well (PG 2), and only ~50% nitrate was found as compared to bromide in the downgradient well (PG 3). The similar nitrate reduction in both sections of the PRB points to no significantly higher reactivity in the rear part than in the front part, assuming a homogeneous iron content in the PRB sections located in front of the central monitoring well (PG 2) and behind it. However, a reduction of approximately 0.5 mM nitrate during one day of residence time within the reactive zone is comparatively small. These findings imply that a lower total reactivity is responsible for the higher contaminant concentration downgradient of the mixture section of the Rheine PRB.

From these results we can not distinguish between an issue of total passivation along single flow pathways and a slight homogeneous passivation of all iron surfaces. However, other investigations, including solid phase characterization (Stender 2002) and column experiments using material from coring (Ebert et al. 2002a), showed only minor passivation of the grey cast iron itself. Therefore, an inhomogeneous mixture of cast iron and gravel caused by construction technique probably combined

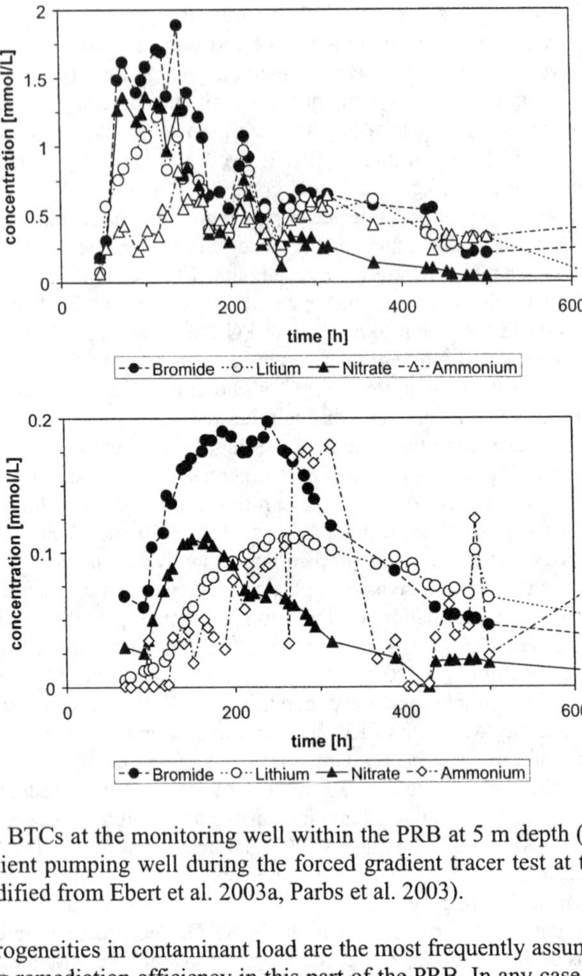

Fig. 16-6. BTCs at the monitoring well within the PRB at 5 m depth (top) and at the downgradient pumping well during the forced gradient tracer test at the Rheine site PRB (modified from Ebert et al. 2003a, Parbs et al. 2003).

with heterogeneities in contaminant load are the most frequently assumed reasons for decreasing remediation efficiency in this part of the PRB. In any case, a flow bypass of the PRB can be ruled out.

16.3.2 Tübingen Site PRB

The funnel-and-gate PRB at the Tübingen site (installed 1998) contains three gates filled with *Gotthart Maier* iron (e.g. Klein and Schad 2000; Schad and Klein 2002; Schad and Teutsch 1998) and a funnel approximately 210 m long (Fig. 16-7). Each gate contains two monitoring transects with wells in the up- and downgradient gravel pack and one or two within the iron-reactive cell. In addition, monitoring wells and compliance wells are installed in the vicinity of the gates. Monitoring results show the proper functioning of the three gates during the first 6 to 12 months after construction (Fig. 16-8). Later, contaminant concentrations measured along the

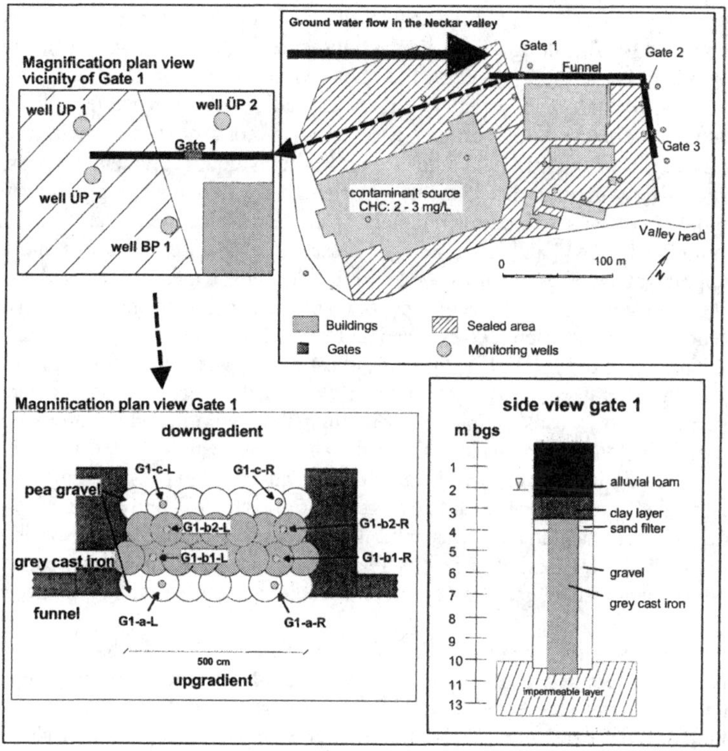

Fig. 16-7. Plan view of the Tübingen site PRB, of gate 1 and side view of gate 1

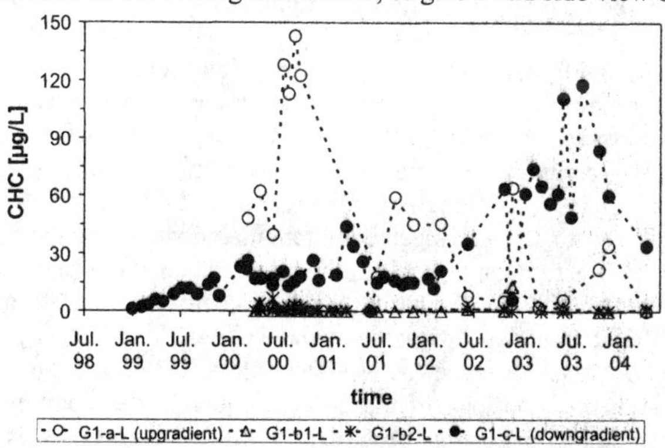

Fig. 16-8. Time series of chlorinated hydrocarbons (CHC, main contaminant cis-DCE) along the left monitoring transect at gate 1 of Tübingen site PRB. Up- and downgradient monitoring wells (G1-a-L and G1-c-L) are located in the gravel zone in ~35 cm distance to the iron-reactive cell.

monitoring transects point to some issues on PRB performance. A typical concentration profile along a compliance transect shows contaminants at the up- and downgradient monitoring points but not within the iron-reactive cell. The composition of the groundwater ~0.5 m downgradient of the reactive cell does not indicate any contact to iron, neither in contaminant concentrations nor in inorganic groundwater constituents. Although a significant hydraulic gradient (approx. 3%) indicates a flow through the gates, the geochemical patterns of the groundwater are in fact nearly equal up- and downgradient of the reactive cell (e.g. Parbs et al. 2003; Purkhold 2001; Schad and Klein 2003). Furthermore, an assessment of bulk flow in the upgradient aquifer is not possible from water-level measurements due to the configuration of the monitoring network in combination with small hydraulic gradients (Schad and Klein 2002; Schad and Klein 2003). Water-level measurements downgradient of the PRB also give no indications of water flow through the gates because the laterally-directed groundwater flow in the Neckar valley is dominant. Furthermore, low background concentrations of chlorinated hydrocarbons is "normal" in this groundwater. Overall, an identification of reasons for the performance limitations observed, i.e. clogging and flow bypass, mixing of treated and untreated groundwater in the downgradient gravel pack, preferential flow along pathways where degradation was inhibited, was not possible on the basis of the standard monitoring results. For this reason, tracer tests were conducted, i.e. two forced gradient tests at gate 1 and a natural gradient tracer test in the vicinity of gate 1.

For the first forced gradient tracer test in November 2002 (Parbs et al. 2003) the hydraulic gradient between wells G1-a-L and G1-c-L was increased by pumping 6.7 L/min (± 0.7) over a time period of 20 days (here, water must be treated by activated carbon). Tracer injection in G1-a-L began after 24 h pumping using 1 kg $NaNO_3$ and LiBr in 0.5 m^3 groundwater at a rate of 1.6 L/min. A recirculation pump in the injection well ensured uniform tracer distribution over the aquifer thickness for six hours. Sample collection in the reactive cell (G1-b1-L and G1-b2-L) at three depth levels (i.e. 5, 7, and 9 m b.s.) using a multi-level low-flow sampling system (Ptak et al. 2000) began three hours after injection and on the following day in the pumping well. The results indicate the existence of at least two flow regions. One is a preferential flow pathway with only minor reactivity and the other is through the reactive iron (Fig. 16-9). Although only the rear part of the BTCs was captured in the pumping well due to the unexpected preferential flow, a comparison of bromide BTCs shows an earlier breakthrough in the downgradient well than that in the wells within the reactive cell. Furthermore, the loss in nitrate suggests only minor Fe^0 reactivity at the preferential flow pathway but a continuous reactivity along the other flow region. A mass balance between the loss of nitrate and recovering of ammonium is difficult because of the later appearance of ammonium or lithium as compared to bromide, indicating a retardation of the cations (Figs. 16-9 and 16-10). Moreover, a partial biotic reduction of nitrate to N_2 (Till et al. 1998) can not be ruled out. In addition to the obvious preferential flow regions, the bromide BTC's in the reactive cell indicate a higher flow velocity in the upper part of the gate (Fig. 16-10). In addition, the retardation of lithium as compared to bromide was variable and less pronounced in the middle part of the reactive cell. This indicates heterogeneities in hydraulics and in geochemical conditions as well within the iron-reactive cell. However, based on these results, preferential flow along less reactive pathways was favoured to cause the performance limitations observed (Diels et al. 2003; Parbs et al. 2005; Parbs et al. 2003).

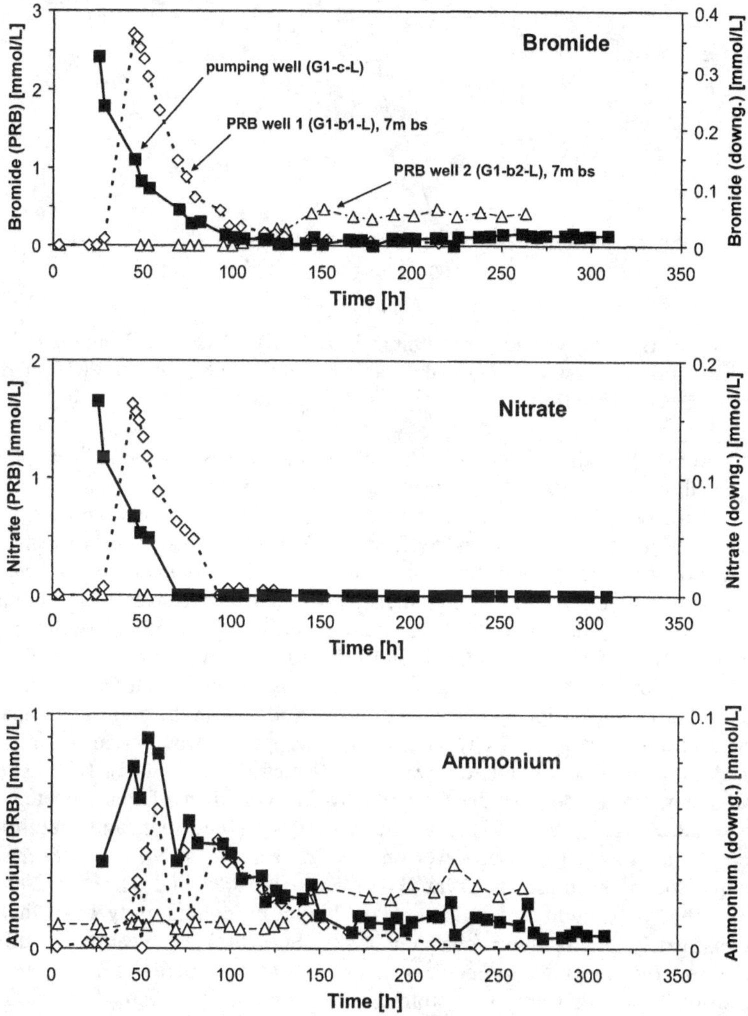

Fig. 16-9. BTC of bromide, nitrate and ammonium in the pumping well (G1-c-L: -■-) and in monitoring wells within the reactive cell at depth level of 7 m b.s. (G1-b1-L: -◊-; G1-b2-L: -Δ-) during the first forced gradient tracer test at the Tübingen site PRB in November 2002.

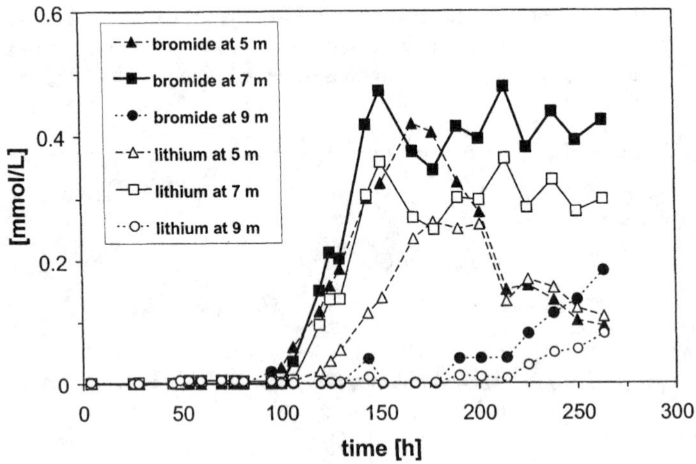

Fig. 16-10. BTC of bromide and lithium in well G1-b2-L in 5, 7, and 9 m below surface during the first forced gradient tracer test at the Tübingen site PRB (modified from Parbs et al. 2003).

A further forced gradient tracer test at the gate 1 was conducted in September 2003 with the objective of localizing the preferential pathways. Therefore the set-up was modified using ten peristaltic pumps to create the drawdown and continuous water sampling by an auto sampler (Ebert 2004). The total pump rate was similar to the first tracer test and the injection procedure, but the recirculation pump ensured a tracer distribution over the aquifer thickness for 24 hours. Overall, the boundary conditions were quite similar for both tests with the exception of an ~1m lower level in natural water table during the second test and extensive sampling begun before tracer injection, in other monitoring wells as well as within the gate. Surprisingly, the tracers did not appear in any of the monitoring wells nor in the multi-level-pumping well over a time span of 12 days but, rather, disappeared from the injection well in two days. Another tracer mixture was then injected (NaCl, $NaNO_3$, LiBr) after the total pump rate was doubled, resulting in a hydraulic gradient of approximately 13%. Nevertheless, only poor BTC's of the tracers were detected, and a significant preferential flow path, as observed in the first tracer test, was not observed. A comparison of bromide BTCs in G1-b1-L as detected in November 2002 and September 2003 indicates a decrease in hydraulic conductivity by a factor of approximately 30, assuming constant porosity (Fig. 16-11). The causes the different results of the tracer tests remain unclear. A few days before the first test was performed, samples were taken from the PRB by coring. A forced release of gas bubbles due to vibrations during coring (Schad 2004) as well as fracturing caused by coring may have been responsible for increased hydraulic conductivity and preferential flow during the first tracer test. However, tracer tests indicated a variable and heterogeneous system, and the results (primarily of the second test) suggest only a minor flow through the gates. This is also supported by the results of a natural gradient test performed temporally between the others. Uranin, an alkaline salt of fluorescein, was injected in a well upgradient of the gate (BP 1) and appeared faster and in higher concentrations in a well lateral of the funnel (ÜP 7) than it did in the monitoring wells directly upgrade of the gate, i.e. G1-a-L and G1-a-R (Fig. 16-12). Furthermore, the appearance of the dye tracer in the monitoring wells outside the funnel-and-gate system (ÜP 1 and ÜP 2) was also faster and the tracer was detected in the downgradient gravel pack (G1-c-L and G1-c-R). This indicates that groundwater without residence in an iron-reactive cell was sampled in downgradient monitoring wells because of the strong reaction between uranin and Fe^0 (Eykholt et

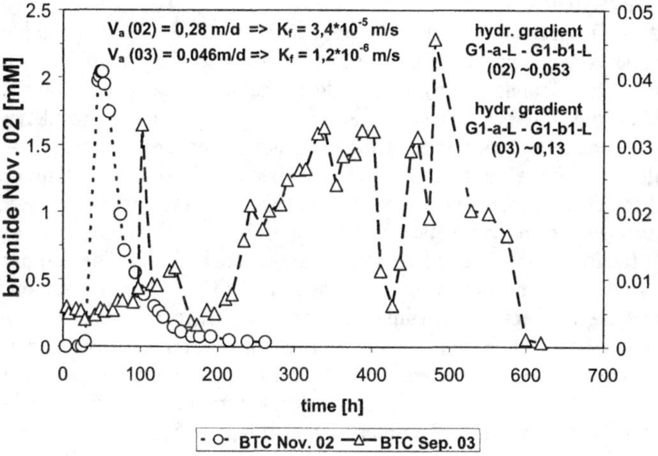

Fig. 16-11. BTCs of bromide in well G1-b1-L in 7 m depth level

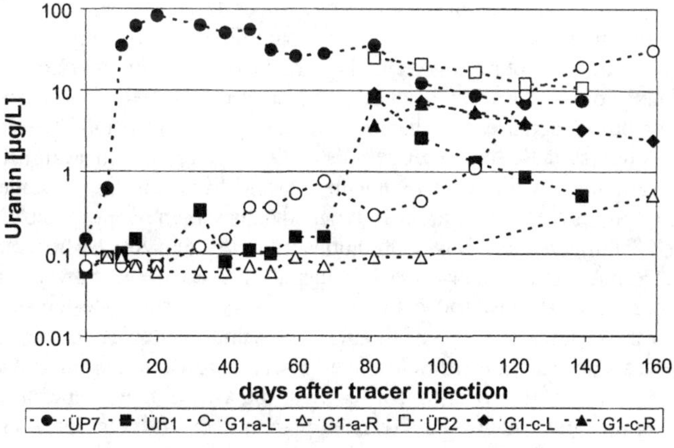

Fig. 16-12. BTCs of uranin during natural tracer test at the Tübingen site PRB (location of the monitoring wells in Fig. 16-7)

al. 1995; Liang et al. 2001). Only a flow bypass of the funnel-and-gate system may sufficiently explain these results of the natural gradient tracer test. Overall, the three tracer tests have shown hydraulic performance limitations and have uncovered a flow bypass of the barrier, which was not able to be recognized from the other monitoring results. While these interpretations are sufficient to demand and to initiate countermeasures (i.e. a further gate using other iron granules is planned), further investigations must be addressed to fully understand the confusing results of the forced gradient tracer test.

16.4 Conclusions

The overall experience from tracer tests conducted at PRB sites is that complex flow patterns can be expected in each PRB and that, in general, other monitoring strategies are not appropriate for uncovering variability in the actual local flow. Furthermore, tracer tests are the only established method for the verification of residence times within an iron-reactive barrier. One of the main issues concerning tracer tests in Fe^0 PRBs is the absence of a tracer which has been proved to be conservatively transported under all geochemical conditions. The variability of residence times in PRBs determined from tracer test results may be partly due to the potential variability of anion and cation retardation in iron-reactive media. The comparison of observed variations of residence times in PRBs and of bromide retardation factors indicates that these effects are of minor importance, because observed retardations varied only by a factor of approximately 1.5. A consideration of an unsteady bromide retardation may be more important when effective pore volumes are determined, particularly in column experiments, because porosity reduction and the variation of bromide retardation is in the same order of magnitude. While tritium tracers are an option in column experiments, a field application of tritium may be difficult at most sites due to its radioactivity. Therefore salt tracers or preferred mixtures of salt tracers are, thus far, the best option at field sites.

It is trivial, but the quality of results from tracer tests depends strongly on monitoring efforts and network configuration, i.e. sampling density, analytical suite, screening intervals of the monitoring wells. Nevertheless, extensive monitoring will not guarantee the capture of tracers because unexpected flow patterns can exist. The forced gradient method increases the chance of capturing tracers but modifies natural bulk flow and is less suitable for investigations of the capture zone or regional flow patterns. In combination with reactive tracers, this method has the potential for uncovering reactivity issues within the reactive media, as shown by tracer tests at the Rheine site and Tübingen site PRBs. Continuous pumping reduces the time needed for tracer tests, but other resource requirements increase, e.g. power supply, analytical suite, reserve pumps, and manpower. An attempt to reduce the resource requirements of tracer tests is the usage of automatic sampling devices or down-hole in-situ ion selective probes, but then additional power supply is required and also a weatherproof installation. However, tracer tests will be always more expensive than other monitoring methods, require an advanced planning, extensive monitoring, and high number of samples. Therefore, tracer tests will not be a standard option for Fe^0 PRB performance monitoring at most sites, but in some cases tracer tests will be the only option for determining the reasons for performance limitations and for designing countermeasures. Furthermore, if a detailed examination of flow patterns, residence times, or capture zone is required, there is no other method available.

Acknowledgments

Research at the Rheine site and Tübingen site was funded by the German Federal Ministry of Education and Science (02WR0208) as a part of the RUBIN research network. Investigations at the Tübingen site were performed in cooperation with IMES GmbH, Amtzell, Germany, and at the Rheine site in cooperation with M&P GmbH, Hanover, Germany. Monitoring results were partly produced by the consultants.

16.5 References

Agrawal, A., Ferguson, W. J., Gardner, B. O., Christ, J. A., Bandstra, J. Z., and Tratnyek, P. G. (2002). "Effects of carbonate species on the kinetics of dechlorination of 1,1,1-Trichloroethane by zero-valent iron." *Environ. Sci. & Technol.*, 36(20), 4326-4333.

Agrawal, A., and Tratnyek, P. G. (1996). "Reduction of nitroaromatic compounds by zero-valent iron metal." *Environ. Sci. & Technol.*, 30(1), 153-160.

Beck, P., Harries, N., and Sweeney, R. (2001). *Design, installation and performance assessment of a zero valent iron permeable reactive barrier in Monkstown, Northern Ireland*, CL:AIRE: Contaminated Land: Applications in Real Environments, London.

Blowes, D. W., Ptacek, C. J., and Jambor, J. L. (1997). "In-situ remediation of chromate contaminated groundwater using permeable reactive walls: Laboratory studies." *Environ. Sci. & Technol.*, 31(12), 3348-3357.

Blowes, D. W., Puls, R. W., Gillham, R. W., Ptacek, C. J., Bennett, T. A., Bain, J. G., Hanton-Fong, C. J., and Paul, C. J. (1999). *An in situ permeable reactive barrier for the treatment of hexavalent chromium and trichloroethylene in ground water: Volume 2, performance monitoring*, EPA/600/R-99/095b, Office of Research and Development, United States Environmental Protection Agency, Washington DC, United States.

Cheng, I. F., Muftikian, R., Fernando, Q., and Korte, N. (1997). "Reduction of nitrate to ammonia by zero-valent iron." *Chemosphere*, 35(11), 2689-2695.

Davis, B. M., Istok, J. D., and Semprini, L. (2002). "Push-pull partitioning tracer tests using radon-222 quantify non-aqueous phase liquid contamination." *J. Cont. Hydrol.*, 58(1), 129-146.

Dethlefsen, F., Bliss, F., Wachter, T., and Dahmke, A. (2004). "Oxidadtionskapazitätsbestimmung durch reaktive Tracer." *Grundwasser*, 9(1), 12-20.

Diels, L., Bastiaens, L., O'Hannesin, S. F., Cortiona, J. L., Alvarez, P. J., Ebert, M., and Schad, H. (2003). "Permeable reactive barriers: A multidisciplinary approach of a new emerging sustainable groundwater treatment technology." In *Proceedings of the Consoil 2003: 8th International FZK/TNO Conference on Contaminated Soil* (CD-Rom), Gent, Belgium, 3607-3618.

Ebert, M. (2004). *Elementares Eisen in permeablen reaktiven Barrieren zur in-situ Grundwassersanierung - Kenntnisstand nach zehn Jahren Technologieentwicklung*, Habilitation, Christian-Albrechts-Universität zu Kiel, Kiel, Germany.

Ebert, M., and Dahmke, A. (2001). *Voruntersuchung zur Materialeignung und Dimensionierung einer Fe^0-basierenden in-situ-Sanierungsmaßnahme am Standort "Bernau"*, Christian-Albrechts-Universität zu Kiel, Kiel, Germany.

Ebert, M., Dahmke, A., Köber, R., Parbs, A., and Schäfer, D. (2003a). "Ionic tracers for the reactivity assessment of Fe(0) PRBs." In *Proceedings of the ConSoil 2003: 8th International FZK/TNO Conference on Contaminated Soil* (CD-Rom), Gent, Belgium, 1582-1590.

Ebert, M., Dahmke, A., Köber, R., Silva-Send, N., and Parbs, A. (2002a). *Bewertung und Weiterentwicklung von Voruntersuchungs- Monitoring- und Qualitätsmanagementansätze für Reaktionswände - eine vergleichende Labor- und Standortstudie unter besonderer Berücksichtigung vollständiger, stoffspezifische Bilanzen des Schadstoffumsatzes beim Einsatz reaktiver Wände*, Christian-Albrechts-Universität zu Kiel, Kiel, Germany.

Ebert, M., Dahmke, A., and Silva-Send, N. (2002b). "Inhibiting substances as tracers for the reactivity assessment of Fe(0) PRBs." In *Proceedings of the 3rd International Conference on Remediation of Chlorinated and Recalcitrant Compounds, Monterey*, California, May 20-23, *2002* (CD-Rom), Battelle Press, Columbus, Ohio, 2A-10.

Ebert, M., Möller, W., and Wegner, M. (1999). "F+E-Vorhaben Reaktive Wand in Rheine -aktuelle Ergebnisse-." *altlasten spektrum*, 2(99), 109-112.

Ebert, M., Wegner, M., Parbs, A., Plagentz, V., Schäfer, D., Köber, R., and Dahmke, A. (2003b). "Prognostizierte und tatsächliche Langzeitstabilität von Fe(0)-Reaktionswänden - Am Beispiel der Reaktionswand am Standort Rheine nach 5-jähriger Betriebszeit." *Grundwasser*, 3, 157-168.

Eykholt, G. R., Baghel, S. S., Sivavec, T. M., Mackenzie, P. D., Haitko, D., and Horney, D. (1995). "Conservative flow tracers for iron column studies." In *Proceedings of the 209th American Chemical Society National Meeting*, Anaheim, CA, 818-821.

Farrell, J., Kason, M., Melitas, N., and Li, T. (2000). "Investigation of the long-term performance of zero-valent iron for reductive dechlorination of trichloroethylene." *Environ. Sci. & Technol.*, 34(3), 514-521.

Focht, R. M., Vogan, J., and O'Hannesin, S. F. (1997). "Hydraulic studies of in-situ permeable reactive barriers." In *Proceedings of the 2nd International Containment Technology Conference*, St. Petersburg, FL., 975-981.

FRTR (2002). *Evaluation of permeable reactive barrier performance*, EPA/542-R/04/004, Prepared for the Federal Remediation Technologies Roundtable (FRTR) by the Tri-Agency Permeable Reactive Barrier Initiative, National Service Centre for Environmental Publications, Cincinnati, Ohio.

Gavaskar, A., Gupta, N., Sass, B., Janosy, R., and Hicks, J. (2000a). *Design guidance for application of permeable reactive barriers for groundwater remediation*, Battelle, Columbus, Ohio.

Gavaskar, A., Gupta, N., Sass, B., Yoon, W.-S., Janosy, R., Drescher, E., and Hicks, J. (2000b). *Design, construction, and monitoring of the permeable reactive barrier in Area 5 at Dover Air Force Base*, Battelle, Columbus, Ohio.

Gavaskar, A., Sass, B., Gupta, N., Drescher, E., Yoon, W.-S., Sminchak, J., Hicks, J., and Condit, W. (2002). *Evaluating the longevity and hydraulic performance of permeable reactive barriers at department of defence sites*, Battelle, Columbus, Ohio.

Gavaskar, A., Sass, B., Gupta, N., Hicks, J., Yoon, S., Fox, T., and Sminchak, J. (1998). *Performance evaluation of a pilot-scale permeable reactive barrier at Former Naval Air Station Moffett Field, Mountain View, California*, Battelle, Columbus, Ohio.

Gillham, R. W., Ritter, K., Zhang, Y., and Odziemkowski, M. S. (2001). "Factors in the long-term performance of granular Iron PRBs." In *Proceedings of 2001 International Containment & Remediation Technology Conference and Exhibition* (CD-Rom), Orlando, Florida, Paper-ID 514.

Gu, B., Phelps, T. J., Liang, L., Dickey, M. J., Roh, Y., Kinsall, B. L., Palumbo, A. V., and Jacobs, G. K. (1999). "Biochemical dynamics in zero-valent iron columns: implications for permeable reactive barriers." *Environ. Sci. & Technol.*, 33(13), 2170-2177.

Gu, B., Watson, D. B., Phillips, D. H., and Liang, L. (2002). "Biogeochemical, mineralogical, and hydrological characteristics of an iron reactive barrier used for treatment of uranium and nitrate." In *Handbook of groundwater remediation using permeable reactive barriers: Applications to radionuclides, trace metals, and nutrients*, Academic Press, New York, 306-343.

Gupta, N., Sass, B. M., Gavaskar, A. R., Sminchak, J. R., Fox, T. C., Snyder, F. A., O'Dwyer, D., and Reeter, C. (1998). "Hydraulic evaluation of a permeable barrier using tracer tests, velocity measurements, and modeling." In *Designing and applying treatment technologies: Remediation of chlorinated and recalcitrant compounds (C1-6)*, Battelle Press, Columbus, Ohio, 157-162.

Hein, P., Freygang, M., and Vigelan, L. (2002). "Horizontaler Eisenganulatreaktor zur in-situ-Abreinigung von LHKW-kontaminiertem GW in Bernau bei Berlin." *altlasten spektrum*, 2, 89-90.

Imhoff, P. T., Pirestani, K., Jafarpour, Y., and Spivey, K. M. (2003). "Tracer interaction effects during partitioning tracer tests for NAPL detection." *Environ. Sci. & Technol.*, 37(7), 1441-1447.

Istok, J. D., Field, J. A., Schroth, M. H., Davis, B. M., and Dwarakanath, V. (2002). "Single-well "push-pull" partitioning tracer test for NAPL detection in the subsurface." *Environ. Sci. & Technol.*, 36(12), 2708-2716.

ITRC (2005). *Permeable reactive barriers: Lessons learned/new directions. PRB-4*, Interstate Technology & Regulatory Council, Permeable Reactive Barriers Team, Washington, D.C.

Juhasz, A. (2001). *Field and numerical experiments to determine the hydraulic properties of gates applied at the funnel and gate system at the Beka site in Tübingen*, Master Thesis, Universtität Tübingen, Tübingen, Germany.

Kamolpornwijit, W., Liang, L., West, O. R., Moline, G. R., and Sullivan, A. B. (2003). "Preferential flow path development and its influence on long-term PRB performance column study." *J. Cont. Hydrol.*, 66, 161-178.

Kjeldsen, P., Locht, T., and Karvonen, A. P. (2001). "Removal of TCE and chromate in a permeable reactive barrier using zero-valent iron." In *Proceedings of 2001 International Containment & Remediation Technology Conference and Exhibition* (CD-Rom), Orlando, Florida, Paper-ID 121.

Klausen, J., and Tröber, S. P. (1995). "Reduction of substituted nitrobenzenes by Fe (II) in aqueous mineral suspensions." *Environ. Sci. & Technol.*, 29(9), 2396-2404.

Klausen, J., Vikesland, P. J., Kohn, T., Burris, D. R., Ball, W. P., and Roberts, A. L. (2003). "Longevity of granular iron in groundwater treatment processes: Solution composition effects on reduction of organohalides and nitroaromatic compounds." *Environ. Sci. & Technol.*, 37(6), 1208-1218.

Klein, R., and Schad, H. (2000). "Results from a full scale funnel-and-gate system at the BEKA site in Tübingen (Germany) using zero-valent iron." In *Proceedings of ConSoil 2000*, Leipzig, Germany, 917-923.

Laase, A. D., Korte, N. E., Baker, J. L., Dieckmann, P., Vogan, J. L., and Focht, R. L. (2000). "Evaluation of the Kansas City Plant iron wall." In *Chemical oxidation and reactive barriers: Remediation of chlorinated and recalcitrant compounds (C2-6)*, Battelle Press, Columbus, Ohio, 417-424.

Lai, K. C.-K., Lo, I. M.-C., and Kjeldsen, P. (2005a). "Remediation of chlorinated aliphatic hydrocarbons in groundwater using Fe^0 PRB." In *Permeable reactive barriers*, IAHS Publication 298, IAHS Press, Wallingford, UK, 12-22.

Lai, K. C.-K., Lo, I. M.-C., and Kjeldsen, P. (2005b). "Investigation of flow pattern of groundwater through a Fe^0 PRB." In *Remediation of chlorinated and recalcitrant compounds - 2004* (CD-Rom), Battelle Press, Columbus, Ohio, 3A-11.

Li, L., Benson, C. H., and Mergener, E. A. (2005). "Impact of mineral fouling on the hydraulic behavior of continuous-wall permeable reactive barriers." *Ground Water*, 43(4), 582-592.

Liang, L., Korte, N. E., Moline, G. R., and West, O. R. (2001). *Long-term monitoring of permeable reactive barriers - progress report,* Oak Ridge National Laboratory, Oak Ridge, Tennessee.

Lo, I. M.-C., Lai, K. C.-K., and Kjeldsen, P. (2003). "Field monitoring of the performance of the permeable reactive barrier at Vapokon site, Denmark." In *RTDF permeable reactive barriers (PRBs) action team meeting*, Niagra Falls, New York, 95-112, <http://www.rtdf.org/public/permbarr/minutes/101603/index.htm> (Aug., 17, 2005).

Mackenzie, P. D., Baghel, S. S., Eykholt, G. R., and Horney, D. P. (1995). "Pilot-scale demonstration of chlorinated ethene reduction by iron metal: Factors affecting iron lifetime." *I&EC Special Symposium*, Atlanta, GA.

Mackenzie, P. D., Horney, D. P., and Sivavec, T. M. (1999). "Mineral precipitation and porosity losses in granular iron columns." *J. Hazard. Mater.*, 68, 1-17.

Matheson, L. J., and Tratnyek, P. G. (1994). "Reductive dehalogenation of chlorinated methanes by iron metal." *Environ. Sci. & Technol.*, 28, 2045-2053.

Mayer, K. U., Blowes, D. W., and Frind, E. O. (2001). "Reactive transport modeling of an in situ reactive barrier for the treatment of hexavalent chromium and trichloroethylene in groundwater." *Wat. Res. Res.*, 37(12), 3091-3108.

Möller, W. (1998). "Bau einer "Reaktiven Wand" als F+E-Vorhaben in Rheine." *TerraTech*, 4.

Morrison, S. J. (2003). "Performance evaluation of a permeable reactive barrier using reaction products as tracers." *Environ. Sci. & Technol.*, 37(10), 2302-2309.

Morrison, S. J., Carpenter, C. E., Metzler, D. R., Bartlett, T. R., and Morris, S. A. (2002). "Design and performance of a permeable reactive barrier for containment of uranium, arsenic, selenium, vanadium, molybdenum, and nitrate at Monticello, Utah." In *Handbook of groundwater remediation using permeable reactive barriers: Applications to radionuclides, trace metals, andnNutrients*, Academic Press, New York, 372-401.

Morrison, S. J., Metzler, D. R., and Carpenter, C. E. (2001). "Uranium precipitation in a permeable reactive barrier by progressive irreversible dissolution of zerovalent iron." *Environ. Sci. & Technol.*, 35(2), 385-390.

Moylan, J. (2003). "Evaluation of the permeable reactive wall at the Northeast Corner Operable Unit, Lake City Army Ammunition Plant (LCCAP), Missouri." In *RTDF permeable reactive barriers (PRBs) action team meeting*, Niagara Falls, New York, 25-41, <http://www.rtdf.org/public/permbarr/minutes/101603/index.htm> (Aug., 17, 2005).

Muza, R., Mushovic, P., and Morrison, S. J. (2003). "Evaluating performance of the Monticello PRB treating uranium and metals." *Technology News and Trends*, (7), 1-2.

Naftz, D. L., Fuller, C. C., Davis, J. A., Morrison, S. J., Feltcorn, E. M., Rowland, R. C., Freethey, G. W., Wilkowske, C., and Piana, M. (2002). "Field demonstration of three permeable reactive barriers to control uranium contamination in groundwater, Fry Canyon, Utah." In *Handbook of groundwater remediation using permeable reactive barriers: Applications to radionuclides, trace metals, and nutrients*, Academic Press, New York, 402-435.

Odziemkowski, M. S., and Gillham, R. W. (1997). "Surface redox reactions on commercial grade granular iron (steel) and their influence on the reductive dechlorination of solvent. Micro raman spectroscopic studies." In *Proceedings of the 213th American Chemical Society National Meeting*, San Francisco, Ca., 177-180.

Odziemkowski, M. S., Schuhmacher, T. T., Gillham, R. W., and Reardon, E. J. (1998). "Mechanism of oxide film formation on iron in simulating groundwater solutions: Raman spectroscopic studies." *Corros. Sci.*, 40(2/3), 371-389.

Parbs, A., Ebert, M., Köber, R., Plagentz, V., and Dahmke, A. (2005). "Field application of reactive tracers for performance assessment of Fe^0 PRBs." In *Remediation of chlorinated and recalcitrant compounds - 2004* (CD-Rom), Battelle Press, Columbus, Ohio, 3A-31.

Parbs, A., Ebert, M., Köber, R., Plagentz, V., Schad, H., and Dahmke, A. (2003). "Einsatz reaktiver Tracer zur Bewertung der Langzeitstabilität und Reaktivität von Fe(0)-Reaktionswänden." *Grundwasser*, 3, 146-156.

Phillips, D. H., Gu, B., Watson, D. B., Roh, Y., Liang, L., and Lee, S. Y. (2000). "Performance evaluation of a zerovalent iron reactive barrier: Mineralogical characteristics." *Environ. Sci. & Technol.*, 34(19), 4169-4176.

Phillips, D. H., Watson, D. B., Roh, Y., and Gu, B. (2003). "Mineralogical characteristics and transformations during long-term operation of a zerovalent iron reactive barrier." *J. Environ. Qual.*, 32, 2033-2045.

Powell, R. M., Puls, R. W., Blowes, D. W., Vogan, J. L., Gillham, R. W., Powell, P. D., Schultz, D., Landis, R., and Sivavec, T. (1998). *Permeable reactive barrier technology for contaminant remediation*, EPA/600/R-98/125, Office of Research and Development, United States Environmental Protection Agency, Washington, DC, United States.

Powell, R. M., Puls, R. W., Hightower, S. K., and Sabatini, D. A. (1995). "Coupled iron corrosion and chromate reduction: Mechanisms for subsurface remediation." *Environ. Sci. & Technol.*, 29(8), 1913-1922.

Ptak, T., Schirmer, M., and Teutsch, G. (2000). "Development and performance of a new multilevel groundwater sampling system." In *Risk, regulatory, and monitoring considerations: Remediation of chlorinated and recalcitrant compounds (C2-1)*, Battelle Press, Columbus, Ohio,95-102.

Puls, R. W., Paul, C. J., Powell, R. M., and Kerr, R. S. (1996). "Remediation of chromate-contaminated ground water using zero-valent iron: Field test at USCG Support Centre, Elizabeth City, North Carolina." In *Proceedings of the 1996 HSRC/WERC Join Conference on the Environment,* Albuquerque, New Mexico.

Purkhold, C. D. (2001). *Hydrochemische Untersuchungen am Funnel and Gate System auf dem BEKA-Gelände in Tübingen*, Diplomarbeit, Universität Tübingen, Tübingen, Germany.

Rahman, A., and Agrawal, A. (1997). "Reduction of nitrate and nitrite by iron metal: Implications for groundwater remediation." In *Proceedings of the 213th American Chemical Society National Meeting*, San Francisco, Ca., 157-159.

Rao, P. S. C., Annable, M. D., and Kim, H. (2000). "NAPL source zone characterization and remediation technology performance assessment: Recent developments and applications of tracer techniques." *J. Contam. Hydrol.*, 45, 63-78.

Reardon, E. J. (1995). "Anaerobic corrosion of granular iron: Measurement and interpretation of hydrogen evolution rates." *Environ. Sci. & Technol.*, 29(12), 2936-2945.

Reeter, C., Gavaskar, A., Gupta, N., and Sass, B. (1999). "Performance evaluations at the Moffett Field and Department of Defense permeable barrier sites." *Hydrological science and technology: short papers*, 15, 24-30.

Roh, Y., Lee, S. Y., and Elless, M. P. (2000). "Characterization of corrosion products in the permeable reactive barriers." *Environ. Geol.*, 40(1-2), 184-194.

Schad, H. (2004). "Lessons learned after 5 years application of PRBs in Germany." In *Proceedings of the 1st International Symposium on Permeable Reactive Barrier*, Belfast, Northern Ireland, 14-16 March, 2004.

Schad, H., and Klein, R. (2002). *Funnel and gate am standort BEKA in Tübingen: 4. Bericht zu laufenden kontrolluntersuchungen zur Überwachung der sanierungsmaßnahme am standort*, IMES GmbH, Amtzell, Germany.

Schad, H., and Klein, R. (2003). *Funnel and gate am standort BEKA in Tübingen: 5. Bericht zu laufenden kontrolluntersuchungen zur Überwachung der sanierungsmaßnahme am standort*, IMES GmbH, Amtzell, Germany.

Schad, H., and Teutsch, G. (1998). "Reaktive wände - Aktueller stand der praxisanwendung." *Geotechnik*, 21(2), 73-82.

Schlicker, O., Ebert, M., Fruth, M., Weidner, M., Wüst, W., and Dahmke, A. (2000). "Degradation of TCE with iron: The role of competing chromate and nitrate reduction." *Ground Water*, 38(3), 403-409.

Sorel, D., Warner, S. D., Longino, B. L., Honniball, J. H., and Hamilton, L. A. (2001). "Dissolved hydrogen measurements at a permeable zero-valent iron reactive barrier." In *Proceedings of the 221st National Meeting American Chemical Society, Division of Environmental Chemistry*, San Diego, Ca., 1175-1180.

Stender, T. (2002). *Vergleichende untersuchung von graugusseisengranulat- und eisenschwammproben aus der reaktionswand Rheine*, Diplomarbeit, Christian-Albrechts-Universität zu Kiel, Kiel, Germany.

Stephens, D. B., Hsu, K.-C., Prieksat, M. A., Ankeny, M. D., Blandford, N., Roth, T. L., Kelsey, J. A., and Whitworth, J. R. (1998). "A comparison of estimated and calculated effective porosity." *Hydrogeol. J.*, 6, 156-165.

Till, B. A., Weathers, L. J., and Alvarez, P. J. (1998). "Fe(0)-supported autotrophic denitrification." *Environ. Sci. & Technol.*, 32(5), 634-639.

Veehmayer, M. K. (2000). *Ein neuer ansatz zur ermittlung des eisenoxid-gehaltes von aquiferen*, Dissertation, Christian-Albrechts-Universität zu Kiel, Kiel, Germany.

Vikesland, P. J., Klausen, J., Zimmermann, H., Roberts, A. L., and Ball, W. P. (2003). "Longevity of granular iron in groundwater treatment process: Changes in solute transport properties over time." *J. Contam. Hydrol.*, 64, 3-33.

Vogan, J. (2003). "Summary of field performance of permeable reactive barrier." In *RTDF permeable reactive barriers (PRBs) action team meeting*, Niagara Falls, New York, 9-22, <http://www.rtdf.org/public/permbarr/minutes/101603/index.htm> (Aug., 17, 2005).

Vogan, J. L., Focht, R. M., Clark, D. K., and Graham, S. L. (1999). "Performance evaluation of a permeable reactive barrier for remediation of dissolved chlorinated solvents in groundwater." *J. Hazard. Mater.*, 68, 97-108.

Watson, D., B., Gu, B., Phillips, D. H., and Lee, S. Y. (1999). *Evaluation of permeable reactive barriers for removal of uranium and other inorganics at the Department of Energy Y-12 Plant, S-3 disposal ponds*, Oak Ridge National Laboratory, Oak Ridge, Tennessee.

Watson, D., Phillips, D. H., and Gu, B. (2002). *Performance evaluation of in-situ iron reactive barriers at the Oak Ridge Y-12 Site*, Environmental Sciences Division, Oak Ridge National Laboratory, Oak Ridge, Tennessee.

Westerhoff, P., and James, J. (2003). "Nitrate removal in zero-valent iron packed columns." *Wat. Res.*, 37, 1818-1830.

Wilkin, R. T., and Puls, R. W. (2003). *Capstone report on the application, monitoring, and performance of permeable reactive barriers for ground-water remediation: Volume 1 - Performance evaluation at two sites,* EPA/600/R-03/045a, Environmental Protection Agency, Cincinnati, Ohio, United States.

Wilkin, R. T., Puls, R. W., and Sewell, G. W. (2003). "Long-term performance of permeable reactive barriers using zero-valent iron: Geochemical and microbiological effects." *Ground Water*, 41(4), 493-503.

Wüst, W. (2000). *Geochemische untersuchungen zur sanierung CKW-kontaminierter aquifere mit Fe(0)-reaktionswänden*, Dissertation, Universität Stuttgart, Stuttgart, Germany.

Zhang, B. P., Johnson, W. P., Piana, M. J., Fuller, C. C., and Naftz, D. L. (2001). "Potential artifacts in interpretation of differential breakthrough of colloids and dissolved tracers in the context of transport in a zero-valent iron permeable reactive barrier." *Ground Water*, 39(6), 831-840.

CHAPTER 17

Hydraulic Studies of Zero-valent Iron in Permeable Reactive Barriers Using Tracer Experiment

Keith C. K. Lai[1], Irene M. C. Lo[2], and Peter Kjeldsen[3]

Abstract: Tracer experiment is the most direct method for the evaluation of the hydraulic performance of zero-valent iron (Fe^0)-based permeable reactive barriers (PRBs). The implementation of tracer experiment involves (1) selection of suitable tracer materials; (2) groundwater modeling; (3) design and installation of tracer injection and groundwater sampling systems; (4) collection and chemical analysis of groundwater samples; and (5) data analysis. The findings obtained from the funnel-and-gate Fe^0 PRB at Vapokon site, Denmark indicated that the natural gradient tracer experiment was able to identify and delineate the preferential flow of contaminated groundwater through the Fe^0 reactive medium of the PRB. Besides, spatial moments analysis of the tracer plume was capable of giving the arrival times and locations of the centre of mass of the tracer plume inside the upgradient and downgradient parts of the Fe^0 reactive medium, thereby resulting in the groundwater velocity of 99.5 m/yr within the Fe^0 reactive medium.

CE Database subject headings: Barriers; Ground-water flow; Hydraulic properties; Iron; Numerical analysis; Tracers; Velocity.

17.1 Introduction

Hydraulic performance of zero-valent iron (Fe^0)-based permeable reactive barriers (PRBs) on groundwater remediation is prone to be affected by seasonal variation in groundwater flow direction and velocity, the potential influence from construction artifacts (ITRC 2005), and mineral precipitates (Wilkin et al. 2002; Andrea et al. 2005), hydrogen gas film (Mackenzie et al. 1999) and biofilms (Gu et al. 2002) forming on the Fe^0 surface. For the sake of evaluating how well the emplaced Fe^0 PRBs meet design specifications, estimating the longevity of Fe^0 PRBs, and evaluating the frequency of maintenance and rejuvenation activities required, monitoring of the hydraulic performance of Fe^0 PRBs after emplacement is necessary. The hydraulic performance of Fe^0 PRBs which should be monitored periodically throughout the operation period includes: (1) the flow pattern of groundwater through the Fe^0 reactive media of PRBs; (2) the residence time available for contaminants inside the Fe^0 reactive media; and (3) the extent of hydraulic capture

[1] Research Associate, Dept. of Civil Engineering, The Hong Kong University of Science and Technology, Clear Water Bay, Kowloon, Hong Kong.
[2] Associate Professor, Dept. of Civil Engineering, The Hong Kong University of Science and Technology, Clear Water Bay, Kowloon, Hong Kong (corresponding author). E-mail: cemclo@ust.hk
[3] Associate Professor, Environment & Resources DTU, Technical University of Denmark, Lyngby, Denmark.

zone created by Fe^0 PRBs (Gavaskar et al. 2000). Monitoring of the groundwater flow pattern through the Fe^0 reactive media can help evaluate whether there is preferential, bypassing or even reverse flow of contaminated groundwater and clogging of the Fe^0 reactive media (RTDF 2000). Moreover, determination of the residence time and extent of hydraulic capture zone aim at checking if there is sufficient residence time inside the Fe^0 reactive media for complete removal of contaminants and generated by-products, and complete capture of contaminant plume by Fe^0 PRBs for the remediation. So far, groundwater flow direction inside Fe^0 PRBs is commonly evaluated using the methods such as water level measurement and tracer experiment (Battelle 1998; Gavaskar et al. 2000; Lai et al. 2006a). The velocity of groundwater water flowing through Fe^0 PRBs is also frequently estimated using Darcy's law (Bedient et al. 1999), by *in situ* groundwater velocity sensor (Ballard 1996), and by implementing tracer experiment (Focht et al. 1997; Vogan et al. 1999; Lai et al. 2006a).

Groundwater direction or flow pattern evaluated by the method of water level measurement requires preparation of water level maps so that groundwater flow pattern can be readily obtained by plotting flow lines along the gradients indicated by the map. However, detailed delineation of the groundwater flow pattern and capture zone is usually difficult if the hydraulic gradient inside and in the vicinity of the Fe^0 reactive media of PRBs is too small for measurement (Gavaskar et al. 2000). Groundwater velocity determination using Darcy's law requires hydraulic gradient, hydraulic conductivity and porosity in the portion of flow field in which velocity estimate is desired (Freeze and Cheery 1979). However, in field situation, hydraulic conductivity measurement often possesses large uncertainty. Combination of this uncertainty and the errors associated with hydraulic gradient and porosity determination may result in a considerable error on the computed groundwater velocity (RTDF 2000). *In situ* permeable flow sensor uses a thermal perturbation technique to measure the magnitude of 3-D groundwater flow velocity in unconsolidated and saturated porous media (Ballard 1996). This technique, however, is still experimental in nature and thus requires further experimental verification.

Rather than investigating the groundwater flow pattern roughly by the approach of the water level measurement, and estimating groundwater velocity roughly using Darcy's law and indirectly from the temperature distribution over the surface of the *in situ* flow sensor, tracer experiment is the most direct method for the hydraulic performance monitoring although it is expensive, time-consuming and difficult to implement (Gavaskar et al. 2000; RTDF 2000). At present, there are two main types of tracer experiment: natural gradient and forced-gradient tracer experiments. The former involves injection of tracers into subsurface and monitoring of the movement of tracer plume under a natural hydraulic gradient (Freyberg 1986; Leblanc et al. 1991; Garabedian et al. 1991). On the other hand, forced-gradient tracer experiment includes both injection of tracers and extraction of groundwater so as to monitor the movement of tracer plume under the hydraulic gradient induced by extraction wells (Mackay et al. 1994; Thorbjarnarson and Mackay 1994). In fact, tracer experiments not only help evaluate the flow pattern and velocity of groundwater inside the Fe^0 reactive media, but also can be used to investigate the reactivity issues or underlying reasons leading to the limited reactivity of Fe^0 PRBs (Ebert et al. 2003; Parbs et al. 2005).

In this chapter, the details of the experimental (laboratory and field) procedure and setup, and modeling study required for the tracer study of the hydraulic performance of Fe^0 PRBs are first introduced through the description and discussion of the natural gradient tracer experiment specifically implemented for the Fe^0 PRB installed at Vapokon site in Denmark. The numerical approaches conventionally used to analyze the experimental data from tracer experiments for the determination of the mass and velocity of conservative tracers in porous media are also covered.

17.2 Vapokon Site Description and Fe^0 PRB Emplacement

Vapokon site is located in the northern part of Fuen Island in Denmark. The unsaturated zone consists of a layer of loam and soil filings extending about 1.5 m below the ground surface, which has an altitude of about 22.2 meters above sea level (masl). The unconfined sandy aquifer underlain by a relatively impermeable clay layer has a thickness of about 9 m. Vapokon groundwater, which is nearly neutral in pH, is categorized as very hard groundwater since it contains about 500 to 700 mg/L as $CaCO_3$ of hardness. In addition, substantial amounts of total alkalinity (TAL) and sulfate ion (SO_4^{2-}), certain concentrations of sodium (Na^+) and chloride ions (Cl^-), and trace quantity of ferrous and ferric ion (Fe_T) could also be detected (Lai et al. 2006b).

Between 1976 and 1996, because of the spill from an underground storage tank of a solvent recycling factory, Vapokon groundwater was heavily contaminated by chlorinated aliphatic hydrocarbons (CAHs), such as tetrachloroethylene (PCE), trichloroethylene (TCE) and cis-dichloroethylene (cis-DCE), and a contaminant plume with an area of 200 × 80 m^2 was created in the downgradient aquifer (Lai et al. 2006b). To remediate the CAH contaminated groundwater, a funnel-and-gate Fe^0 PRB, which consists of two 110 and 130 m long impermeable sheet pilings (funnel) and a permeable treatment zone, was installed at Vapokon site in September 1999 (Fig. 17-1). Inside the permeable treatment zone, there is a Fe^0 reactive medium with

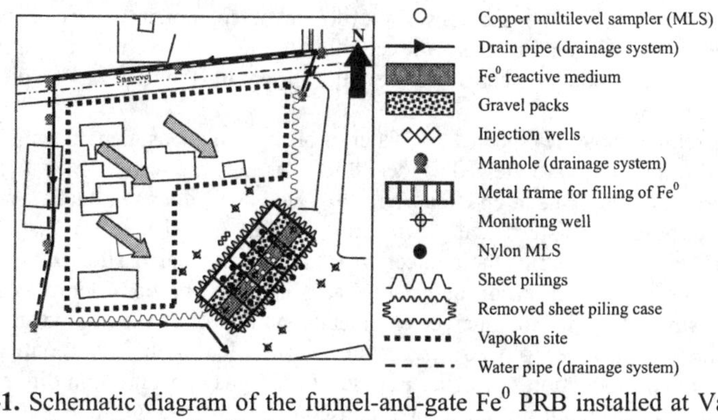

Fig. 17-1. Schematic diagram of the funnel-and-gate Fe^0 PRB installed at Vapokon site, Denmark, and the distributions of the monitoring wells, nylon and copper multilevel samplers over the Fe^0 PRB (adapted from Lai et al. 2006a)

dimensions of 14.5 × 9 × 0.8 m (W × D × L) and initial porosity of 0.6 flanked by 1 m thick layer of gravel pack on both upgradient and downgradient sides. Both impermeable funnel and permeable treatment zone were keyed into the till underneath the aquifer to ensure a contact between the Fe^0 PRB and the clay layer beneath. In addition, a drainage system was also installed in the upgradient of the Fe^0 PRB to prevent any uncontaminated groundwater from flowing into the capture zone of the funnel-and-gate system, thereby lengthening the residence time inside the Fe^0 reactive medium of the PRB (Lai et al. 2006b).

17.3 Natural Gradient Tracer Experiment for the Hydraulic Performance Monitoring of the Fe^0 PRB at Vapokon Site

The natural gradient tracer experiment implemented for the funnel-and-gate Fe^0 PRB at Vapokon site aimed to monitor the flow pattern of groundwater through the Fe^0 reactive medium of the PRB and determine the groundwater velocity or residence time inside the Fe^0 reactive medium. It was conducted from 1 November 2002 to 8 August 2003 and in total, 70 times of groundwater samplings corresponding to about 13000 samples had been conducted over a period of 280 days, thereby obtaining breakthrough curves (BTCs) from 48 discrete sampling points. Numerical analysis of the BTCs ultimately resulted in the groundwater velocity within the Fe^0 reactive medium. The overall methodology for the hydraulic performance monitoring of the Fe^0 PRB at Vapokon site using natural gradient tracer experiment involved following steps: selection of suitable tracer materials; hydrologic and hydrogeologic modelings; design and installation of injection and monitoring systems; collection and chemical analysis of groundwater samples; and data analysis.

17.3.1 Selection of Suitable Tracer Materials

Past experience from various column studies (Andrea et al. 2005; Lo et al. 2006) and field applications (McMahon et al. 1999; O'Hannesin and Gillham 1998; Vogan et al. 1999; Wilkin and Puls 2003; Lai et al. 2006b) indicated that Fe^0 reactivity could significantly alter the inorganic composition of groundwater such as calcium, magnesium and bicarbonate ions. In order to select tracer materials, which are chemically and physically conservative within the Fe^0 reactive medium, a laboratory column experiment is needed to conduct prior to the natural gradient tracer experiment.

Currently, isotopic species of water molecule such as deuterium (D_2O) and tritiated water (T_2O) are viewed as ideal tracers as they are water molecule and travel with groundwater (Bitsch and Jensen 1990). However, their radioactive propensity, and the possible disturbance of natural tritium and deuterium distribution in subsurface highly restrict their injection into a contaminated site. Dissolved noble gases such as helium, argon and neon gases can also be considered as near-ideal tracers since they do not interact chemically with porous media and groundwater constituents. However, physical tracer problems including diffusion, outgassing and underground production occurring in case of high gas concentration differences and high uranium contents in subsurface consequently discourage their extensive uses (Moser 1995). Conventionally, tracer salts including bromide (Battelle 1998; Vogan et al. 1999; Fiorenza et al. 2000; Gavaskar et al. 2000), chloride (Freyberg 1986;

Mackay et al. 1986), iodide (Mackay et al. 1994; Thorbjarnarson et al. 1994), lithium (Leblanc et al. 1991; Garabedian et al. 1991), sodium (Thierrin and Kitanidis 1994) and potassium (Moser 1995) are the materials frequently used in field tracer experiments because of their conservative and relatively environmental-friendly propensity, and their commonly low background concentration in subsurface. Nevertheless, with respect to Fe^0, Ebert et al. (2003) reported the possible retardation of bromide within a Fe^0 reactive medium depending upon the groundwater characteristics. Furthermore, obvious retardation of lithium was also observed within a Fe^0 reactive medium, which had been completely exhausted for the remediation of chromate-contaminated groundwater.

To select conservative tracers for the natural gradient tracer experiment conducted at Vapokon site, Vapokon groundwater spiked with lithium bromide (LiBr), potassium chloride (KCl) and T_2O was fed into 100% Fe^0 packed column. Although the conservative propensity of Li^+ and Br^- with respect to Fe^0 is in doubt, they were still selected as candidate tracer materials for further study because of their past application for the tracer study of the hydraulic performance of Fe^0 PRBs (Battelle 1998; Vogan et al. 1999; Fiorenza et al. 2000; Gavaskar et al. 2000; Ebert et al. 2002; Ebert et al. 2003). The Fe^0 used in the laboratory column should be the same as that used in the funnel-and-gate Fe^0 PRB. Besides, running the laboratory column experiment using Vapokon groundwater was to take account of the influence of groundwater characteristic on the transport of various tracer materials within the Fe^0 reactive medium. Sampling of water was implemented hourly at the effluent end of the column whereto approximately 3 mL of sample was collected for lithium (Li^+), bromide (Br^-), potassium (K^+), chloride (Cl^-) and T_2O assays until complete BTCs were achieved.

Fig. 17-2 illustrates the BTCs of T_2O, Li^+, Br^-, K^+ and Cl^- obtained from the laboratory Fe^0 packed column. Since T_2O is an ideal tracer, the transport behavior of other cations and anions within the column can be readily deduced by comparing their BTCs with that of T_2O. As can be seen, the breakthrough time of T_2O was approximately at a pore volume of one, thereby further corroborating its conservative within the Fe^0 reactive medium. Regarding the BTCs of Li^+, Br^-, K^+ and Cl^-, it is readily to observe that K^+ was significantly retarded, whereas there were only minor retardations of Li^+, Br^- and Cl^-. Besides, the good agreement between the influent and effluent concentrations for the T_2O, Li^+, Br^-, K^+ and Cl^- after the breakthrough also indicated the absence of any chemical transformations within the Fe^0 packed column. With the application of CXTFIT program (Toride et al. 1999), the BTCs were fitted by the mathematical steady-state solution of 1-D advection-dispersion equation, thereby giving a retardation factor of 1.37 for K^+ and 1.12, 1.13, and 1.17 for Li^+, Br^- and Cl^-, respectively. It is widely accepted that tracer materials work well in the Fe^0 reactive medium for the investigation of groundwater flow pattern and velocity if their retardation factors are 1.2 or less (Gavaskar et al. 2000). In accordance with this rule of thumb, Li^+, Br^- and Cl^- were classified as conservative tracers with respect to the Fe^0 PRB at Vapokon site. After taking account of their background concentration ($Li^+ < 0.1$ mg/L, $Br^- < 0.05$ mg/L, $Cl^- \sim 43$ to 50 mg/L) and the susceptibility of measurement, Li^+ injected as LiBr was finally selected as a main tracer for the natural gradient tracer experiment at Vapokon site (APHA/AWWA/WEF 1998).

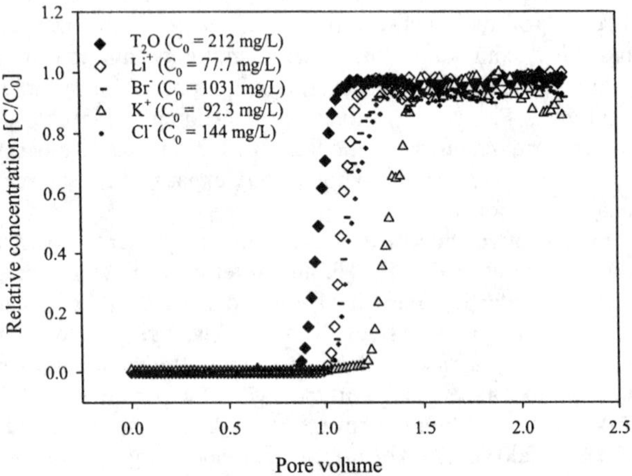

Fig. 17-2. T_2O, Li^+, Br^-, K^+ and Cl^- breakthrough curves (BTCs) obtained from the laboratory column experiment for the selection of suitable conservative tracer materials with respect to the Fe^0 PRB installed at Vapokon site, Denmark (adapted from Lai et al. 2006a)

17.3.2 Groundwater Modeling

After selecting the suitable tracer material, another question was the design parameters for tracer injection and groundwater sampling systems. The design parameters for the Li^+ injection system included the location of Li^+ injection wells, Li^+ injection rate and duration, and the Li^+ concentration of injected stock solution; while the location and spacing of sampling network, and groundwater sampling rate were the design parameters for the sampling system. To optimize these design parameters, groundwater modeling was implemented so that Li^+ plume created in the subsurface could be readily captured by the sampling system, and was able to provide enough information to determine the groundwater flow pattern and velocity inside the Fe^0 reactive medium of the PRB.

17.3.2.1 Design Parameters for the Tracer Injection System

Basically, the location of the Li^+ injection wells, Li^+ injection rate and duration, and the Li^+ concentration of the injected stock solution are the keyed factors influencing the spreading, size and concentration of the Li^+ plume created. Since the prime objective of this natural gradient tracer experiment was to monitor the hydraulic performance of the Fe^0 reactive medium of the PRB, the location of the Li^+ injection wells should be situated in the vicinity of the Fe^0 reactive medium. Generally, injection wells are recommended to place apart on a line perpendicular to groundwater flow direction and screen approximately 2 to 5 m below groundwater table, and should be able to create a tracer plume with a width of about 1.5 m and a height of approximately 1 m around injection wells. In our tracer experiment, if the injection wells are situated too close to the Fe^0 reactive medium, spreading or dispersion of the Li^+ plume may be minimal and thereby may not be captured by the downgradient sampling network. Moreover, too low Li^+ injection rate and Li^+ concentration in the injected stock solution may result in non-detectable level of the Li^+ plume inside the Fe^0 reactive medium. On the other hand, if the injection wells are located too far away from the Fe^0 reactive medium, the Li^+ plume created may be overspread. Under this situation, the created Li^+ plume may suffer boundary effects from the Fe^0 reactive medium and the impermeable funnel wall interface or the Fe^0 reactive medium and the aquifer interface, which can disturb the natural movement or spreading of the Li^+ plume through the Fe^0 reactive medium.

Another important point is that the high concentration of Li^+ in the injected stock solution may result in high Li^+ concentration in the tracer plume, which may possess significantly high density in comparison to the native groundwater at Vapokon site. Therefore, the movement of the Li^+ plume may be subject to the density gradient as it travels through the aquifer or the Fe^0 reactive medium. It may not be the same as the resulting path of the natural groundwater. In addition, to minimize the disturbance from the stock solution injection to the natural groundwater flow, the Li^+ injection rate cannot be too high. Conventionally, the volume of stock solution being injected should between 2% and 5% of the groundwater passing naturally through injection wells (Rügge et al. 1999; Broholm et al. 2001). Besides, in our case, the duration for the Li^+ injection should not be too long in order to minimize the experimental time required for obtaining complete Li^+ BTCs.

17.3.2.2 Design Parameters for the Groundwater Sampling System

In addition to the Li^+ injection system, the location and spacing of the sampling network, and the groundwater sampling rate are also crucial in capturing the injected Li^+ plume. In our tracer experiment, sampling network should be located along the simulated flow path of the Li^+ plume in order to explore its trajectory and flowing velocity within the Fe^0 reactive medium. Moreover, the horizontal and vertical spacing of the sampling network should be close enough to delineate the Li^+ concentration distribution within the plume, thereby allowing identification of the centre of mass of the Li^+ plume. Generally, the horizontal spacing of sampling network is between 0.5 and 2.0 m and the vertical spacing usually ranges from 0.25 to 1.0 m (Rügge et al. 1999; Broholm et al. 2001). Certainly, the horizontal spacing of sampling network should be increased gradually as a tracer plume spreads (Leblanc et al. 1991). As regards the groundwater sampling rate from the sampling network at Vapokon site, the criteria for the optimization were that the pumping rate or the drawdown created should not be able to disturb the natural movement or spreading of the Li^+ plume and the amount of the Li^+ removed from the sampling process should be insignificant in comparison to the total injected mass. Traditionally, the mass of a tracer material removed during each sampling round should only be about 0.1% of total injected mass (Leblanc et al. 1991).

17.3.2.3 Groundwater Flow Model and Solute Transport Program

Groundwater modeling is an important tool for incorporating site-specific hydrogeology into the design of Fe^0 PRBs. Once a detailed groundwater flow model for Fe^0 PRBs is established, various design configurations and setting, performance, monitoring and longevity scenarios can be readily evaluated. Thus the combined effects of several critical design parameters on the PRB design and the various scenarios can be simultaneously incorporated into a single model (Gupta and Fox 1999; Gavaskar et al. 2000). Generally, groundwater modeling is used at the several stages of Fe^0 PRB application including initial feasibility assessment, site selection, design optimization, design of performance monitoring system and longevity assessment (Gupta and Fox 1999). The groundwater modeling conducted for the natural gradient tracer experiment at Vapokon site was at the stage of the design of performance monitoring system in which the design parameters of the Li^+ injection system and groundwater sampling system were optimized.

To simulate the groundwater flow and Li^+ transport through the Fe^0 reactive medium of the PRB at Vapokon site, a 3-D groundwater flow model, which is able to simulate the sharp hydraulic conductivity contrast at the Fe^0 reactive medium and the aquifer interface or the impermeable funnel wall and the aquifer interface, was required. Furthermore, such 3-D groundwater flow model should be capable of handling the aquifer and the Fe^0 reactive medium heterogeneity, simulating the effects of internal sinks or sources induced by streams, drains, tunnels or wells on the groundwater flow through the Fe^0 PRB, and should be amenable to use with other solute transport or particle-tracking programs. Generally, commercially available model MODFLOW, which had been used to simulate particle movement through the pilot-scale PRBs at Naval Air Station (NAS) Moffett Federal Airfield (Battelle 1996) and Dover Air Force Base (Battelle 1997), is sufficient to fulfill the above requirements. In our study, MODFLOW coupled with multi-species contaminant transport model, namely MT3DMS, was applied for simulating the 3-D movement of the Li^+ plume along the Fe^0 PRB. MODFLOW was first used to develop a steady-state numerical approximation of the groundwater flow field through the Fe^0 PRB. This simulated system was established using the available site hydrologic and hydrogeologic data, and the characterization data of the Fe^0 PRB, such as porosity, hydraulic conductivity, level of water table and Darcy velocity or specific discharge. Then MT3DMS was applied in conjunction with MODFLOW for the simulation of the advection and dispersion of Li^+ in the simulated groundwater system.

Fig. 17-3 illustrates the simulated Li^+ plume through the upgradient and downgradient gravel packs, and the Fe^0 reactive medium at Vapokon site after optimizing the Li^+ injection and groundwater sampling systems. As can be seen from Figs. 17-3a and 17-3c, injection of the stock solution containing 3000 mg/L of Li^+ into the aquifer via the three injection wells at a rate of 1.44 mL/min was simulated resulting in the Li^+ plume with an initial width of 0.80 m, a height of 1.2 m and a Li^+ concentration of about 200 mg/L around the injection wells after 1 day of the injection. Moreover, in accordance with the simulated results obtained after 40 days of the Li^+ injection (i.e., Figs. 17-3b and 17-3d), it was believed that the optimized groundwater sampling network in practice had a high possibility of capturing and delineating the Li^+ plume movement through the Fe^0 reactive medium in which the maximum Li^+ concentration within the Fe^0 reactive medium was simulated ranging between 40 and 60 mg/L. After implementing the groundwater modeling, the Li^+ injection system and groundwater sampling system were then installed at Vapokon site accordingly.

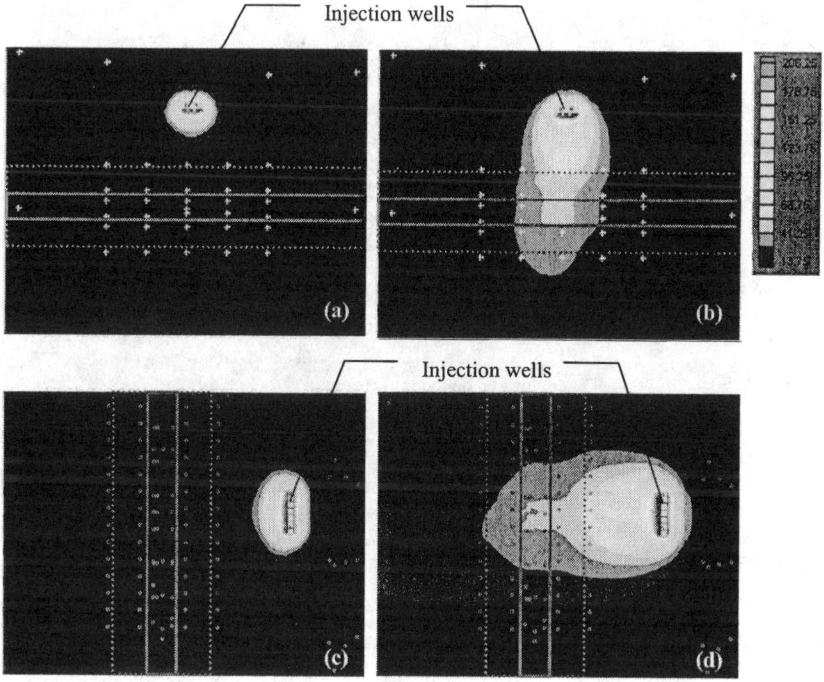

Fig. 17-3. Groundwater modeling results from MODFLOW and MT3DMS simulations conducted for the natural gradient tracer test at Vapokon site. Plan view of the simulated Li^+ plume after (a) 1 day and (b) 40 days of the Li^+ injection. Side view of the simulated Li^+ plume after (c) 1 day and (d) 40 days of the Li^+ injection. ▭ represents the Fe^0 reactive medium, ⬚ refers to the upgradient or downgradient gravel packs, + or **o** represents the sampling points of the sampling network.

17.3.3 Injection Wells and System

The injection wells and system used in this natural gradient tracer experiment basically were similar to those applied in the Vejen and Grindsted, Denmark (Rügge et al. 1999; Broholm et al. 2001). Vapokon groundwater collected from the upgradient monitoring well (i.e., M11 in Fig. 17-4a and spiked with about 3000 mg/L of Li^+ as LiBr was prepared as the stock solution for the injection. The stock solution was injected into the aquifer via the three injection wells (i.e., IW1, IW2 and IW3) installed approximately 3.5 m from the front of the Fe^0 reactive medium (Fig. 17-4a) and screened from 16.8 to 17.8 masl (Fig. 17-4b). The well screens were wrapped

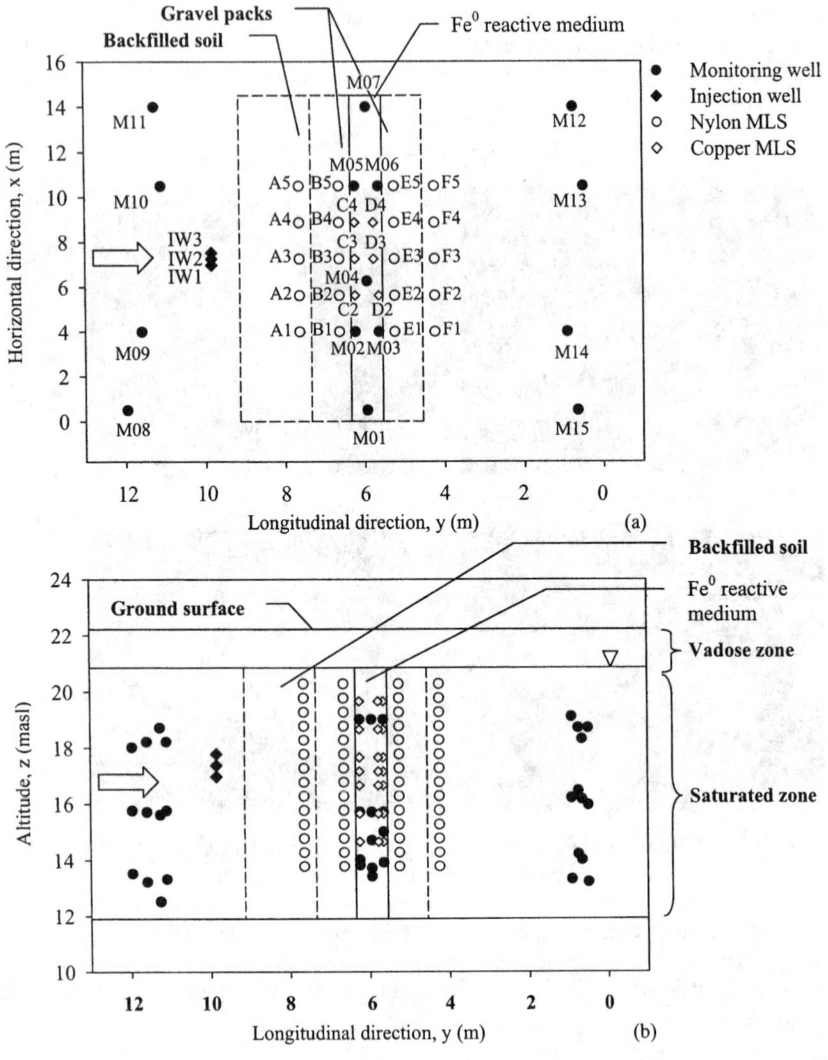

Fig. 17-4. Distribution of the monitoring wells, injection wells, nylon and copper multi-level samplers over the permeable treatment zone of the Fe^0 PRB at Vapokon site: (a) plan distribution and (b) vertical distribution (adapted from Lai et al. 2006a)

with a nylon mesh preventing the blockage of the screens by native soil (Lai et al. 2006a). During the emplacement, the three injection wells were located 0.275 m (centre to centre) apart on a line perpendicular to the groundwater flow direction (Fig. 17-4a). Three vertical sampling points located in the upper, middle and lower positions of the well screens were placed on each injection well to monitor the concentration level and homogeneity of the injected Li^+ in the groundwater (Fig. 17-5a).

As illustrated in Fig. 17-5, each injection well consisted of a PVC pipe (OD 125 mm, ID 117.5 mm) in which there was a second inner PVC pipe (OD 63 mm, ID 57 mm and perforated over 1 m at the bottom) housing a stainless steel (SS) rotary mixer and SS tube (OD 4 mm, ID 2 mm) for the supply of the stock solution. The SS rotary mixer, which was connected by a PVC rod to a motor above the well and seated in an oil-free Teflon seat at the bottom, assured complete mixing of the injected stock solution with groundwater within the screened section of each well. Approximately 127 L of the stock solution was continuously injected into the aquifer between 1 November 2002 and 1 January 2003. The injection system sequentially injected 1.44 mL of the stock solution over 60 sec once every 180 sec in each of the three injection wells. The SS rotary mixer was operated continuously during the injection period to ensure a complete mixing in each well.

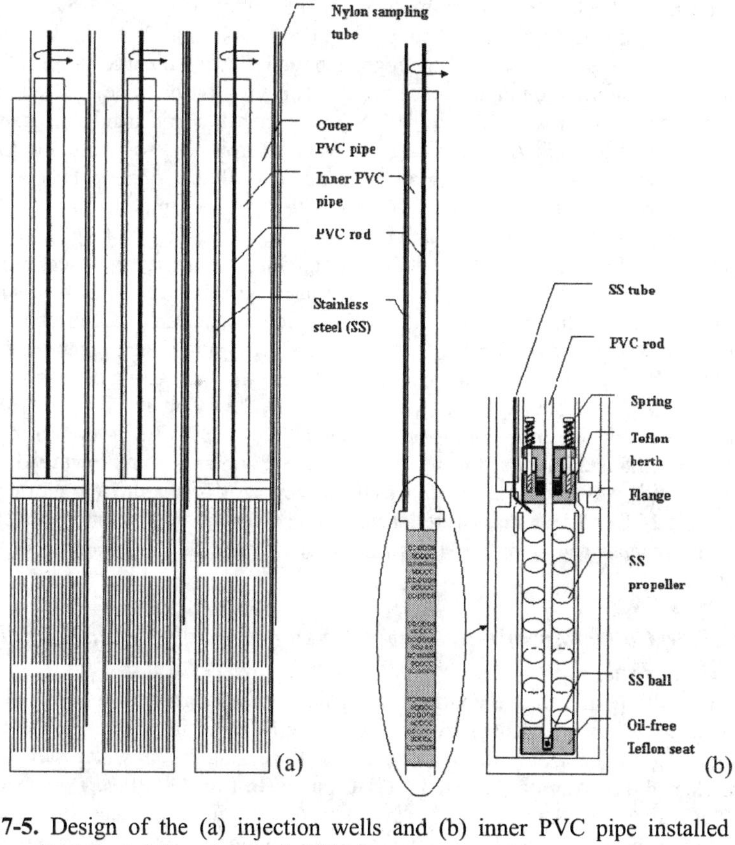

Fig. 17-5. Design of the (a) injection wells and (b) inner PVC pipe installed at Vapokon site (adapted from Lai et al. 2006a)

17.3.4 Groundwater Sampling Network and Systems

Groundwater sampling network installed across the permeable treatment zone of the PRB at Vapokon site consisted of two main types of well or sampler. During the time of the Fe^0 PRB installation, 45 piezometers each ended by 1 m long screen had been installed in the 15 monitoring wells (3 piezometers in a well) locating in the upgradient (M08 to M11) and downgradient (M12 to M15) locations, as well as in the middle (M01, M04 and M07), upgradient (M02 and M05) and downgradient (M03 and M06) parts of the Fe^0 reactive medium (Fig. 17-4a). The 3 piezometers in each of the monitoring well were usually screened at depths of 12.5 to 19.1 masl so as to collect groundwater vertically along the aquifer (Fig. 17-4b).

In addition to the monitoring wells already installed at Vapokon site, a dense network of multilevel sampler (MLS) consisting of 20 nylon MLSs (280 screens) and 6 copper MLSs (42 screens) was also installed downgradient from the three injection wells by Geoprobe technique (Figs. 17-4a and 17-4b). The nylon MLS consisted of a hollow PVC pipe (OD 20 mm, ID 10 mm) with 14 nylon sampling tubes (OD 4 mm, ID 2.5 mm) attached outside with a vertical spacing of 0.5 m. Each nylon tube was screened by connecting to a slotted Teflon tube (OD 4 mm, ID 3 mm) with a length of 7.5 cm. It was installed into the upgradient and downgradient locations of the Fe^0 reactive medium through the casing (i.e., galvanized iron pipe) driven into the subsurface in advance. The nylon MLSs were then lowered inside the galvanized iron pipe and leveled. Afterwards, the galvanized iron pipe was pulled out and the sediment was allowed to collapse around the nylon MLSs. Copper MLS consisted of a hollow iron pipe (OD 28 mm, ID 14 mm) with 7 copper sampling tubes (OD 5 mm, ID 3 mm) attached inside with a vertical spacing of 0.5 or 1 m (Fig. 17-6). It was specifically designed for the groundwater sampling from the Fe^0 reactive medium. Since complete collapse of the Fe^0 filings around the MLSs could not be assured, the copper MLS was directly driven into the Fe^0 reactive medium by Geoprobe without a casing to ensure a close contact between the screens and the Fe^0 reactive medium, and thereby prevented vertical crossflow of the groundwater along the copper MLSs. Across the permeable treatment zone of the PRB, the nylon and copper MLSs were installed in a fence formation. Fences of the nylon MLSs with 14 monitoring depths between 13.8 and 20.3 masl were situated 2.2, 3.2, 4.6 and 5.6 m downgradient from the injection wells; whereas the fences of the copper MLSs with 7 monitoring depths between 14.8 and 19.8 masl were located 3.6 and 4.1-4.2 m from the injection wells. The horizontal spacing of the MLSs within each fence was about 1.63 m (Figs. 17-4a and 17-4b).

17.3.5 Collection of Groundwater Samples and Chemical Analysis of Tracers

During the period of the continuous injection of Li^+, groundwater sampling was conducted every 1 to 4 days. After ceasing the Li^+ injection on 1 January 2003, the sampling frequency reduced gradually and the last groundwater sampling was conducted on 8 August 2003. A 12 V DC pump (In-Line 991, *Whale*) connected to a 10 m long nylon tube was used for groundwater sampling from the monitoring wells; while a sampling manifold for simultaneous sampling of six sampling points by the suction provided by a vacuum pump was applied for collecting groundwater from

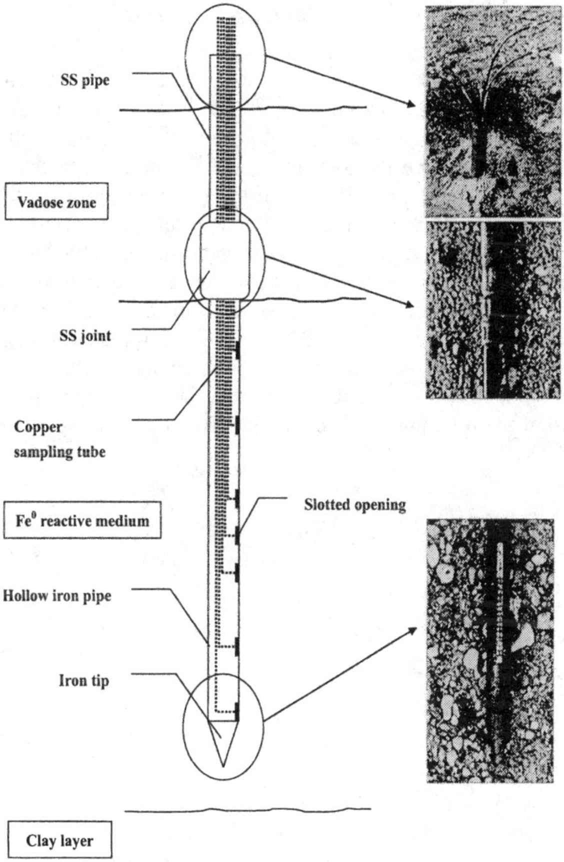

Fig. 17-6. Design of the copper MLS specifically installed into the Fe^0 reactive medium for groundwater sampling (adapted from Lai et al. 2006a)

both nylon and copper MLSs. Prior to the collection of groundwater sample, 50 mL and 5 to 15 L of groundwater were first removed from the MLSs and monitoring wells, respectively, so as to prevent the collection of stagnant water. Collected groundwater sample was then immediately stored into a 20-mL PE bottle with a screw cap. After undergoing 4 hours of gravitational settling, the supernatant was extracted from the PE bottle for Li^+ measurement with a flame photometer (PFP 7, *Jenway*). Br^- concentration in the groundwater samples collected from the sampling points located within the Li^+ plume was also measured using ion chromatograph (LC 20 Chromatography, *Dionex*) in laboratory so as to compare the transport behavior of the Li^+ with that of Br^- across the permeable treatment zone of the PRB.

17.3.6 Data Analysis – Spatial Moments Analysis for the Lithium Plume

Spatial moments analysis for the Li^+ distribution within the tracer plume can result in the mass and velocity of the Li^+ plume inside the Fe^0 reactive medium (Freyberg 1986; Garabedian et al. 1991; Rügge et al. 1999). Theoretically, the results from the zeroth moment analysis can give the total mass of Li^+ within the plume. The flow velocity of the Li^+ plume can be obtained from the first moment analysis in which the locations of the centre of mass of the Li^+ plume at different periods of time are identified. The relationship between the spatial moments, and the mass and the velocity of the Li^+ plume can be derived based upon the approach developed by Aris (1956). Concerning the velocity of the Li^+, when the time derivative of the equation for the first spatial moment of the Li^+ distribution [Eq. (17.1)] is used in conjunction with the 3-D advection-dispersion equation (ADE) for a conservative solute modified by Bear (1979) [Eq. (17.3)], it can be illustrated that the Li^+ velocity is equal to the temporal variation of the first moment [Eq. (17.4)] (Freyberg 1986; Garabedian et al. 1991).

$$\frac{d\bar{x}_i}{dt} = \iiint_{-\infty}^{\infty} n \frac{x_i}{M} \frac{\partial c}{\partial t} dx_1\, dx_2\, dx_3 \tag{17.1}$$

where

$$M = \iiint_{-\infty}^{\infty} nc\, dx_1\, dx_2\, dx_3 \tag{17.2}$$

$$\frac{\partial c}{\partial t} = -V_i \frac{\partial c}{\partial x_i} + D_{ij} \frac{\partial^2 c}{\partial x_i \partial x_j} \tag{17.3}$$

$$V_i = \frac{d\bar{x}_i}{dt} \tag{17.4}$$

where c is the Li^+ concentration; D_{ij} is the dispersion coefficient tensor, $i, j = 1, 2$ and 3; M is total Li^+ mass; n is the porosity; t is time; V_i is the Li^+ velocity in the x_i direction, $i = 1, 2$ and 3; and $\bar{x}_i$ is the first moment of the Li^+ distribution, $i = 1, 2$ and 3.

Computations of the spatial moments were conducted using a numerical integration of the Li^+ distribution in which the values of the concentration at the sampling points were linearly interpolated over the domain and then integrated to obtain the moment estimates (Garabedian et al. 1991). For this analysis, the 3-D spatial moments of the following form were applied:

$$M_{ijk} = \iiint_{\Omega} (x_1 - u\bar{x}_1)^i (x_2 - u\bar{x}_2)^j (x_3 - u\bar{x}_3)^k\, nc\, dx_1\, dx_2\, dx_3 \tag{17.5}$$

where i, j and k are exponents with values 1 and 2; u is a flag for the central moment with value 0 or 1, depending on the moment being evaluated; and Ω is the test domain.

Since the Li^+ concentration was sampled at points rather than continuously in space, it was required to interpolate the Li^+ concentrations as part of the numerical integration procedure. Integration over the vertical was the first step to be performed at each sampler or monitoring well location because the horizontal positions are the same for each set of concentrations at a MLS or monitoring well. The concentration was hypothesized varying linearly between each pair of adjacent sampling points in the vertical. Afterwards, integration over the horizontal plane was conducted in which a linear interpolation between three sampler locations (i.e., a triangular subdomain) of the vertically integrated values was assumed and substituted into Eq. (17.5), thereby establishing Eq. (17.6).

$$M_{ijk} = \sum_{p=1}^{N} \iint_{\Omega_p} n\left(A_1 + A_2 x_1 + A_3 x_2\right)_k \left(x_1 - u\bar{x}_1\right)^i \left(x_2 - u\bar{x}_2\right)^j dx_1\, dx_2 \quad (17.6)$$

where A_1, A_2 and A_3 are the coefficients defining the vertically integrated concentration distribution; k is the exponent of the vertical spatial integration; N is the number of triangular subdomains in the test domain; and Ω_p is the triangular subdomain.

A set of equations from finite element analysis were used to solve Eq. (17.6). In fact, each of the expressions within parentheses in Eq. (17.6) is a linear function over the triangular subdomain and can be represented as a sum of linear basis functions, thereby resulting in Eq. (17.7).

$$M_{ijk} = \sum_{p=1}^{N} \iint_{\Omega_p} n\left(\sum_{L=1}^{3} I_L \phi_L\right)_k \left(\sum_{L=1}^{3} \left(x_1 - u\bar{x}_1\right)_L \phi_L\right)^i \left(\sum_{L=1}^{3} \left(x_2 - u\bar{x}_2\right)_L \phi_L\right)^j dx_1\, dx_2 \quad (17.7)$$

where I_L is the vertically integrated value at point L; and ϕ_L (x_1, x_2) is the linear basis function assuming the value of 1 at L and 0 at the other two points.

A formula, which is useful for the integration of linear basis functions over a triangular area, is shown in Eq. (17.8) (Zienkiewicz 1977).

$$\iint_{\Omega_p} \phi_1^i \phi_2^j \phi_3^k dx_1\, dx_2 = \frac{2\, A_t\, i!\, j!\, k!}{(i + j + k + 2)!} \quad (17.8)$$

where A_t is the area of the triangular subdomain.

By expanding the polynomials and sequentially applying Eq. (17.8), Eq. (17.7) can be solved, then summing the N triangular subregions until the total integral is computed (Garabedian et al. 1991).

For the sake of investigating the movement of the centre of mass of the Li^+ plume along the Fe^0 reactive medium, the study area shown in Fig. 17-4a (i.e., x = 0.5 to 14 m; y = 5.24 to 6.64 m) was divided into two domains. These are upgradient part of the Fe^0 reactive medium and downgradient part of the Fe^0 reactive medium (Fig. 17-7), and numerical integration of the Li^+ plume for each sampling round was conducted in each domain. By carrying out the spatial moments analysis of the Li^+ distribution for each round of the sampling, the temporal change of the Li^+ mass within the domain can be obtained. The arrival time and location of the centre of mass of the Li^+ plume in the domain thereupon can be simply identified and computed from the Li^+ distribution in the sampling round, giving the maximum mass of Li^+ within the domain (Bedient et al. 1999). After determining the arrival times and locations of the centre of mass of the Li^+ plume in each domain, the flow velocity of the Li^+ plume or groundwater can be readily obtained from the temporal change of the first moment of the Li^+ distribution [i.e., Eq. (17.4)]. In this study, the programs "Triangulation" and "Moment", which were written by Garabedian et al. (1988), were used for automatic establishment of the triangular subdomains and the numerical integration of the Li^+ distribution in these two domains, respectively.

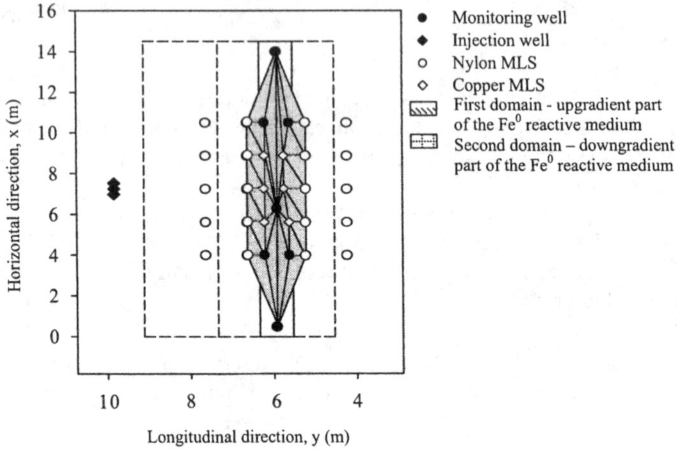

Fig. 17-7. The two domains (i.e., upgradient part and downgradient part of the Fe^0 reactive medium), and the corresponding triangular subdomains along the Fe^0 reactive medium of the PRB at Vapokon site

17.4 Results and Discussion

17.4.1 Flow Pattern of the Lithium Plume

The plan and side views of the movement of the Li^+ plume across the permeable treatment zone (i.e., from the fence of MLSs A1-A5 to that of F1-F5) of the PRB at Vapokon site are illustrated in Figs. 17-8a and 17-8b, respectively. As shown in Fig. 17-8a, before entering the upgradient gravel pack, the Li^+ plume moved exactly along the centerline of the permeable treatment zone without pronounced lateral displacement since the three injection wells were also situated in the centerline (Fig. 17-4a). However, there was sudden change of the flow path in the immediate vicinity of the front edge of the Fe^0 reactive medium, probably due to the clogging of the central or northeast parts of the Fe^0 reactive medium caused by mineral precipitates. Sudden longitudinal diminution of the hydraulic conductivity from the highly permeable gravel pack to the Fe^0 reactive medium and spatial variability of the hydraulic conductivity of the gravel pack (Freeze and Cherry 1979) may also exert certain contributions to the sudden change of the flow direction. However, the absence of a sudden change of the Li^+ flow path between the Fe^0 reactive medium and the downgradient gravel pack indicated the insignificance of their influences.

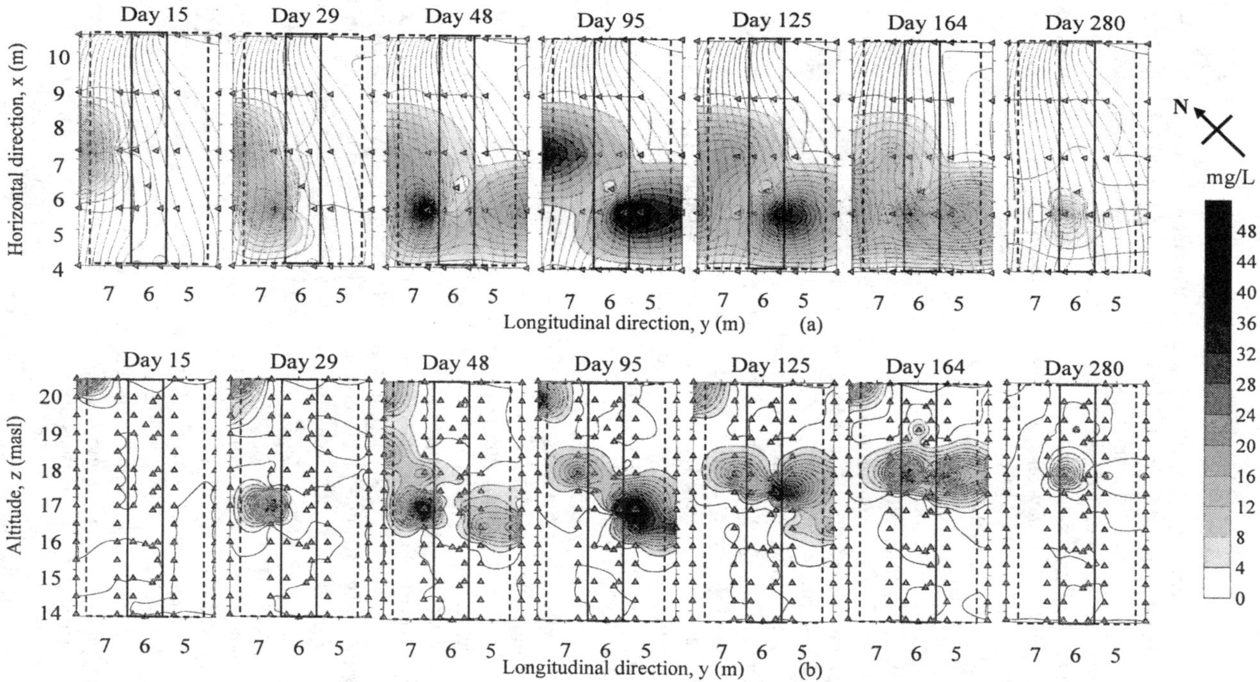

Fig. 17-8. Movement of the Li^+ plume across the permeable treatment zone (i.e., from the fence of MLSs A1-A5 to that of F1-F5) of the PRB at Vapokon site: (a) plan view and (b) side view of the Movement. Δ represents the screens of the monitoring wells, nylon and copper MLSs. — means water table contour. ▭ represents the Fe^0 reactive medium and ⬚ refers to the upgradient and downgradient gravel packs (adapted from Lai et al. 2006a).

After Day 29 (i.e., 29 days after the Li^+ injection), it can be readily observed that the Li^+ plume moved preferentially across the Fe^0 reactive medium.

Fig. 17-8b shows the vertical distribution of the Li^+ plume across the permeable treatment zone. It is clear to examine that the Li^+ plume was first detected from the sampling point in the fence of MLSs A1-A5 located just below the water table (i.e., A3-20.3 masl – the sampling point in MLS A3 with an altitude of 20.3 masl). Since the injection wells were screened from 16.8 to 17.8 masl, an upward movement of the Li^+ plume from the injection wells to the front edge of the upgradient gravel pack was highly expected. This upward movement of the Li^+ plume can most likely be ascribed to the presence of clay lens between the injection wells and upgradient gravel pack since clay soil could be collected from an altitude of about 16 to 19 masl during the installation of the injection wells. It is believed that this clay lens hindered the Li^+ plume from moving towards the gravel packs and Fe^0 reactive medium at altitudes the same as those of the well screens and thereby forced the upward movement of the Li^+ plume. However, within the upgradient gravel pack, the Li^+ plume dropped significantly from an altitude of 20.3 to about 16.5 masl on Day 29. Once the Li^+ plume entered the Fe^0 reactive medium, it traveled at a depth between 16 and 18 masl with minor drop of the altitude. Such an insignificant drop of the altitude within the Fe^0 reactive medium was most likely due to the influences of the density contrast between the Li^+ plume and the native groundwater, as observed at the Borden, Ontario (Mackay et al. 1986) and Cape Cod, Massachusetts (LeBlanc et al. 1991) as well as the recharge from precipitation, which added a layer of water on the groundwater surface and thereby suppressed the altitude of the Li^+ plume.

The noticeable drop of the Li^+ plume within the upgradient gravel pack, however, was not merely caused by the sinking of the denser Li^+ plume into the native groundwater and rainfall precipitation. Although heavy rainfall precipitation had been recorded between Days 6 and 28 (Fig. 17-9), it would not permanently induce the downward movement of the Li^+ plume within the upgradient gravel pack. Analogous to the reason causing the lateral change of the Li^+ flow path in the immediate vicinity of the Fe^0 reactive medium (Fig. 17-8a), clogging of the upper part of the Fe^0 reactive medium, most likely caused by mineral precipitates, is believed to be the main underlying reason leading to the significant drop of the Li^+ plume. The absence of the Li^+ in the upper part of the Fe^0 reactive medium can further verify the insignificance of the contribution from the sudden longitudinal diminution of the hydraulic conductivity from the upgradient gravel pack to the Fe^0 reactive medium on the downward movement of the Li^+ plume within the upgradient gravel pack. Besides, the consistent movement of the Li^+ plume between the Fe^0 reactive medium and the downgradient gravel pack can also indicate the insignificant influence from the spatial variability of the hydraulic conductivity of the gravel pack on the pronounced sinking of the Li^+ plume within the upgradient gravel pack.

According to the observation from Fig. 17-8b, the Li^+ plume traveled across the Fe^0 reactive medium at two altitudes. At the initial stage of the tracer experiment (i.e., from Days 15 to 48), the Li^+ plume entered and passed through the Fe^0 reactive medium at a lower altitude, which was about 17 masl. However, commencing from Day 48, the Li^+ plume started entering the Fe^0 reactive medium at a higher altitude (i.e., 17.5 to 18 masl) and subsequently traveled across the Fe^0 reactive medium at

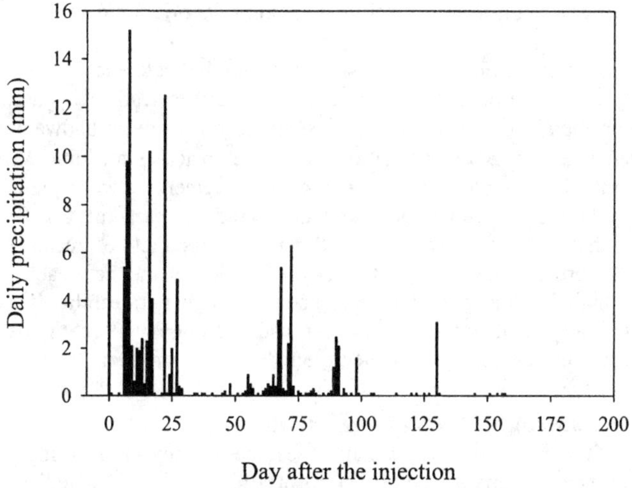

Fig. 17-9. Daily rainfall precipitation obtained from the station located approximately 5 km to the east of Vapokon site during the period of the natural gradient tracer experiment

this depth. The vertical distribution of the Li^+ plume on Day 95 clearly illustrated the change of the traveling altitude of the Li^+ plume from about 17 to 18 masl. Although the underlying reasons causing the lower traveling altitude of the Li^+ plume between Days 15 and 48 is not fully understood, it is believed that the suppression caused by the areal recharge from precipitation between Days 6 and 28 is the main culprit (see Fig. 17-9). The heavy rainfall precipitation recorded in that period of time substantially added a thick layer of water on the groundwater surface and thereby caused nearly a 1 m drop of the Li^+ plume. Based upon the meteorological data obtained from a station located approximately 5 km to the east of Vapokon site, the total precipitation between Days 6 and 28 was about 78 mm. Similar trends of variation in the daily precipitation could also be observed from the measurement conducted at Vapokon site where 112 mm of total precipitation was recorded in the same period of time. This is equivalent to a water layer with thickness in a range of 329 to 400 mm in the upgradient gravel pack (assuming the porosity of the gravel pack was in a range of 0.28 to 0.34) (Bedient et al. 1999). The high possibility of clogging of the upper surface of the Fe^0 reactive medium, most likely due to the intermittent exposure of the upper surface to air during the drought season, may also encourage the areal recharge on the Fe^0 reactive medium to flow to the top of the upgradient gravel pack, which at maximum can add another 320 mm of water in the upgradient gravel pack. Depending on the heterogeneity of the vadose zone, more areal recharge may be directed to the top of the upgradient gravel pack, which consequently aggravates the suppression of the Li^+ plume by precipitation. After Day 28, the rainfall was comparatively low so that vertical movement of groundwater induced by precipitation was minimal. As a consequence, the Li^+ mass injected between Days 29 and 61 did not suffer the suppression caused by precipitation, thereby flowing through the Fe^0 reactive medium at its original depth, about 17.5 to 18 masl.

The observations from Figs. 17-8a and 17-8b indicate that the groundwater conclusively passed through the Fe^0 reactive medium instead of bypassing, either over or under, the reactive medium. Besides, reversed flow of groundwater could not be examined in the upgradient location of the Fe^0 reactive medium. Virtually, as discussed in the latter section, there was a good agreement between the mass of the Li^+ passing through the upgradient and downgradient parts of the Fe^0 reactive medium so that there was no loss of the Li^+ across the Fe^0 reactive medium. However, the preferential flow of the Li^+ plume across the Fe^0 reactive medium proposed a high likelihood of the existence of a high permeability zone therein, which may greatly localize the contaminant load and thereby affect the performance of the Fe^0 PRB on treating contaminated groundwater (Ebert et al. 2002).

17.4.2 Breakthrough Curves of Bromide

Despite the fact that Li^+ was the main tracer being measured during the natural gradient tracer experiment, the Br^- concentration in the collected groundwater samples were also measured selectively in a laboratory in order to compare the transport behavior of Li^+ with that of Br^- inside the Fe^0 reactive medium. Figs. 17-10a to 17-10d illustrate the typical Li^+ and Br^- BTCs obtained from the sampling points in MLSs B2, C3, D2 and E2. It is clear to see that similar to the phenomenon observed from the laboratory column experiment (Fig. 17-2), the Li^+ BTCs were conclusively consistent with those of the Br^- wherever in the gravel pack (B2 and E2)

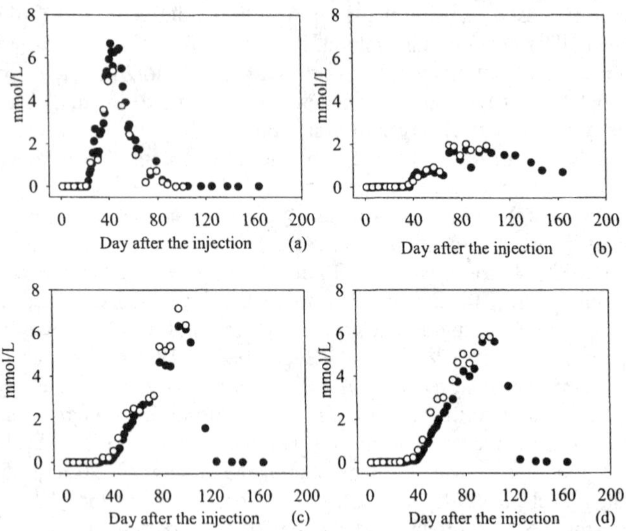

Fig. 17-10. Breakthrough curves (BTCs) of Li^+ and Br^- obtained from the sampling points (a) B2-16.8 masl, (b) C3-17.7 masl, (c) D2-16.7 masl, and (d) E2-16.8 masl (B2-16.8 masl means the sampling point in MLS B2 with an altitude of 16.8 masl) (adapted from Lai et al. 2006a)

and the Fe^0 reactive medium (C3 and D2) having a molar ratio of nearly 1 to 1. Although Ebert et al. (2003) had reported the importance of groundwater characteristics on the retardations of Li^+ and Br^- within the Fe^0 reactive medium, the influence of the characteristics of the Vapokon groundwater on the transports of the Li^+ and Br^- across the Fe^0 reactive medium was observed to be insignificant in this experiment. In addition, the analogy of the Li^+ and Br^- BTCs also indicates that Li^+ and Br^- were conservative within the Fe^0 reactive medium.

17.4.3 Mass of Lithium Flowing through the Fe^0 Reactive Medium

Fig. 17-11 illustrates the temporal variation of the Li^+ mass within the domains of upgradient and downgradient parts of the Fe^0 reactive medium. As can be seen, the mass of Li^+ in each domain first increased to a maximum and afterwards decreased back to the background level. This variation in the Li^+ mass is seemingly based upon an instantaneous source of Li^+ injection (Bedient et al. 1999). The Li^+ and Br^- BTCs obtained from the upgradient and downgradient gravel packs, and the Fe^0 reactive medium also followed the instantaneous source pattern of the injection (Fig. 17-10). However, as aforesaid, the Li^+ was continuously injected into the three injection wells over the first 61 days (i.e., from 1 November 2002 to 1 January 2003) of the tracer experiment. Therefore, the temporal variation of the Li^+ mass in the domains or the Li^+ concentration in the sampling points located downgradient from the injection wells is expected to be fairly steady within the injection period. Fig. 17-12 indicates the temporal change of the Li^+ concentration of the injected cloud created around the injection wells. It can be readily observed that, although there was fluctuation in the Li^+ concentration (especially in IW3), the concentration generally maintained a steady value throughout the injection period and the Li^+ concentration quickly dropped back to a non-detectable level after ceasing the injection on Day 61.

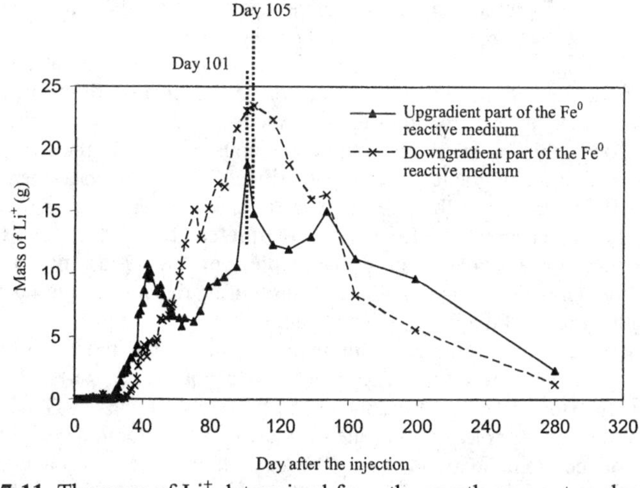

Fig. 17-11. The mass of Li^+ determined from the zeroth moment analysis within the domains of the upgradient part of the Fe^0 reactive medium and downgradient part of the Fe^0 reactive medium during the period of the natural gradient tracer experiment

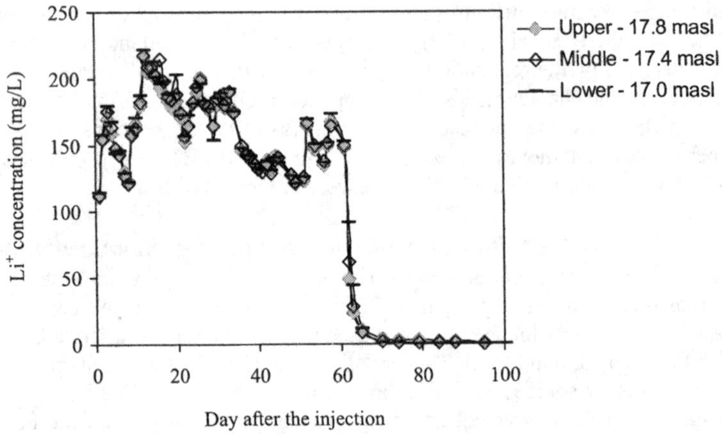

Fig. 17-12. Average Li^+ concentration measured from the upper, middle and lower positions of the screens of the three injection wells during the natural gradient tracer experiment

The transformation from the continuous source pattern to the instantaneous source pattern of the injection for the temporal variation of the Li^+ mass or Li^+ concentration in the downgradient of the injection wells is believed to be mainly due to the accumulation of the Li^+ within the upgradient gravel pack (Das and Nassehi 2003). It is believed that the clogging of the upgradient interface of the upper part of the Fe^0 reactive medium created complex pressure distribution in the upper part of the upgradient gravel pack, which accordingly established the flow circulation therein. When the Li^+ plume entered thc upgradient gravel pack at an altitude just below the groundwater table (Fig. 17-8b), the clogged section in the upgradient interface of the upper part of the Fe^0 reactive medium encouraged a downward movement of the Li^+ plume towards the centre of the created flow circulation. Subsequently, the Li^+ in the groundwater accumulated in the flow circulation (Das and Nassehi 2003), thereby resulting in the continuous increase in the Li^+ mass or concentration within the upgradient gravel pack (Fig. 17-10a). The movement of the centre of the flow circulation from the upgradient gravel pack to the Fe^0 reactive medium brought the accumulated Li^+ mass to the preferential path through the Fe^0 reactive medium, thereby resulting in the pattern of the instantaneous source injection for the temporal changes of the Li^+ mass in the domains of the upgradient and downgradient parts of the Fe^0 reactive medium.

Based upon the results from the zeroth moment analysis, the maximum Li^+ mass determined in the domains of the upgradient and downgradient parts of the Fe^0 reactive medium were 18.69 g and 23.42 g, respectively (Table 17-1). The good agreement of the maximum Li^+ mass or the Li^+ mass of the centre of the Li^+ plume in the domains of the upgradient and downgradient parts of the Fe^0 reactive medium indicates the successful capture of the flow path of the centre of the Li^+ plume along the Fe^0 reactive medium by the monitoring network. This also verifies the conservative nature of the Li^+ within the Fe^0 reactive medium and further shows the absence of any bypassing flow over the Fe^0 reactive medium.

Table 17-1. Maximum Li^+ Mass, and Arrival Times and Locations of the Centre of Mass of the Li^+ Plume in the Domains of Upgradient and Downgradient Parts of the Fe^0 Reactive Medium

Domain	[1]Maximum mass of Li^+ (g)	Arrival time of the maximum Li^+ mass (d)
Upgradient part of the Fe^0 reactive medium	18.69	101
Downgradient part of the Fe^0 reactive medium	23.42	105

Domain	[2]$\bar{x}$ (m)	[2]$\bar{y}$ (m)	[2]$\bar{z}$ (masl)	[3]Travel distance (m)
Upgradient part of the Fe^0 reactive medium	6.62	6.32	17.3	1.09
Downgradient part of the Fe^0 reactive medium	5.92	5.51	17.0	

[1]Zeroth moment analysis – maximum Li^+ mass within the domains
[2]First moment analysis – location of the centre of mass of the Li^+ plume in each domain
[3]Travel distance = $\sqrt{(\bar{x}_{Up} - \bar{x}_{Down})^2 + (\bar{y}_{Up} - \bar{y}_{Down})^2 + (\bar{z}_{Up} - \bar{z}_{Down})^2}$

17.4.4 Groundwater Velocity

After computing the locations of the centre of mass of the Li^+ plume in the two domains through the first moment analysis of the Li^+ distribution, the corresponding travel distance of the Li^+ plume from the upgradient part to the downgradient part of the Fe^0 reactive medium could be determined, which is approximately 1.09 m (Table 17-1). In accordance with Fig. 17-11, it is easily to observe that the centre of mass of the Li^+ plume reached the upgradient part of the Fe^0 reactive medium on Day 101. Then it took 4 days to travel to the downgradient part of the Fe^0 reactive medium, which corresponded to a groundwater velocity of approximately 99.5 m/yr and a residence time of about 70.4 h inside the Fe^0 reactive medium. Basically, the groundwater velocity determined from the spatial moment analysis is more reliable than that computed from a single BTC in upgradient and downgradient locations. This is because the former method conclusively determines the groundwater velocity from the flow path of the centre of a tracer plume; whereas the upgradient and downgradient BTCs analyzed in the latter method cannot ensure being collected from the flow path of the centre of a tracer plume.

17.5 Conclusions

Monitoring of the hydraulic performance of the Fe^0 PRBs, including the groundwater flow pattern and residence time inside the Fe^0 reactive media, and the extent of the hydraulic capture zone created, after the installation is obligatory so as to evaluate whether the emplaced Fe^0 PRBs meet their design specifications. Among various approaches for the hydraulic performance monitoring, tracer experiment is the most reliable method although it is expensive, time-consuming and difficult to implement. The overall methodology required for implementing the tracer experiment involves following steps: selection of suitable tracer materials; groundwater modeling; design and installation of the tracer injection and groundwater sampling systems; collection and chemical analysis of groundwater samples; and data analysis. Before beginning the tracer experiment, a tracer material which is chemically and physically conservative within the Fe^0 reactive media should be selected first. Then groundwater modeling is conducted to optimize the design parameters of the tracer injection and groundwater sampling systems for the subsequent installation. These design parameters include the location of injection wells, the tracer injection rate and duration, the tracer concentration in the injected stock solution, the location and spacing of sampling networks, and the groundwater sampling rates. Generally, the 3-D groundwater flow model required for the modeling should be able to simulate a sharp hydraulic conductivity contrast in porous media, handle the heterogeneity of the aquifers and Fe^0 reactive media, simulate the effects of internal sinks or sources on the groundwater flow through the Fe^0 PRBs, and is amenable to use with other solute transport or particle-tracking programs. Groundwater sampling and chemical analysis processes are implemented to delineate the movement and spreading of the injected tracer through the Fe^0 reactive media. Numerical analysis of the tracer distribution over the Fe^0 reactive media is to determine the flow velocity of the tracer plume through the Fe^0 reactive media.

The monitoring results obtained from the funnel-and-gate Fe^0 PRB at Vapokon site, Denmark showed the success of the natural gradient tracer experiment for the identification and delineation of the preferential flow of the contaminated groundwater through the Fe^0 reactive medium in which Li^+ was selected as a conservative tracer. Furthermore, the absence of bypassing flow over the Fe^0 reactive medium and revered flow of groundwater in the upgradient location of the Fe^0 reactive medium could also be verified from the observed movement of the Li^+ plume. Besides, spatial moments analysis of the Li^+ distribution was able to identify the arrival times and locations of the centre of mass of the Li^+ plume in the upgradient and downgradient parts of the Fe^0 reactive medium, thereby resulting in the groundwater velocity of 99.5 m/yr and residence time of 70.4 h inside the Fe^0 reactive medium.

17.6 References

Andrea, P. D., Lai, K. C.-K., Kjeldsen, P., and Lo, I. M.-C. (2005). "Effect of groundwater inorganics on the reductive dechlorination of TCE by zero-valent iron." *Water, Air and Soil Pollut'n*, 162, 401-420.

APHA, AWWA and WEF (1998). *Standard methods for the examination of water and wastewater*, 20th Ed., American Public Health Association, American Water Works Association and Water Environment Federation, Washington DC, United States.

Aris, R. (1956). "On the dispersion of a solute in a fluid flowing through a tube." *Proc R Soc London*, Ser A, 235, 67-78.

Ballard, S. (1996). "The in situ permeable flow sensor: A ground-water flow velocity meter." *Ground Water*, 34(2), 231-240.

Battelle (1996). *Draft evaluation of funnel-and-gate pilot study at Moffett Federal Airfield with groundwater modeling*, Environmental Security Technology Certification Program and Naval Facilities Engineering Services Center, United States Department of Energy, Port Hueneme, CA, United States.

Battelle (1997). *Design/test plan: Permeable barrier demonstration at area 5, Dover AFB*, Air Force Research Laboratory, Tyndall AFB.

Battelle (1998). *Performance evaluation of a pilot-scale permeable reactive barrier at Former Naval Air Station Moffett Field Mountain View, California*, Technical Report TR-2093-ENV, Naval Facilities Engineering Service Center, Port Hueneme, California. Batelle, Columbus, Ohio.

Bear, J. (1979). *Hydraulics of groundwater*, McGraw-Hill, New York.

Bedient, P. B., Rifai, H. S., and Newell, C. J. (1999). *Ground water contamination transport and remediation*, 2nd Ed., Prentice Hall PTR, Upper Saddle River, NJ, United States.

Bitsch, K., and Jensen, K. H. (1990). "Natural gradient dispersion test in a sandy aquifer using tritium as tracer." In *Proceedings of the International Conference and Workshop on Transport and Mass Exchange Processes in Sand and Gravel Aquifers*, Ontario, Canada.

Broholm, M. M., Rügge, K., Tuxen, N., Højberg, A. L., Mosbæk, H., and Bjerg, P. L. (2001). "Fate of herbicides in a shallow aerobic aquifer: A continuous field injection experiment (Vejen, Denmark)." *Water Resour. Res.*, 37(12), 3163-3176.

Das, D. B., and Nassehi, V. (2003). "Modeling of contaminants mobility in underground domains with multiple free/porous interfaces." *Water Resour. Res.*, 39(3), 1072, doi: 10.1029/2002WR001506.

Ebert, M., Dahmke, A., and Silva-Send, N. (2002). "Inhibiting substances as tracers for the reactivity assessment of Fe(0) PRB's." In *Proceedings of the 3rd International Conference on Remediation of Chlorinated and Recalcitrant Compounds*, Monterey, United States.

Ebert, M., Dahmke, A., Köber, R., Parbs, A., and Schäfer, D. (2003). "Ionic tracers for the reactivity assessment of Fe(0) PRB's." In *Proceedings of the 8th International FZK/TNO Conference on Contaminated Soil, ConSoil 2003* (CD-Rom), Gent, Belgium, 1582-1590.

Fiorenza, S., Oubre, C. L., and Ward, C. H. (2000). *Sequenced reactive barriers for groundwater remediation*, Lewis Publishers, Boca Raton, Florida, United States.

Freeze, R. A., and Cherry, J. A. (1979). *Groundwater*, Prentice-Hall, Inc., Englewood Cliffs New Jersey, United States.

Freyberg, D. L. (1986). "A natural gradient experiment on solute transport in a sand aquifer 2. Spatial moments and the advection and dispersion of nonreactive tracers." *Water Resour. Res.*, 22(13), 2031-2046.

Focht, R. M., Vogan, J. L., and O'Hannesin, S. F. (1997). "Hydraulic studies of in-situ permeable reactive barriers." In *Proceedings of the International Containment Technology Conference and Exhibition*, St. Petersburgh, Florida, United States.

Garabedian, S. P., Gelhar, L. W., and Celia, M. A. (1988). *Large-scale dispersive transport in aquifers: Field experiments and reactive transport theory*, Rep. 315, 290. Ralph M. Parson Lab., Department of Civil Engineering, Mass. Inst. Technol., Cambridge.

Garabedian, S. P., LeBlanc, D. R., Gelhar, L. W., and Celia, M. A. (1991). "Large-scale natural gradient tracer test in sand and gravel, Cape Cod, Massachusetts 2. Analysis of spatial moments for a nonreactive tracer." *Water Resour. Res.*, 27(5), 911-924.

Gavaskar, A., Gupta, N., Sass, B., Janosy, R., and Hicks, J. (2000). *Design guidance for application of permeable reactive barriers for groundwater remediation*, Strategic Environmental Research and Development Program, Battelle Columbus, Ohio.

Gu, B., Watson, D. B., Wu, L. Y., Phillips, D. H., White, D. C., and Zhou, J. Z. (2002). "Microbiological characteristics in a zero-valent iron reactive barrier." *Environ. Monit. Assess.*, 77, 293-309.

Gupta, N., and Fox, T. C. (1999). "Hydrogeologic modeling for permeable reactive barriers." *J. Hazard. Mater.*, 68, 19-39.

ITRC (2005). *Permeable reactive barriers: Lessons learned/new directions, PRB-4*, Interstate Technology & Regulatory Council, Permeable Reactive Barriers Team, Washington DC, United States.

Lai, K. C.-K., Lo, I. M.-C., and Kjeldsen, P. (2006a). "Natural gradient tracer test for the permeable reactive barrier in Denmark 1. Field study of tracer movement." *J. of Hazardous, Toxic and Radioactive Waste Management, ASCE*, in press.

Lai, K. C.-K., Lo, I. M.-C., Birkelund, V., and Kjeldsen, P. (2006b). "Field monitoring of a permeable reactive barrier for removal of chlorinated organics." *J. Environ. Eng. ASCE*, 132(2), 199-210.

LeBlanc, D. R., Garabedian, S. P., Hess, K. M., Gelhar, L. W., Quadri, R. D., Stollenwerk, K. G., and Wood, W. W. (1991). "Large-scale natural gradient tracer test in sand and gravel, Cape Cod, Massachusetts 1. Experimental design and observed tracer movement." *Water Resour. Res.*, 27(5), 895-910.

Lo, I. M.-C., Lam, C. S.-C., and Lai, K. C- K. (2006). "Hardness and carbonate effects on the reactivity of zero-valent iron for Cr(VI) removal." *Wat. Res.*, 40, 595-605.

Mackay, D. M., Freyberg, D. L., Roberts, P. V., and Cherry, J. A. (1986). "A natural gradient experiment on solute transport in a sand aquifer 1. Approach and overview of plume movement." *Water Resour. Res.*, 22(13), 2017-2029.

Mackay, D. M., Bianchi-Mosquera, G., Kopania, A. A., and Kianjah, H., and Thorbjarnarson, K. W. (1994). "A forced-gradient experiment on solute transport in the Borden aquifer 1. Experimental methods and moment analyses of results." *Water Resour. Res.*, 30(2), 369-383.

Mackenzie, P. D., Horney, D. P., and Sivavec, T. M. (1999). "Mineral precipitation and porosity losses in granular iron columns." *J. Hazard. Mater.*, 68(1-2), 1-17.

McMahon, P. B., Dennehy, K. F., and Sandstrom, M. W. (1999). "Hydraulic and geochemical performance of a permeable reactive barrier containing zero-valent iron, \Denver Federal Center." *Ground Water*, 37(3), 396-404.

Moser, H. (1995). "Groundwater tracing." In *Tracer technologies for hydrological systems*, International Association of Hydrological Sciences (IAHS) Publ. No. 229, 119-124.

O'Hannesin, S. F., and Gillham, R. W. (1998). "Long-term performance of an in-situ 'iron wall' for remediation of VOCs." *Ground Water*, 36(1), 164-170.

Parbs, A., Ebert, M., Köber, R., Plagentz, V., and Dahmke, A. (2005). "Field application of reactive tracers for performance assessment of Fe^0 PRBs." In *Proceedings of the Remediation of Chlorinated and Recalcitrant Compounds - 2004* (CD-Rom), Battelle Press, Columbus, Ohio, 3A-31.

RTDF (2000). *Summary of the remediation technologies development forum permeable reactive barriers action team meeting*, Remediation Technologies Development Forum, Feb 16-17, Melbourne, Florida, United States.

Rügge, K., Bjerg, P. L., Pedersen, J. K., Mosbæk, H., and Christensen, T. H. (1999). "An anaerobic field injection experiment in a landfill leachate plume, Grindsted, Denmark 1. Experimental setup, tracer movement, and fate of aromatic and chlorinated compounds." *Water Resour. Res.*, 35(4), 1231-1246.

Thorbjarnarson, K. W., and Mackay, D. M. (1994). "A forced-gradient experiment on solute transport in the Borden aquifer 2. Transport and dispersion of the conservative tracer." *Water Resour. Res.*, 30(2), 385-399.

Thierrin, J., and Kitanidis, P. K. (1994). "Solute dilution at the Borden and Cape Cod groundwater tracer tests." *Water Resour. Res.*, 30(11), 2883-2890.

Toride, N., Leij, F. J., and van Genuchten, M. Th. (1999). *The CXTFIT code for estimating transport parameters from laboratory or field tracer experiments version 2.1*, research report no. 137. United States Salinity Laboratory Agricultural Research Service, United States Department of Agriculture, Riverside, California, United States.

Vogan, J. L., Focht, R. M., Clark, D. K., and Graham, S. L. (1999). "Performance evaluation of a permeable reactive barrier for remediation of dissolved chlorinated solvents in groundwater." *J. Hazard. Mater.*, 68, 97-108.

Wilkin, R. T., Puls, R. W., and Sewell, G. W. (2002). *Environmental research brief - Long-term performance of permeable reactive barriers using zero-valent iron: An evaluation at two sites*, EPA/600/S-02/001, Office of Research and Development, United States Environmental Protection Agency, Washington DC 20460, United States.

Wilkin, R. T., and Puls, R. W. (2003). *Capstone report on the application, monitoring, and performance of permeable reactive barriers for ground-water remediation: Volume 1 – performance evaluations at two sites*, EPA/600/R-03/045a, Office of Research and Development, United States Environmental Protection Agency, Cincinnati, OH 45268, United States.

Zienkiewicz, O. C. (1977). *The finite element method*, 3rd Ed., McGraw-Hill, New York.

About the Editors

Dr. Irene M. C. Lo is currently Associate Professor of the Department of Civil Engineering and Associate Director of Environmental Engineering Program at the Hong Kong University of Science and Technology (HKUST). She had been Visiting Associate Professor of Technical University of Denmark and the University of Wisconsin at Madison in 1999 and 2000, respectively. She received her M.S. and Ph.D. in Civil Engineering from the University of Texas at Austin, USA. Dr. Lo is a chartered engineer of Hong Kong Institute of Engineers (HKIE) in Environmental discipline. She was bestowed the 2003 Peter H K Chan Best Paper Award by HKIE in recognition of her outstanding applied research achievement. Dr. Lo was the recipient of the 2002 Wong Kuan Cheng Scholar Award and the recipient of the 2003 Best Paper award from the International Conference on Soil and Groundwater Contamination and Clean-up. In 2004, she had been selected by the Award Committee of American Society of Civil Engineers (ASCE) as the recipient of the 2004 James Croes Medal. The highly coveted and prestigious James Croes Medal was awarded to Dr. Lo for her research paper on innovative waste containment barriers for subsurface pollution control that is judged worthy of special commendation for its merit as a contribution to engineering science. Dr. Lo's research includes remediation technologies for contaminated soils and groundwater, pollutant migration and transport behaviors in saturated and unsaturated soils, and innovative liner materials for waste containment facilities. She has published over 120 refereed journal and conference papers.

Dr. Rao Surampalli, an Engineer Director, has been working in the United States Environmental Protection Agency for 20 years. He received his Ph.D. in Environmental Engineering from Iowa State University in 1985 and M.S in Environmental Engineering from Oklahoma State University in 1978. He is a Registered Professional Engineer in the branches of Civil and Environmental Engineering, and also an Adjunct Professor of Environmental Engineering in four well-known universities in United States including University of Nebraska at Lincoln, University of Missouri at Columbia, University of Missouri at Rolla, and Iowa State University. Since 1984, Dr. Surampalli has published/presented over 300 papers in refereed journals or at national and international conferences. He is also the author of five books and twenty-two book chapters. Dr. Surampalli is the Editor-in-chief of the American Society of Civil Engineers "Hazardous, Toxic, and Radioactive Waste Management Journal" and Editor of the Water Environment Federation's Water Environment Research Journal. He also serves on the editorial boards of three other refereed journals. In 2001, he was awarded the "Founders Gold Medal" and also was named as the "National Federal Engineer of the Year" by the National Society of Professional Engineers (NSPE). In 2005, he was elected as Fellow of the American Association for the Advancement of Science (AAAS), a prestigious honor in America. He is also a Diplomate of the American Academy of Environmental Engineering and Fellow of the American Society of Civil Engineers. Oklahoma State and Iowa State Universities have honored him with "Distinguished Engineering Alumni" awards in 2001 and 2002, respectively. His environmental specialization and area of interest mainly involve hazardous and solid waste management, bioremediation, groundwater pollution control and treatment, clean technologies, design of water and wastewater treatment systems, biological waste treatment processes, and sludge treatment and disposal.

Dr. Keith C. K. Lai is currently Post-doctoral Fellow of the National Institute of Scientific Research (INRS) at University of Quebec. He had been Research Associate of the Department of Civil Engineering at the Hong Kong University of Science and Technology (HKUST). He received his M.Phil. in Chemical Engineering in 2001 and his Ph.D. in Civil Engineering in 2004 from HKUST. Dr. Lai's Ph.D. study involved a laboratory investigation of the influence of groundwater velocity and temperature on the performance of zero-valent iron (Fe^0) for the remediation of chlorinated aliphatic hydrocarbons (CAHs) in groundwater, field-oriented researches on the long-term performance of a Fe^0-based permeable reactive barrier (PRB) on groundwater remediation, and the 3-D transport of conservative tracers through a Fe^0 PRB. His post-doctoral works mainly focus on the influence of groundwater geochemistry and the competitive effect of CAHs on the Fe^0 performance for the remediation of chromate-contaminated groundwater in which the surface morphology of Fe^0 was examined by X-ray photoelectron spectroscope, Raman spectroscope, scanning electron microscope, X-ray diffractometer and Fourier transform infrared spectroscope. In April 2006, Dr. Lai was awarded the fellowship by the Croucher Foundation in Hong Kong for supporting a research study in Department of Civil, Architectural and Environmental Engineering at University of Texas at Austin.

Index

1,1,1-trichloroethane (TCA). *See* TCA
1,1,2,2-tetrachloroethane (TCE). *See* TCE
acid mine drainage, arsenic removal 121–122
algae blooms 77
alkalinity 276*f*
aluminum oxide, arsenic removal 140–141
ammonium, as byproduct of nitrate removal 90, 91*f*, 103–104, 104*f*; removal of 104–106, 106*f*, 107*f*
ammonium chloride 98
anions, effect on arsenic removal 122–127; effect on nitrate removal 87–89, 88*f*
arsenic 2, 112; in acid mine drainage 121–122
arsenic removal, aluminum oxide 140–141; analysis of iron oxides 138*f*; batch tests 112, 113*t*, 114–120, 115*f*, 116*f*, 117*f*, 118*f*, 119*t*; chemical processes 132*t*; field tests 112, 113*t*; hydraulic retention time 163; ion effects 154–159; iron corrosiveness 163–164; iron oxides 140–141; laboratory tests 112, 113*t*; modeling 161–163, 162*f*, 163*f*; oxidation in Fe^0 systems 139, 139*f*, 154–157; permeable reactive barrier 112; phosphate concentration 160*f*; removal mechanisms by Fe^0 154–155, 156*f*; removal rates 156–157; research needs 141–142
auger electron spectroscopy, TCE removal 176; comparison of palladized iron samples 181*f*, 181*t*

Bangladesh, groundwater, arsenic removal 127–128, 130
batch tests, arsenic removal 112, 113*t*, 114–120, 115*f*, 116*f*, 117*f*, 118*f*, 119*t*; contaminant treatability 250; halogenated methanes 191
Belfast, Northern Ireland, case study 238–239, 238*f*
bentonite 36, 233
bimetalic reductants 249; deactivation 216–217; first uses 207; literature on 208*t*; loss of reactivity 216; poisoning by inorganic species 217; preparation of 209–210
biopolymer slurry 273, 275, 274*f*
Bordon site, Ontario, Canada, permeable reactive barrier 224–225, 225*f*
breakthrough curves, tracer tests 19*f*, 20*f*, 284–285, 286*f*, 287*f*, 288*f*, 297*f*, 298*f*, 299*f*, 314*f*, 328–329
bromate 2
bromated methanes 192–193, 193*f*
bromide 284, 285, 286–287, 299*f*, 328–329; transport behavior 289

CAHs 9–11; concentration 13*t*, 21–22, 252–253, 253*f*; dechlorination 12, 17–19, 23, 24*f*, 25, 48; groundwater remediation 224–225, 235, 238–239; rate of 51; with Fe^0 52
calcium carbonate 275–276, 277*f*
carbon dioxide, effect on nitrate removal 84–85, 85*f*
carbon-halogen bond 194, 196*f*
Carbonate green rust (CGR) 133–137
caisson permeable reactive barriers 227, 227*f*; emplacement 231, 231*f*
capital investment, estimating 259–260, 259*t*
case studies, groundwater remediation 12
cations, effect on nitrate removal 87–89, 88*f*
clinoptilolite 104–106

chlorate 2
chlorinated aliphatic hydrocarbons. *See* CAHs
chlorinated hydrocarbons 2
chlorinated methanes 193–194; parallel dehalogenation reactions 198–200, 200*f*; sequential dehalogenation reaction 196–197, 198*f*;
chloroform 48; degradation rate 255*f*; parallel dehalogenation reactions 198–200, 200*f*
chromate, effect on TCE removal 41; effluent concentrations 40–41, 40*f* 42*f*, 43*f*, 44*f*; measurement of 38; reduction rate 41*t*; reduction with surfactant-modified zeolite/Fe^0 36; removal from aqueous solutions 35–36
chromium 35. *See also* hexavalent chromium
clinoptilolite 98, 99
column experiments 12, 14*f*, 49*f*, 285–289; arsenic removal 127–131, 127*f*, 128*f*, 129*f*; contaminant treatability 250–252, 250*f*; materials 49; TCE removal 175
Comprehensive Environmental Response, Compensation, and Liability Act 2; sites 246*t*, 247
continuous permeable reactive barriers 226, 226*f*
copper, effect of 215
cylindrical reactor, nitrate removal 80*f*

DDT 1
dechlorination rate 13*t*
denitrification 95
Denmark, Kolding site 72
Denmark, Vapokon site 12, 13*t*, 14*f*, 15*t*, 311–312, 311*f*; performance monitoring 312; permeable reactive barrier 235–237, 236*f*; rainfall 327, 327*f*; tracer test 290–291
dichlorodiphenyl-trichloroethane| (DDT). *See* DDT
Dover Air Force Base, PRB cost estimates 259*t*, 260, 260*t*
drinking water quality 78
drinking water standards 2–5, 61, 190

Elizabeth City, North Carolina, chromium removal site 70–72; arsenic removal column tests 128*f*, 129*f*, 134*f*, 137; permeable reactive barrier construction 237–237, 237*f*, 238*f*

field tests, arsenic removal 112, 113*t*, 157–161; comparison of 158*f*
finite element analysis, tracer test 323
full scale studies 12
funnel-and-gate permeable reactive barriers 226, 226*f*, 296*f*, 311*f*; emplacement 229, 229*f*, 232–233, 236*f*, 238–239, 238; hydraulic performance 272; tracer tests 289–290; width determination 254, 256, 256*f*

geochemistry, influence on hydraulic performance 275–277; modeling 254, 256–257
GeoSiphon/GeoFlow 228, 228*f*
groundwater contaminants, half-life 248; monitoring 257; treatability testing 250–253; treatable by reactive materials 245–246*t*
groundwater flow 254–256, 272, 273*f*, 310, 316–317, 324–328, 325*f*,
groundwater modeling 314, 316–317, 317*f*
groundwater remediation 13*t*, 35, 238–239
groundwater, samples 51, 257, 315, 320, 321*f*; CAH concentration 16, 251–252; chemical analysis 320 321; sources of arsenic contamination 151–152
groundwater standards 2–5, 3*t*
groundwater, synthetic, composition of 50*t*

groundwater temperature, seasonal variations 10
groundwater velocity 246–247, 246*t*, 254, 255*f*, 268–270, 271, 283, 310, 331

halogenated contaminants 173; batch experiments 191; headspace analysis 191; influence of surface intermediates 201–202, 202*f*; products from nanoscale iron treatment 195*t*;
HDTMA-bentonite. *See* organo bentonite
Hexadecyltrimethylammonium (HDTMA)-bentonite. *See* organo-bentonite
hexavalent chromium (Cr(VI)) 2, 55–58, 57*f*, 67; aqueous 67*f*; contaminant level in drinking water 61; passivation effects on removal 68; removal by Fe^0 62–63, 237–238; removal rates 63–66, 63*f*, 66*f*. *See also* chromium
humic acid, effect on nitrate removal 86–87, 87*f*
hydraulic fracturing 234, 235*f*
hydraulic retention time, arsenic removal 163
hydrochloric acid 98
hydrogen, adsorption of 211–213, 212*f*
hydrogen gas, formation during TCE reduction 179*t*
hydrogeologic modeling 254, 256–257
hydroxyl groups 83–84

in situ redox manipulation 69–71
injection wells, tracer experiments 317–319, 318*f*, 319*f*
ion exchange 95, 99
iron consumption 103–104, 104*f*
iron corrosion, anodic control 65, 66, 66*f*
iron oxides 140–141
iron foam 249–250

ISRM. *See in situ* redox manipulation

jetting 234, 234*f*

laboratory Fe^0 packed columns 10–11
laboratory tests, arsenic removal 112, 113*t*; barrier system 37*f*; groundwater remediation 12, 16; materials preparation 36–38; materials preparation 38*t*; nitrate reduction 98; reducing agents 71; surface characteristics 39*t*; tracers 285–289
liners 36
lithium bromide 16–17; tracer injection 314–315; analysis 322–324; temporal variation of flow 329–331

mackinawite precipitation 71
metal hydrides 211
microbial activities 10
mineral precipitates, effect on Fe^0 10–11
modeling, arsenic removal 161–163, 162*f*, 163*f*
monitoring wells 12, 294, 297*f*, 317, 318*f*

nanoparticles, benefits of 217–218; cost effectiveness 189, 189*t*; effect on human health 190; in treatment of common environmental contaminants 188*t*
nickel 214, 214*f*; effect of 215
nitrate 2, 77; initial concentration 85–86, 86*f*
nitrate levels, measurement 80; Taiwan 78*t*
nitrate loading, effect on rate of nitrate reduction 101–103, 102*f*
nitrate removal, anion effects 87–89, 88*f*; biological denitrification 98; byproducts 90–92, 90*f*, 96, 103–104, 104*f* carbon dioxide effects 84–85, 85*f*; cation effect 87–89, 88*f*; chemical reduction 96; Fe^0-based

processes 78; Fe^0 dosages 82*f*; humic acid effects 86–87, 87*f*; hydrogen peroxide 84*f*; ion exchange 95; laboratory tests 98–99; nitrogen gas formation 91*f*, 96; operating modes, effects 89–90, 90*f*; pH conditions 82*f*, 83, 96; propanol 83*f*, 84*f*, 96; prototype for complete removal 92*f*
nitrite 2, 77
nitroaromatic compounds 2
noble metals, catalytic activity 207–209; effect on reaction rate 215–216

operating modes, effect on nitrate removal 89–90, 90*f*
organo-bentonite 35, 36–37; organic carbon contests 39*t*; sorption or chromate 42. *See* also bentonite
oxo-anion removal rates 97

palladium, effect of 215; iron enhancement 173, 210*f*; plating efficiency 176, 177*f*; acid washed and unwashed, compared 180–184, 181*f*
PCE 2, 10; concentration 251–252, 251*f*, 252*f*; dechlorination 17–18; degradation rate 253, 254*f*; reduction with surfactant-modified zeolite/Fe^0 36
pentachlorophenol 2
permeable reactive barrier technology (PRB) 48; Bordon site, Ontario, Canada 224–225, 225*f*; caisson 227, 227*f*, 231, 231*f*; case studies 235–239; construction methods 274*f*, 283; contaminant concentration 6; continuous PRB 226, 226*f*; core sampling 258; costs 259–260, 259*t*, 260*t*; design methodology 234–244, 244*f*; Elizabeth City, North Carolina, case study 70–72, 237–238, 237*f*, 238*f*; emplacement techniques 229, 229*t*, 232–233, 233–235, 234*f*, 235*f*; first field application 2; funnel-and gate 226, 226*f*; hydraulic performance 11, 268*f*, 271, 275–276, 282–283, 309–310; mandrel-based emplacement 231; monitoring 257–258, 283–284, 209–310; performance evaluation 267–268, 268*f*; reactive material selection 247–250, 248*t*; thickness 27–28, 28*f*; treatment zones, emplacement 230; trench 227*f*, 228; types of 225–228
pH, effect on nitrate removal 78, 79, 81, 82*f*, 85*f*, 97–98, 100–101, 100*f*
phosphate, effect on arsenic removal 123–124, 124*f*, 125*f*, 130
plume hydrogeology 271
pump-and-treat systems 5*f*; contaminant concentration 5*f*; permeable reactive barriers, compared 5–6, 243

RCRA. *See* Resource Conservation and Recovery Act
research needs, arsenic removal 141–142
Resource Conservation and Recovery Act (RCRA) 2
Rhein site permeable reactive barrier, tracer test 291–294, 291*f*, 292*f*, 294*f*

selenium 2
sewer lines, influence on groundwater flow 272–273, 273*f*
sheet piles 232–233, 233*f*, 236*f*, 272
silicate, effect on arsenic removal 125–127, 126*f*
site investigation 244–247
slurry injection 234–235, 234*f*
slurry wall construction 233, 238–239, 238*f*
sodium carbonate 87, 88*f*
sodium hydroxide 98
sodium hypochlorite 98
sodium nitrate 98
sodium nitrite 98
sodium sulfate 98
storage tanks 36

soil mixing 234–235
sulfate, effect on arsenic sorption 123
sulfate reducing bacteria 69, 217
sulfuric acid 98
Superfund Law. *See* Comprehensive Environmental Response, Compensation, and Liability Act
surface intermediates, influence on reaction mechanisms 201–202
surface water standards 2–5; 3*t*
surfactant-modified zeolite 36

TCA 48; dechlorination 17–18; degradation rate 255*f*; rate of dechlorination 53*t*
TCE 1, 10, 35; concentration 25*f*, 52*f*; dechlorination 17–18, 48, 237–238, 237*f*; byproducts analysis 175; degradation rate 253, 254*f*, 255*f*; effect of chloroform 54–55; effect of TCA 52–54, 53*t*; effluent concentrations 40–41, 40*f*, 42*f*, 43*f*, 44*f*, 177, 178*f*, 180*f*; hydrogen gas 179*f*; measurement of 38; rate of 53*t*; reduction rate 41, 41*t*; tracer test results 285, 286*f*; with palladized granular iron 174–175
technetium 2
temperature effects 54–55, 55*f*
tetrachloroethylene. *See* PCE
tracer tests advantages and disadvantages 283–284; design parameters 314–315; laboratory 285–289; material selection 312–314, 314*f*; plume analysis 323, 324–328, 324*f*; resource requirements 300
trench excavation, permeable reactive barriers 230–231, 230*f*, 236*f*
trench stabilization, permeable reactive barriers 230
trenching, continuous 232, 232*f*, 237–238, 238*f*, 273, 274*f*
trichloroethylene. *See* TCE
Tübingen site permeable reactive barrier, tracer test 294–300, 295*f*, 297*f*, 298*f*

uranium 2
utilities, buried 272

Vapoken site, Denmark, groundwater remediation 12, 13*t*, 14*f*, 15*t*, 311–312, 311*f*; performance monitoring 312; permeable reactive barrier 235–237, 236*f*; rainfall 327, 327*f*; tracer test 290–291

water chemistry, effect on Cr(VI) removal 68–69
water level measurements 283, 310

x-ray photoelectron spectroscopy, T removal 176; comparison of palladized iron samples 182–183, 183*f*

zero-valent iron (Fe^0), arsenic removal 131–133, 132*t*, 154–156; conduit of electrons 214; corrosion 83, 84, 131, 133, 134*f*, 135*f*, 136*f*, 163–164; dechlorination rate 9–11; disposal of filter media 164; dosages 82*f*; effect of carbon dioxide 84–85, 85*f*; first uses 224; field-scale applications 69; filings 36; flow-through thickness 268; grain size 268, 269*f*, 270; hydraulic characteristics 268–271; lithium mass flow 329–331, 329*f*, 330*f*, 331*f*; longitudinal dispersivity 19–21; media longevity 164; mineralogical changes 131, 133*f*, 134*f*, 135*f*, 136*f*; nanoscale 121, 187–188; palladium 173–175; particle size 79; porosity 269*t*, 276, 285; primary roles 213–215; reaction with chromate 41*f*; thickness 10
zero-valent iron (Fe^0) permeable reactive barriers, and pump-and-treat systems, compared" 5–6, 243